Fundamentals of Programmable Logic Controllers, Sensors, and Communications

Second Edition

Jon Stenerson
Fox Valley Technical College

Prentice Hall
Upper Saddle River, New Jersey   **Columbus, Ohio**

Library of Congress Cataloging-in-Publication Data

Stenerson, Jon.
    Fundamentals of programmable logic controllers, sensors, and communications / Jon Stenerson. -- 2nd ed.
        p.   cm.
    Includes index.
    ISBN 0-13-746124-0 (hc)
    1. Programmable controllers. 2. Telecommunication systems.
I. Title.
TJ223.P76S74 1999
629.8'9--dc21   97-44151
                                                                    CIP

Cover photo: © Uniphoto
Editor: Charles E. Stewart, Jr.
Assistant Editor: Kate Linsner
Editorial Assistant: Kim Yehle
Production Editor: Patricia S. Kelly
Design Coordinator: Karrie M. Converse
Cover Designer: Katherine Hanley
Production Manager: Deidra M. Schwartz
Marketing Manager: Ben Leonard

This book was set in Times by Jon Stenerson and was printed and bound by Courier/Westford, Inc. The cover was printed by Phoenix Color Corp.

 © 1999, 1993 by Prentice-Hall, Inc.
Simon & Schuster/A Viacom Company
Upper Saddle River, New Jersey 07458

Printed in the United States of America

10 9 8 7 6 5 4 3 2 1

ISBN: 0-13-746124-0

Prentice-Hall International (UK) Limited, *London*
Prentice-Hall of Australia Pty. Limited, *Sydney*
Prentice-Hall of Canada, Inc., *Toronto*
Prentice-Hall Hispanoamericana, S. A., *Mexico*
Prentice-Hall of India Private Limited, *New Delhi*
Prentice-Hall of Japan, Inc., *Tokyo*
Simon & Schuster Asia Pte. Ltd., *Singapore*
Editora Prentice-Hall do Brasil, Ltda., *Rio de Janeiro*

To those administrators who understand the importance of technical education and enable all of us.

Jon Stenerson

Contents

Preface   **vi**

Acknowledgements   **viii**

Chapter 1: Overview of Programmable Logic Controllers   **1**

History of PLCs 2
PLC Components 7
Allen-Bradley Memory Organization and I/O Addressing 30
Logical Addressing 31
I/O addressing 32
GE Fanuc Memory Organization and   I/O Addressing...................33
PLC Applications 35
Questions 39

Chapter 2: Overview of Number Systems   **43**

Decimal..44
Binary Numbering System 45
Binary Coded Decimal System 47
Octal 48
Hexadecimal 49
Questions 54

Chapter 3: Fundamentals of Programming   **57**

Ladder Logic...58
Allen-Bradley Contacts 59
Coils 59
Allen-Bradley Coils 60
Ladder Diagrams 60
Allen-Bradley SLC File Organization and I/O Numbering 65
Allen-Bradley Normally Closed Contacts 72
Multiple Contacts 73
Branching..75

Special Contacts 79
Program Flow Instructions 83
GE Fanuc Program Flow functions 86
GE Fanuc Subroutine Instructions 87
PLC Scanning and Scan Time 88
Questions 91

Chapter 4: Timers and Counters 95

Timers 96
Allen-Bradley PLC-5, SLC 500, and MicroLogix 1000 Timers 101
GE Fanuc Timers 105
Gould Modicon Timers 108
Omron Timers 108
Square D Timers 111
Texas Instruments and PLC Direct Timers 112
Cascading Timers 114
Counters 115
Allen-Bradley PLC-5, SLC 500, and MicroLogix 1000 Counters ... 118
GE Fanuc Counters 121
Gould Modicon Counters 122
Omron Counters 123
Square D Counters 124
Texas Instruments and PLC Direct Counters 126
Programming Hints 127
Questions 131

Chapter 5: Industrial Sensors 137

The Need For Sensors 138
Sensor Types 138
Digital Sensors 140
Electronic Field Sensors (Field Sensors) 152
Sensor Wiring 157
Sourcing and Sinking Sensors 160
Analog Sensors 161
Installation Considerations 178
Typical Applications 180
Questions 188

Chapter 6: Input/Output Modules and Wiring   189

I/O Modules 190
Digital (Discrete) Modules 190
Analog Modules 199
Remote I/O Modules 202
Communication Modules 203
Position Control Modules 205
Vision Modules 208
Bar-Code Modules 210
Proportional, Integral and Derivative Modules (PID) 210
Fuzzy Logic Modules 217
Radio-Frequency Modules 218
Operator Input/Output Devices 219
Questions 223

Chapter 7: Arithmetic Instructions   225

Introduction 226
Allen-Bradley Instructions 227
GE Fanuc Functions 250
Gould Modicon Instructions 261
Omron Arithmetic Instructions 264
Square D Arithmetic Instructions 269
Texas Instruments and PLC Direct Arithmetic Instructions 274
Questions 277

Chapter 8: Advanced Programming   279

Sequential Control 280
Sequencer Instructions 282
Shift Resister Programming 285
Stage Programming 286
Step Programming 291
Fuzzy Logic 294
State Logic 302
Questions 307

Chapter 9: IEC 1131-3 Programming   **309**

   Overview of IEC 1131-3 310
   Structured Text Programming 312
   Function Block Diagram Programming 317
   Ladder Diagramming 320
   Instruction List Programming 322
   Sequential Function Chart Programming 325
   Questions 337

Chapter 10: Overview of Plant Floor Communication   **339**

   Introduction 340
   Levels of Plant Communication 340
   Questions 364

Chapter 11: Industrial Networks   **367**

   Overview of Industrial Networks 368
   Types of Industrial Buses 371
   Process Bus Standards 372
   Device Buses 372
   Process Buses 383
   CONTROLNETtm 386
   Integrating Networks 389
   Questions 392

Chapter 12: Supervisory Control and Data Acquisition   **393**

   Overview Of Supervisory Control and Data Acquisition (SCADA) 394
   Sample Application 397
   Application Development 410
   Questions 430

Chapter 13: PC-Based Control   **431**

   Introduction to PC-Based Control 432
   Flowchart Programming 436
   Application Development 438
   Gello 446
   IOworks 454

SoftPLC 456
Questions 459

Chapter 14: Industrial Automation Controllers   461

Overview of Industrial Automation Controllers 462
Industrial Automation Controller Hardware 471
Questions 479

Chapter 15: Single-Board Controllers   481

Overview 482
Single-Board Controllers 482
Simple Tank Application 490
Indexing, Sealing and Perforating Application 493
Questions 506

Chapter 16: Lockout/Tagout   507

Overview of Lockout/Tagout 508
Lockout 510
Tagout 511
Training 511
Requirements for Lockout/Tagout Devices 512
Application of Control 514
Sample Lockout Procedure 517
Sample Lockout/Tagout Checklist 520
Questions 521

Chapter 17: Installation and Troubleshooting   523

Installation and Troubleshooting 524
Installation 524
Industrial Controller Maintenance 536
PLC Troubleshooting 537
Questions 542

Appendix A: Internet Addresses   543

Appendix B: Common Electrical Symbols   545

Glossary   547

Preface

Programmable control has transformed manufacturing. There is a huge need for trained personnel who can program and integrate industrial controllers and devices.

I began writing this text for my students when I was unable to find a practical, affordable text that used a generic approach to various brands and types of industrial controllers.

Each chapter begins with a generic approach to the topic. Each topic is clearly explained through the use of common, easy to understand, generic examples. Examples are then shown for specific controllers. The brands covered are Allen-Bradley, Gould Modicon, Omron, PLC Direct by Koyo, and Square D Corporation. There are many illustrations and practical examples. Each chapter has objectives and questions. Some chapters have programming exercises. After completing the reading and exercises, the reader will be able to easily learn how to work with any new industrial controller. The text also covers new and emerging technologies such as IEC 1131 programming, industrial automation controllers, embedded controllers, supervisory control and data acquisition, fuzzy logic and radio frequency, as well as new programming techniques such as step, stage, and state logic programming. A chapter is also devoted to the topic of lockout/tagout. The book is not intended to replace the technical manual for the specific controller. The book does, however, explain the common instructions so that the reader will then be able to efficiently learn the use of new instructions from any technical manual.

Chapters 1 through 4 provide the basic foundation for the use of PLCs. Chapter 1 focuses on the history and fundamentals of the PLC. Chapter 2 covers number systems. Chapter 3 covers contacts, coils, and the fundamentals of programming. Chapter 4 focuses on timers, counters, and logical program development.

Chapter 5 covers industrial sensors and their wiring. The chapter focuses on types and uses of sensors. Sensors covered include optical, inductive, capacitive, encoders, resolvers, ultrasonic, and thermocouples. The wiring and practical application of sensors is stressed.

Chapter 6 covers I/O modules and wiring. The chapter covers digital and analog modules, communication modules, position control modules, barcode modules, radio frequency modules, fuzzy logic modules, speech modules, and others.

Chapter 7 covers the common arithmetic instructions including add, subtract, multiply, divide, and compare. In addition, logical operators, average, standard deviation, and number system conversion instructions are covered.

Chapter 8 covers advanced programming. Sequential logic, shift registers, step logic, stage logic, fuzzy logic, and state logic are all covered. Several of these programming techniques are very new, but will rapidly become prevalent.

Chapter 9 is an introduction to IEC 1131-3 programming methods. IEC 1131-3 is the new standard that covers programming languages for industrial controllers.

Chapter 10 covers plant communication. Device communications are sure to increase in importance as companies integrate their enterprises. This chapter provides a foundation for the integration of plant floor devices.

Chapter 11 looks at industrial networks. Industrial networks will radically change the way industry automates and integrates. Industrial networks will change the demands on technicians and maintenance personnel.

Chapter 12 covers supervisory control and data acquisition (SCADA) software. The fundamentals of SCADA are covered and then a practical SCADA application is developed.

Chapter 13 considers PC-based control. PCs are rapidly gaining acceptance as industrial controllers. This chapter looks at some of the various approaches that are being taken by software developers for PC-based control. PC-based hardware is also briefly covered.

Chapter 14 looks at industrial automation controllers. Industrial automation controllers are a new breed of controller aimed at special applications such as motion control. These controllers are very powerful and easy to program to develop machine and motion control applications.

Chapter 15 examines single-board controllers, also called embedded controllers. Single-board controllers are microprocessor-based controllers that are generally programmed in C, BASIC, or assembly language. The programming and use of a C-based controller is then examined.

Chapter 16 is an in-depth look at lockout/tagout procedures for personnel who are working in industrial environments.

Chapter 17 focuses on the installation and troubleshooting of PLC systems. The chapter begins with a discussion of cabinets, wiring, grounding, and noise. The chapter then covers troubleshooting. This chapter will provide the fundamental groundwork for proper installation and troubleshooting of integrated systems.

Appendix A provides Internet addresses for various manufacturers that are covered in the book.

Appendix B shows some common I/O device symbols.

I appreciate the suggestions of the reviewer for this edition, Robert F. Wolf of Waukesha County Technical College.

Acknowledgments

This text would not have been possible without the help I received from corporations, companies, and individuals. We would especially like to thank the following companies for their assistance.

Allen-Bradley

Control Technology Corporation

Omron Electronics Inc.

PLC Direct by Koyo

Z-World Inc.

Wonderware Inc.

Special thanks to Tim Sevener of Control Technology Corporation for his assistance.

Chapter 1

Overview of Programmable Logic Controllers

In a very short period of time, programmable logic controllers (PLCs) have become an integral and invaluable tool in industry. In this chapter we will examine how and why PLCs have gained such wide application. We will also take an overall look at what a PLC is.

OBJECTIVES

Upon completion of this chapter, you will be able to:

> *Explain some of the reasons why PLCs are replacing hardwired logic in industrial automation.*

> *Explain such terms as* **ladder logic, CPU, programmer, input devices,** *and* **output devices.**

> *Explain some of the features of a PLC that make it an easy tool for an electrician to use.*

> *Explain how the PLC is protected from outside disturbances.*

> *Draw a block diagram of a PLC.*

> *Explain the types of programming devices available.*

HISTORY OF PLCS

The programmable logic controller may be the best example ever of taking an existing technology and applying it to meet a need. In the 1960s and 1970s, industry was beginning to see the need for automation. Industry saw the need to improve quality and increase productivity. Flexibility had also become a major concern. Industry needed to be able to change processes quickly to meet the needs of the consumer.

The Old Way

Imagine an automated manufacturing line in the 1960s and 1970s. There was always a huge wiring panel to control the system. The wiring panel could cover an entire wall. Inside the panel were masses of electromechanical relays. These relays were all hardwired together to make the system work. Hardwiring means that an electrician had to install wires between the connections of the relays. An engineer would design the logic of the system and the electricians would be given a blueprint of the logic and would have to wire the components together. There were hundreds of electromechanical relays in a system before programmable logic controllers were developed.

The drawing that the electrician was given was called a ladder diagram. They were called ladders because they resemble a ladder in appearance. The ladder showed all the switches, sensors, motors, valves, relays, etc., that would be in the system. It was the electrician's job to wire them all together (see Figure 1-1).

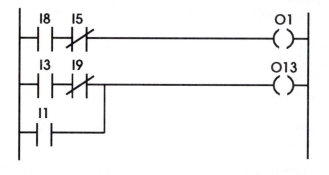

Figure 1-1. PLC ladder diagram. Note the similarity in appearance to a ladder. There are three rungs in this example. Power is represented by the left and right uprights. There are five inputs and two outputs in this example. Inputs are on the left, outputs at the right of each rung.

It is not hard to imagine that the engineer made a few small errors in his/her design. It is also conceivable that the electrician may have made a few errors in wiring the system. It is also not hard to imagine a few bad components. The only way to see if everything was correct was to run the system. Systems are normally not perfect on the first attempt. Troubleshooting was done by running the actual system. This was a very time-consuming process. You must also remember that no product could be manufactured while the wiring was being changed. The system had to be disabled for wiring changes. This means that all of the production personnel associated with that production line were without work until the system was repaired. After the electrician had completed the troubleshooting and repair, the system was ready for production.

Disadvantages of the Old Way

One of the problems with this type of control is that it is based on mechanical relays. Mechanical devices are usually the weakest link in systems. Mechanical devices have moving parts that can wear out. If one relay failed, the electrician might have to troubleshoot the whole system again. The system was down again until the problem was found and corrected.

Another problem with hardwired logic is that if a change must be made, the system must be shut down and the panel rewired. If a company decided to change the sequence of operations (even a minor change), it was a major expense and loss of production time while the system was not producing parts.

The First Programmable Controllers

General Motors saw the need for a replacement for hardwired control panels. Increased competition forced the automakers to improve manufacturing performance in both quality and productivity. Flexibility, rapid changeover, and reduced downtime became very important.

GM realized that a computer could be used for logic instead of hardwired relays. The computer could take the place of the huge, costly, inflexible, hardwired control panels. If changes in the system logic or sequence of operations were needed, the program in the computer could be changed instead of rewiring. Imagine eliminating all the downtime associated with wiring changes. Imagine being able to completely change how a system operated by simply changing the software in a computer.

The problem was how to get an electrician to accept and use a computer. Systems are often very complex and require complex programming. It was out of the question to ask plant electricians to learn and use a computer language in addition to their other duties.

GM's Hydromatic Division realized the need and acted. In 1968, Hydromatic wrote the design criteria for the first programmable logic controller. *Note*: There were companies already selling devices that performed industrial control, but they were simple sequencing controllers—not PLCs as we think of them today (see Figure 1-2).

The specifications required that the new device:

be price competitive with the relay systems it was intended to replace;

be a solid-state device (electronic, not mechanically operated);

have easily replaceable input and output devices;

have the flexibility of a computer;

be able to function in an industrial environment (vibration, heat, dirt, etc.);

be modular so that subassemblies could be easily removed for repair or replacement;

be reprogrammable so it could be used for other tasks (reusable);

be easily programmed and maintained by plant electricians and technicians.

GM solicited interested companies and encouraged them to develop a device to meet the design specifications.

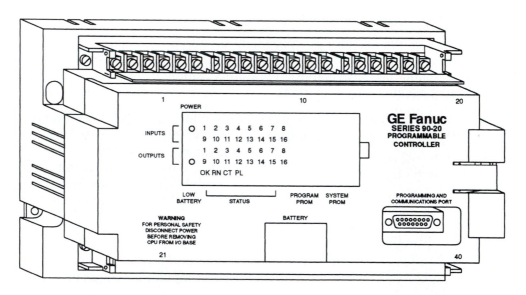

Figure 1-2. A drawing of a GE Fanuc Series 90-20 PLC. Notice the input and output LEDs on the front panel. Notice the status indicators also. Courtesy GE Fanuc Inc.

Figure 1-3. A computer being used to troubleshoot a system. Courtesy of Reiko Iwao.

Bedford Associates (Bedford, MA) proposed something called a Modular Digital Controller (MODICON) to a major US car manufacturer. Other companies at the time proposed computer-based schemes, one of which was based upon the PDP-8. Products were being developed to meet these specifications in 1968, and in 1969 PLC products were available. The MODICON 084 brought the world's first PLC into commercial production. Gould Modicon developed the first device that met the specifications. The key to the programmable logic controllers' success is that the device is programmed using the same language that electricians already used: ladder logic. Electricians and technicians could very easily understand these new devices because the logic looked similar to the old logic they had always worked with. They did not have to learn a new programming language. In fact, the ladder looks almost like the electrical diagrams they used to troubleshoot the systems.

PLCs made troubleshooting a very easy task also. All of the inputs and outputs could be viewed on the screen of the programming unit (see Figure 1-3). Many PLCs also provided I/O and status LEDs on the front of the PLC or modules to assist in troubleshooting. The problem in the system could be found very quickly. If the process needed to be changed, the new ladder could be written off-line. The system could be running while the electrician was making changes in the ladder diagram. The system would only have to be shut down for a few minutes while the new ladder (logic) was downloaded to the PLC (no rewiring).

The ease of change is very important in manufacturing. This capability allows for rapid production or model change. This even makes it possible to customize products through a system to meet individual consumers' needs. Just-in-time and other new manufacturing techniques require smaller lot sizes of production, which mean more changeovers in production. Speed and flexibility are crucial in this environment. PLCs are the perfect control device for this need. The program in the PLC (ladder diagram) controls how the PLC runs the manufacturing process. If the program is changed, the PLC can completely change the process. Today there are many makers of PLCs. See Figures 1-4 and 1-5 for examples of PLCs.

Figure 1-4. Two sizes of GE Fanuc PLCs. Courtesy GE Fanuc Inc.

PLCs were originally called PCs (programmable controllers). This caused some confusion when personal computers became prevalent. To avoid confusion PCs became personal computers and programmable controllers became PLCs (programmable logic controllers). The original PLCs were simple devices. They were simply on/off control devices. They could take input from devices such as switches, digital sensors, etc., and turn output devices on or off. They were very appropriate for simple relay replacement applications. They were not well suited for complex control of temperatures, positions, pressures, etc. Since the early days, manufacturers of PLCs have added numerous features and enhancements. PLCs have also been given the capability to handle extremely complex tasks such as position control, process control, and other difficult applications. Their speed of operation has drastically improved. Their ease of programming has also improved. Special-purpose modules have been developed for such applications as radio frequency communications, vision inspection, and even speech. It would be difficult to imagine a task that a PLC could not handle.

Figure 1-5. An Allen-Bradley SLC 500 PLC and modules.
Courtesy of Rockwell Automation/Allen-Bradley Company Inc.

PLC COMPONENTS

The PLC is really an industrial computer in which the hardware and software have been specifically adapted to the industrial environment and the electrical technician. Figure 1-6 shows the functional components of a typical PLC. Note the similarity to a computer.

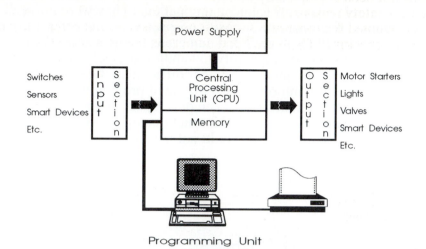

Figure 1-6. Block diagram of the typical components that make up a PLC. Note particularly the input section, the output section, and the central processing unit (CPU). The broken arrows from the input section to the CPU and from the CPU to the output section represent protection that is necessary to isolate the CPU from the real-world inputs and outputs. The programming unit is used to write the control program (ladder logic) for the CPU. It is also used for documentation of the programmer's logic and troubleshooting the system.

Central Processing Unit

The central processing unit (CPU) is the brain of the PLC (see Figure 1-7). It contains one or more microprocessors to control the PLC. The CPU also handles the communication and interaction with the other components of the system. The CPU contains the same type of microprocessor that one could find in a microcomputer. The difference is that the program that is used with the microprocessor is just written to accommodate ladder logic instead of other programming languages. The CPU executes the operating system, manages memory, monitors inputs, evaluates the user logic (ladder diagram), and turns on the appropriate outputs.

The factory floor is a very noisy environment. The big problem is the electrical noise that motors, motor starters, wiring, welding machines, and even fluorescent lights create. PLCs are hardened to be noise immune.

PLCs also have elaborate memory-checking routines to be sure that the PLC memory has not been corrupted by noise or other problems. Memory checking is undertaken for safety reasons. It helps assure that the PLC will not execute if memory is corrupted for some reason. Most computers do not offer noise hardening or memory checking. There are a few industrial computers that do.

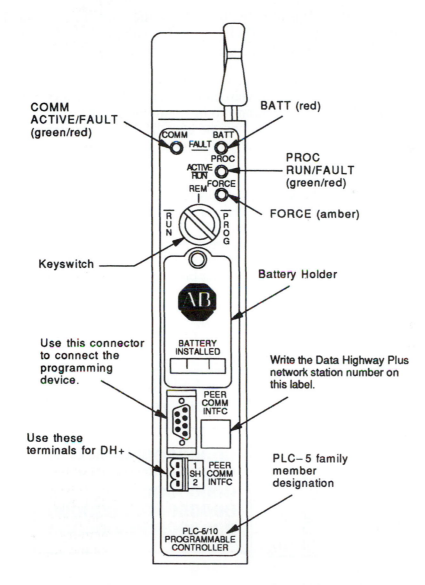

Figure 1-7. PLC-5/10 processor module (CPU). Note the many indicators for error checking. Also note the keyswitch for switching modes between run and program mode. Courtesy of Rockwell Automation/Allen-Bradley Company Inc.

Memory

PLC memory can be of various types. Some of the PLC memory is used to hold system memory and some is used to hold user memory.

Operating System Memory

Read-Only Memory

Read-only memory (ROM) is used by the PLC for the operating system. The operating system is burned into ROM by the PLC manufacturer. The operating system controls functions such as the system software that the user uses to program the PLC. The ladder logic that the programmer creates is a high-level language. A high-level language is a computer language that makes it easy for people to program. The system software must convert the electrician's ladder diagram (high-level language program) to instructions that the microprocessor can understand. ROM is not changed by the user. ROM is nonvolatile memory, which means that even if the electricity is shut off, the data in memory is retained.

User Memory

The memory of a PLC is broken into blocks that have specific functions. Some sections of memory are used to store the status of inputs and outputs. (*Note*: Input/output is typically represented as I/O.) These are normally called I/O image tables. The states of inputs and outputs are kept in I/O image tables. The real-world state of an input is stored as either a 1 or a 0 in a particular bit of memory. Each input or output has one corresponding bit in memory (see Figures 1-8 and 1-9).

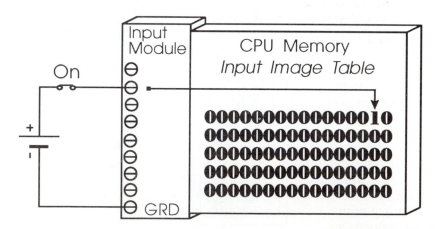

Figure 1-8. How the status of a real-world input becomes a 1 or a 0 in a word of memory. Each bit in the input image table represents the status of one real-world input.

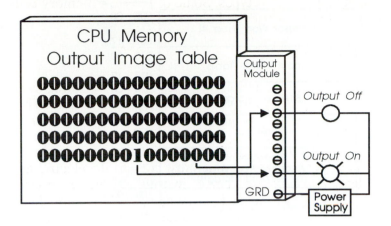

Figure 1-9. How a bit in memory controls one output. If the bit is a 1, the output will be on. If the bit is a 0, the output will be off. (This is active-high logic.)

Other portions of the memory are used to store the contents of variables that are used in a user program. For example, a timer or counter value would be stored in this portion of memory. Memory is also reserved for processor work areas.

Memory Maps
Diagrams that show the layout, uses, and location (addresses) of memory are called memory maps. Figure 1-10 shows an example of an Allen-Bradley memory map for a PLC-2.

Random Access Memory
Random access memory (RAM) is designed so that the user can read or write to the memory. Ram is commonly used for user memory. The user's program, timer/counter values, input/output status, etc., are stored in RAM.

RAM is volatile, which means that if the electricity is shut off, the data in memory is lost. This problem is solved by the use of a lithium battery. The battery takes over when the PLC is shut off. Most PLCs use CMOS-RAM technology for user memory. CMOS-RAM chips have very low current draw and can maintain memory with a lithium battery for an extended period of time, two to five years in many cases.

Figure 1-11 shows an example of battery replacement. Notice that in this case the manufacturer provided two battery connections so that the new battery can be connected without powering down the system. The battery being replaced can then be removed. This minimizes the chance of losing data.

	Word Address ➙ ☐	Bit Address
Processor Work Area #1	000 00 ↓ 007 17	
Output Image Table	010 00 ↓ 017 17	
Bit/Word Storage	020 00 ↓ 026 17	
Reserved	027	
Timer/Counter Accumulated Values (or Bit/Word Storage)	030 00 ↓ 077 17	
Processor Work area #2	100 00 ↓ 107 17	
Input Image Table	110 00 ↓ 117 17	
Bit/Word Storage	120 00 ↓ 127 17	
Timer/Counter Preset Values	130 00 ↓ 177 17	
Expanded Data Table and/or User Program	200 00 ↓ End of Memory	

Figure 1-10. The memory map for an Allen-Bradley PLC-2. This is the factory configuration. Note that the addresses are in octal.

Processor work area is used by the processor for its internal operations. These addresses are not available to the programmer for addressing. Note that there are two work areas in this memory map.

Timer/counter accumulated value storage is used to store the present accumulated values for timers and counters.

Timer/counter preset value storage is used to store the user preset values for timers and counters. If the user wants a counter to count up to five this is the area where the value five is stored.

The input /output image table is an area of memory dedicated to I/O data. During every scan each input controls one bit in the input image table. Individual outputs are also controlled by one bit in the output image table.

Electrically Erasable Programmable Read-Only Memory
Electrically erasable programmable read-only memory (EEPROM) can function
in almost the same manner as RAM. The EEPROM can be erased electrically
instead of by ultraviolet light (see Figure 1-12). The EEPROM is also nonvolatile
memory, so it does not require battery backup. EEPROM modules are available
for many PLCs today. They are often small cartridges that can store several
thousand bytes of memory (see Figure 1-13). They can be used to store user
programs. The program is copied to user RAM for execution of the ladder.

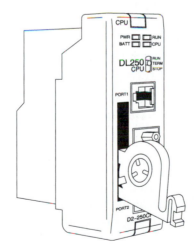

*Figure 1-11. This figure shows how the battery would be
replaced in a PLC Direct 205 PLC. The photo at the top shows
how small and yet incredibly powerful the PLC is. The diagram on
the bottom shows how the battery is changed. Courtesy PLC Direct
by Koyo.*

Figure 1-12. Two kinds of memory: RAM and EPROM. Note the window on the EPROM memory chip. If the window is exposed to ultraviolet light, the memory is erased.

Figure 1-13. Left: A memory cartridge for a Siemens Industrial Automation, Inc. PLC. Right: A SRAM (static random access memory) card for an Omron PLC. Courtesy of Omron Electronics and Siemens Industrial Automation, Inc.

PLC Programming Devices

There are many devices that are used to program PLCs. These devices do not need to be attached to the PLC once the ladder is written. The devices are just used to write the user program for the PLC. They may also be used to trouble-shoot the PLC.

Dumb Terminal

A dumb terminal is a device that has a keyboard and a monitor. There is no intelligence in the device. This requires that all of the brains for operation of the PLC and for the programming system be located in the PLC. The dumb terminal is just used to send information to the PLC and display the information that is returned from the PLC. A dumb terminal sends out the ASCII equivalent of whichever key was pressed. The output from this device is sent by serial communication to the CPU. The CPU can also send ASCII information back to the dumb terminal. The main advantage of the device is that it is inexpensive and can be used with a variety of devices.

Dumb terminals are not widely used anymore because they cannot upload/download or store programs. Users have also demanded more in the way of documentation and troubleshooting, so the dumb terminal seems doomed for most applications.

Dedicated Industrial Terminals

These are terminals that have built-in intelligence. They are dedicated to one brand of PLC (see Figure 1-14). In fact, many are dedicated to only a few models of one brand. In many cases, they must be attached to the PLC to be able to program (on-line programming). Some dedicated terminals allow off-line programming. "Off-line" means that the program can be written without being connected to the PLC. The ladder can then be downloaded to the PLC. Dedicated devices can be used to troubleshoot ladder logic while the PLC is running. They can force inputs and outputs on or off for troubleshooting.

Figure 1-14. Allen-Bradley programming terminal.

The big disadvantage of these terminals is that one is required for each different brand of PLC a company might have. They are also quite expensive. For these reasons they are becoming less prevalent.

Hand-Held Programmers

Hand-held programmers are often used to program small PLCs (see Figure 1-15). They are inexpensive and easy to use. Handheld programmers are basically dumb devices also. They must be attached to a PLC to be used. They are handy for troubleshooting. They can easily be carried out to the manufacturing system and plugged into the PLC. Once plugged in they can be used to monitor the status of inputs, outputs, variables, counters, timers, etc. This eliminates the need to carry a large programming device out onto the factory floor. Handheld programmers can also be used to force inputs and outputs on or off for troubleshooting. Handheld programmers are designed for the factory floor. They typically have membrane keypads that are immune to the contaminants in the factory environment.

One disadvantage is that these programmers cannot show very much of a ladder on the screen at one time. Small operator interfaces are also available to allow input/output to/from PLCs (see Figure 1-16).

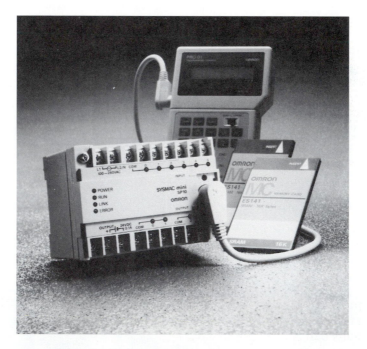

Figure 1-15. A hand-held programmer. Note the SRAM memory card. Courtesy of Omron Electronics.

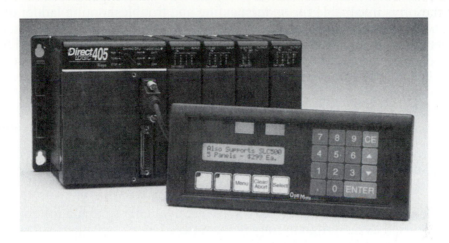

Figure 1-16. Operator interface. Courtesy PLC Direct by Koyo.

Microcomputers

The microcomputer is rapidly becoming the most commonly used programming device. The same microcomputer can program any brand of PLC that has software available for it. This means that a microcomputer can be used to program virtually any PLC. The microcomputer can also be used for off-line programming and storage of programs. One disk can hold many ladders. The microcomputer can also upload and download programs to a PLC. The microcomputer can also force inputs and outputs on and off.

This upload/download capability is vital for industry. Occasionally, PLC programs are modified on the factory floor to get a system up and running for a short period of time. It is vital that once the system has been repaired the correct program is reloaded into the PLC. It is also useful to verify from time to time that the program in the PLC has not been modified. This can help to avoid dangerous situations on the factory floor. Some automobile manufacturers have set up communications networks that regularly verify the programs in PLCs to assure that they are correct.

The microcomputer can also be used to document the PLC program. Notes for technicians can be added and the ladder can be output to a printer for hardcopy so that the technicians can study the ladder diagram.

DirectSOFT is an example of Windows-based microcomputer software for programming PLCs (see Figure 1-17). DirectSOFT is a powerful software package that allows off-line and on-line programming. It allows ladder diagrams to be stored to a floppy disk, uploaded/downloaded from/to the PLC. The software allows monitoring the operation of the ladder while it is executing. The software allows forcing system inputs/outputs (I/O) on and off. This is extremely valuable for troubleshooting. It also allows the programmer to document the ladder. This documentation is invaluable for understanding and troubleshooting ladder diagrams. The programmer can add notes, names of input or output devices, and comments that may be useful for troubleshooting and maintenance. The addition of notes and comments allows any technician to readily understand the ladder diagram. This allows any technician to troubleshoot the system, not just the person who developed it. The notes and/or comments could even specify replacement part numbers if so desired. This would facilitate rapid repair of any problems due to faulty parts.

The old way was that the person who developed the system had great job security because no one else could understand what had been done. A properly documented ladder allows any technician to understand it.

All of the leading brands of PLCs have software available so that a microcomputer can be used as the programming device. PLC Direct calls their software DirectSOFT. Square D calls their programming software SY/MATE. Omron calls their programming software SYSMATE.

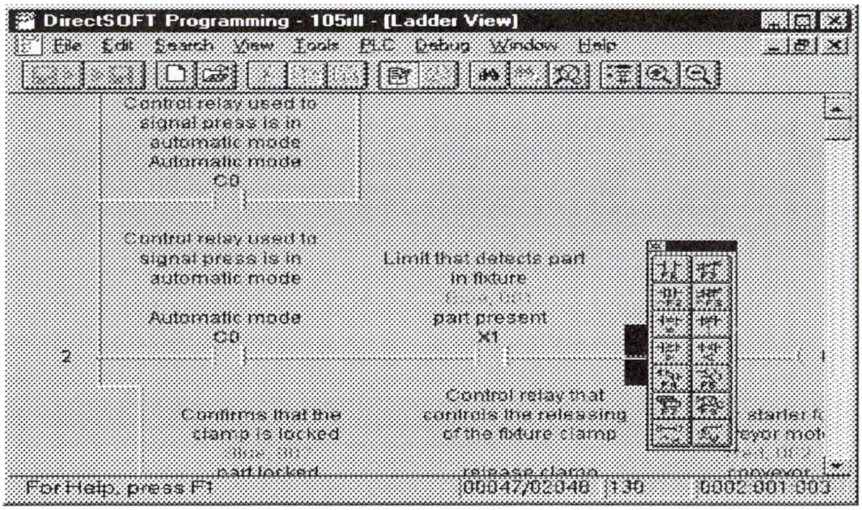

Figure 1-17. Typical PLC programming software. Note the comments and tagnames to make the ladder logic more understandable. Courtesy PLC Direct by Koyo.

IEC 1131-3 Programming
The IEC has developed a standard for PLC programming. The latest IEC standard (IEC 1131-3) has attempted to merge PLC programming languages under one international standard. PLCs can now be programmed in function block diagrams,

instruction lists, C and structured text. The standard is accepted by an increasing number of suppliers and vendors of process control systems, safety-related systems, industrial personal computers, etc. An increasing number of application software vendors offer products based on IEC 1131-3.

Power Supply

The power supply is used to supply power for the central processing unit. Most PLCs operate on 115 VAC. This means that the input voltage to the power supply is 115 VAC. The power supply provides various DC voltages for the PLC components and CPU. On some PLCs the power supply is a separate module. This is usually the case when extra racks are used. Each rack must have its own power supply.

The user must determine how much current will be drawn by the I/O modules to ensure that the power supply will supply adequate current. Different types of modules draw different amounts of current. *Note*: The PLC power supply is not typically used to power external inputs or outputs. The user must provide separate power supplies to power the inputs and outputs of the PLC. Some of the smaller PLCs do supply voltage to be used to power the inputs, however.

Input Section

The input portion of the PLC performs two vital tasks. It takes inputs from the outside world and protects the CPU from the outside world. Inputs can be almost any device. The input module converts the real-world logic level to the logic level required by the CPU. For example, a 250-VAC input module would convert a 250-VAC input to a low level DC signal for the CPU.

Common input devices would include switches, sensors, etc. These are often called field devices. Field devices are gaining extensive capability, especially the ability to communicate over industrial communications networks. Other smart devices, such as robots, computers, and even PLCs, can act as inputs to the PLC.

The inputs are provided through the use of input modules. The user simply chooses input modules that will meet the needs of the application. These modules are installed in the PLC rack (see Figure 1-18).

The PLC rack serves several functions. It is used to physically hold the CPU, power supply, and I/O modules. The rack also provides the electrical connections and communications between the modules, power supply, and CPU through the backplane.

The modules are plugged into slots on the rack (see Figure 1-19). This ability to plug modules in and out easily is one of the reasons PLCs are so popular. The ability to change modules quickly allows very rapid maintenance and repair.

Input/output numbering is a result of which slot the module is plugged into. Omron calls their slots channels. The combination of the channel number and actual input or output point on the module determines the number (see Figure 1-20).

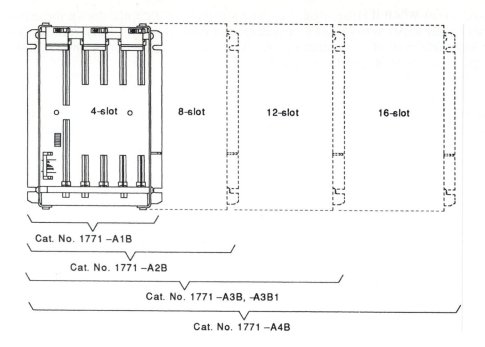

4-slot   8-slot   12-slot   16-slot

Cat. No. 1771 –A1B

Cat. No. 1771 –A2B

Cat. No. 1771 –A3B, –A3B1

Cat. No. 1771 –A4B

Figure 1-18. Various rack sizes. Courtesy of Rockwell Automation/Allen-Bradley Company Inc.

Figure 1-19. Rack filled with modules. Courtesy GE Fanuc Inc.

There are cases when it is necessary to have more than one rack. Some medium-sized and large PLCs allow more than one rack. There are applications that have more I/O points than one rack can handle. There are also some cases when it is desirable to locate some of the I/O away from the PLC. For example, imagine a very large machine. Rather than run wires from every input and output to the PLC, an extra rack might be used. The I/O on one end of the machine is wired to modules in the remote rack. The I/O on the other end of the machine is wired to the main rack. The two racks communicate with one set of wiring rather than running all of the wiring from one end to the other.

I/O Channel Number + Bit Number (0-15)			I/O Address
Channel 150, bit 3	150	03	15003
Channel 1, bit 12	001	12	00112

Figure 1-20. How I/O is numbered in the Omron C200H PLC. Two examples are shown. If bit number 3 needs to be used on channel number 150, the I/O number would be 15003. If bit number 12 needs to be used on channel 1, the I/O number would be 00112.

Many PLCs allow the use of multiple racks (see Figure 1-21). When more than one rack is used it is necessary to identify which rack the I/O is in. Omron handles this through the channel number. Allen-Bradley requires that the I/O number include the rack number (see Figures 1-22 to 1-25).

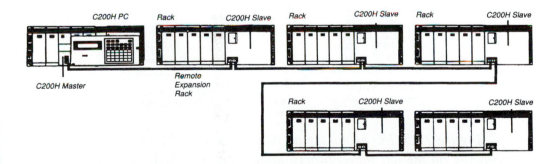

Figure 1-21. Use of multiple racks. In this case five additional racks were connected to the main PLC. This allows more I/O to be used. It also allows for remote mounting of the racks. Courtesy of Omron Electronics.

112
07

Figure 1-22. Numbering of an Allen-Bradley input. This input would correspond to input terminal 7 of group 2 of rack 1. The first 1 means that it is an input. Courtesy of Rockwell Automation/Allen-Bradley Company Inc.

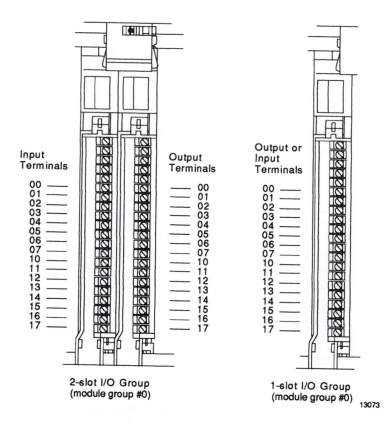

Input
Terminals

00
01
02
03
04
05
06
07
10
11
12
13
14
15
16
17

Output
Terminals

00
01
02
03
04
05
06
07
10
11
12
13
14
15
16
17

Output or
Input
Terminals

00
01
02
03
04
05
06
07
10
11
12
13
14
15
16
17

2-slot I/O Group
(module group #0)

1-slot I/O Group
(module group #0)   13073

Figure 1-23. I/O group for an Allen-Bradley PLC-5. The drawing on the left shows two 16-point modules which comprise one group. They are in module group 0 in this case. One word will be used for the 16 inputs and one word will be used for the 16 outputs. Note the octal numbering. The drawing on the right shows a one-slot I/O group (module group 0). An I/O group can contain up to 16 input terminals and 16 output terminals and can occupy 1/2, one, or two slots. Courtesy of Rockwell Automation/Allen-Bradley Company Inc.

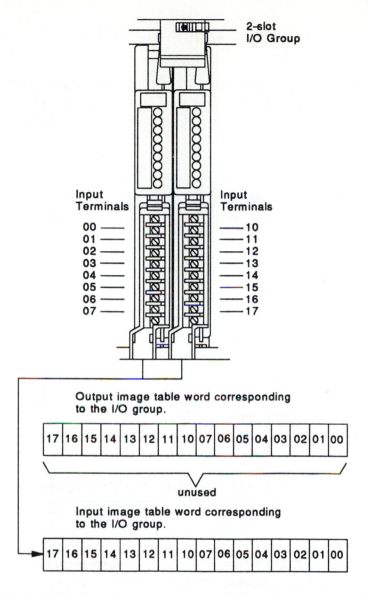

Figure 1-24. Two 8-point input modules using one word of the input image table for a PLC-5. Note that the inputs are numbered 00 to 17. Remember that octal numbering is used. This is 16 inputs or one word. Note that two slots are used to form this one I/O group. Courtesy of Rockwell Automation/Allen-Bradley Company Inc.

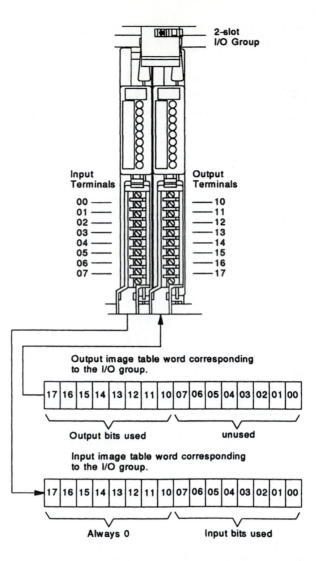

Figure 1-25. Use of one 8-point input module and one 8-point output PLC-5 module. In this case two slots are used to create one group. Because input and output modules have both been used in this group, two words of memory must be used. The first eight bits of the input image table word are used for inputs 00-07 and the last eight bits of the word from the output image table are used for outputs 10 to 17. If two 16-point modules are used, one must be an input module and one must be an output module. In that case both modules use a full word of memory. Courtesy of Rockwell Automation/Allen-Bradley Company Inc.

Optical Isolation

The other task that the input section of a PLC performs is isolation. The PLC CPU must be protected from the outside world and at the same time be able to take input data from the outside world. This is typically done by opto-isolation (see Figure 1-26). "Opto-isolation" is short for "optical isolation." This means that there is no electrical connection between the outside world and the CPU. The two are separated optically. The outside world supplies a signal which turns on a light in the input card. The light shines on a receiver and the receiver turns on. There is no electrical connection between the two.

The light separates the CPU from the outside world up to very high voltages. Even if there was a large surge of electricity, the CPU would be safe. (Of course, if the voltage is too large, the opto-isolator can fail and may cause a circuit failure.) Optical isolation is used for inputs and outputs.

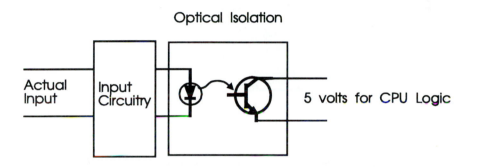

Figure 1-26. Typical optical isolation circuit. The arrow represents the fact that only light travels between the input circuitry and the CPU circuitry. There is no electrical connection.

Input modules provide the user with various troubleshooting aids. There are normally light emitting diodes (LEDs) for each input. If the input is on, the CPU should see the input as a high (or a 1).

Input modules also provide circuits that debounce the input signal. Many input devices are mechanical and have contacts. When these devices close or open there are unwanted "bounces" that close and open the contacts. Debounce circuits make sure that the CPU sees only debounced signals. The debounce circuit also helps eliminate the possibility of electrical noise from firing the inputs.

Inputs to Input Modules

Sensors are commonly used as inputs to PLCs. Sensors can be purchased for a variety of purposes. They can sense part presence, count pieces, measure temperature, pressure, or size, sense for proper packaging, and so on.

There are also sensors that are able to sense any type of material. Inductive sensors can sense ferrous metal objects, capacitive sensors can sense almost any material, and optical sensors can detect any type of material.

Other devices can also act as inputs to a PLC. Smart devices such as robots, computers, vision systems, etc., often have the ability to send signals to a PLC's input modules (see Figure 1-27). This can be used for handshaking during operation. A robot, for example, can send the PLC an input when it has finished a program.

Sensor types and use are covered in detail in Chapter 5.

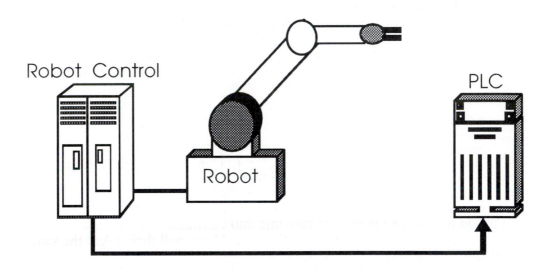

Figure 1-27. A "smart" device can also act as an input device to a PLC. The PLC can also output to the robot. Robots and some other devices typically have a few digital outputs and inputs available for this purpose. The use of these inputs and outputs allows for some basic handshaking between devices. Handshaking means that the devices give each other permission to perform tasks at some times during execution to assure proper performance and safety. When devices communicate with a digital signal it is called primitive communication.

Output Section

The output section of the PLC provides the connection to real-world output devices. The output devices might be motor starters, lights, coils, valves, etc. These are often called field devices. Field devices are gaining extensive capability, especially the ability to communicate over industrial communications networks. Field devices can be either input or output devices. Output modules can be purchased to handle DC or AC voltages. They can be used to output analog or digital signals. A digital output module acts like a switch. The output is either energized or deenergized. If the output is energized, the output is turned on, just like a switch.

The analog output module is used to output an analog signal. An example of this is a motor whose velocity we would like to control. An analog module puts out a voltage that corresponds to the desired speed.

Output modules can be purchased with various output configurations. They are available as modules with 8, 16, and 32 outputs. Modules with more than eight outputs are sometimes called high-density modules. They are generally the same size as the eight-output modules. They just have many more components within the module. For that reason high-density modules will not handle as much current for each output. This is because of the size of components and the heat generated by them.

Current Ratings

Module specifications will list an overall current rating and an output current rating. For example, the specification may give each output a current limit of 1 ampere. If there are eight outputs, we would assume that the output module overall rating would be 8A. This is a poor leap in logic. The overall rating of the module current will normally be less than the total of the individuals. The overall rating might be 6A. The user must take this into consideration when planning the system. Normally, each of the eight devices would not pull their 1A at the same time. Figure 1-28 shows an example of I/O wiring.

Output Image Table

The output image table is a part of CPU memory (see Figure 1-29). The user's logic determines whether an output should be on or off. The CPU evaluates the user's ladder logic. If it determines that an output should be on it stores a one in the bit that corresponds to that output. The one in the output image table is used to turn on the actual output through an isolation circuit (see Figure 1-30).

The outputs of small PLCs are often relays. This allows the user to mix and match output voltages or types. For example, some outputs could then be AC and some DC. There are relay output modules available for some of the larger PLCs too. The other choices are transistors for DC outputs and triacs for AC outputs. There are many types of field devices that can be connected to outputs. Figure 1-31 shows a few examples.

Output modules are covered in greater detail in Chapter 6.

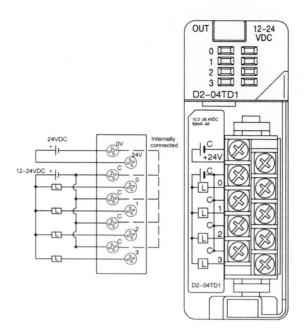

Figure 1-28. PLC Direct I/O wiring courtesy PLC Direct.

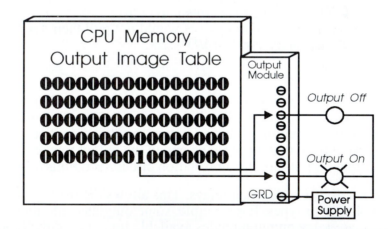

Figure 1-29. How a typical PLC handles outputs. The CPU memory contains a section called the output image table. The output image table contains the desired states of all outputs. If there is a 1 in the bit that corresponds to the output, the output is turned on. If there is a 0, the output is turned off. This is called active-high logic.

Chapter 1: Overview of Programmable Logic Controllers

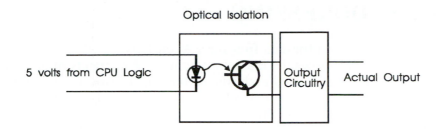

Optical Isolation

5 volts from CPU Logic

Output Circuitry

Actual Output

Figure 1-30. How PLC output isolation works. The CPU provides a five-volt signal which turns on the LED. The light from the LED is used to fire the base of the output transistor. There is no electrical connection between the CPU and the outside world.

Figure 1-31. A few possible output devices including a contactor, AC motor, starter and a valve.

ALLEN-BRADLEY MEMORY ORGANIZATION AND I/O ADDRESSING

The memory organization for data files for a Allen-Bradley PLC-5 is shown in Figure 1-32. The chart shows the maximum number of elements of each type, the file type, the number that is used to designate the type of file, and the number of words per element. The programmer designates the type of file according to the type of data that will be stored in it.

Maximum # of Elements	File Type	File #	Words/element
32*	Output Image	0	1
32*	Input Image	1	1
32	Status	2	1
1000	Bit	3	1 (16 bits)
1000	Timer	4	3
1000	Counter	5	3
1000	Control	6	3
1000	Integer	7	1
1000	Floating Point	8	2
1000	File Type as Needed	9-999	
* Up to 64 for the PLC-5/25 processor			

Figure 1-32. Memory organization for a PLC-5.

Most files use one-word elements. For these each bit can be addressed individually. The range of values that can be stored in one word of memory is -32,768 to +32,767. Floating-point files use two words of memory. The floating-point element must be addressed as a whole. Individual bits cannot be accessed. Timer and counter files use three words of memory. The words (or elements) are used to store the status bits, .PRE (preset) values, and .ACC (accumulated) values. These are covered in more detail later.

Chapter 1: Overview of Programmable Logic Controllers

LOGICAL ADDRESSING

The Allen-Bradley (AB) system of addressing allows you to address bits, elements (words), or data files. The format varies slightly depending on the type of file (see Figure 1-33). There are occasions when the status of a particular bit must be accessed. There are also occasions when a whole word must be accessed.

#Xf:e.s/b

 # File-address identifier

X File Type

B=bit	C=Counter
F=Floating point	I=Input
N=Integer	O=Output
R=Control	S=Status
T=Timer	A=ASCII-For display only
D=BCD	

File Number

0=Output	1=Input
2=Status	3=Bit
4=Timer	5=Counter
6=Control	7=Integer
8=Floating point	9-999 for additional files

: Separates file and element numbers

e Element number

0-37 Octal for I/O files
0-31 Decimal for the status file
0-999 Decimal for all other file types

s Use only with counter, timer, and control files

.PRE(preset) .ACC(accumulated)	.LEN(length) .POS(position)

/ Delimiter separates bit numbers from element or subelement numbers

b Bit number

0-17 Octal for I/O files	0-15 for all other file types

Figure 1-33. The addressing scheme for AB PLCs.

I/O ADDRESSING

I/O addressing varies from the general format. Figure 1-34 shows the method in which I/Os are addressed. I/Os are addressed by the programmer specifying an I or an O for input or output, the rack number, the I/O group number, and the terminal (bit) number.

O:rg/00-17   I:rg/00-17

I = Input

O = Output

r = Assigned Rack Number

g = I/O Group Number

00-17 = Bit (terminal) Number

Figure 1-34. How AB I/O addressing is done.

Status files are addressed by the programmer specifying the element number and the bit number (see Figure 1-35). Bits or elements (words) can be addressed. Note that individual bits can be addressed and used in ladder logic for most file types (see Figure 1-36).

S:e/b

S = Status

e = Element Number (0-31)

b = Bit Number (0-15)

Figure 1-35. How AB status files are addressed.

B3:15/0	This is binary file 3, element 15, bit 0.
I:02/7	This is input 7 of I/O group 2, rack 0.
C5:6.DN	This is counter file 5, element 6, bit 13 (bit 13 is the done bit).
N7:3/4	This is the bit address in an integer file. It is integer file 7, element 3, bit number 4.

Figure 1-36. Allen-Bradley bit addressing examples.

It is often desirable to access a whole element (word) rather than a single bit. For example, if the programmer needed to access the accumulated count of a counter or the accumulated time of a timer to make a logic decision, a whole element would be accessed. Elements (words) can be accessed by specifying the file type and element to be addressed. Figure 1-37 shows examples of element addressing.

S:3	This is the format for addressing element 3 in a status file.
N7:3	This is the format for addressing element 3 in integer file number 7.
F8:53	This is the format for addressing element 53 in floating point file number 8.

Figure 1-37. Examples of Allen-Bradley element addressing.

GE FANUC MEMORY ORGANIZATION AND I/O ADDRESSING

GE Fanuc uses the following letters to represent different types of contacts, coils and memory. Real-world, discrete inputs are identified as %I. %I0003 would be the name of real-world input 3 in a ladder diagram. Real-world outputs are identified as %Q. %Q0005 would be the name of real-world output 5. Discrete internal coils are identified as %M. See Figure 1-38 for a list of the types of discrete types that are available.

Label	Description
%I	Discrete Input
%Q	Discrete Output
%M	Discrete Internal Coil
%T	Discrete Temporary Coil
%G	Discrete Global Data
%S, %SA, %SB, %SC	Discrete System Status References

Figure 1-38. GE Fanuc programming labels for discrete type data and I/O.

GE Fanuc uses the following letters to represent analog I/O data and register data. Real-world, analog inputs are identified as %AI. %AI0003 would be the name of real-world analog input 3 in a ladder diagram. Real-world outputs are identified as %AQ. %AQ0005 would be the name of real-world output 5. System registers are identified as %R. See Figure 1-39 for a list of the types of analog and register types that are available.

Label	Description
%R	System Register
%AI	Analog Inputs
%AQ	Analog Outputs

Figure 1-39. GE Fanuc label names for registers and analog type data.

GE Fanuc Contacts

There are several types of contacts available (see Figure 1-40). The normally open and normally closed contacts function just as we have seen before. The continuation contact is a little different. The continuation coil passes power if the preceding contact is on (see Figure 1-41).

GE Fanuc Coils

There are several types of coils available (see Figure 1-42). There is, of course, the normally open coil that we are familiar with (-()-). This coil is on if it receives power from the contacts to the left on the rung.

GE Fanuc Contacts		
Instruction	Function	Description
<+>---	Continuation Contact	The continuation contact passes power to the right if the preceding continuation coil is set on.
-| |-	Normally Open Contact	A normally open contact passes power if the associated reference is is on.
-|/|-	Normally Closed Contact	A normally closed contact passes power if the associated reference is is off.

Figure 1-40. GE Fanuc contacts.

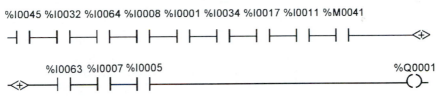

Figure 1-41. Sample GE Fanuc ladder diagram.

The set (-(S)-) coil is used to latch a coil on. If the preceding contacts on the rung are true the set contact turns on. It stays on even if the contacts to its left become deenergized. The set coil can only be turned off if a reset (-(R)-) coil is used.

The negated coil (-(/)-) is a normally closed coil. If the coil receives power from the rung, it is off. If the coil does not receive power from the rung, it is on.

A retentive coil (-(M)-) is like a normally open coil. The difference is that it retains its state through a power failure and a stop-to-run transition. This can be very important if the state of a coil needs to be retained for a system to operate safely and correctly through a power failure or stop-to-run transition.

There are two types of transition coils available. The positive transition coil turns on for only one scan if it senses a low-to-high transition of the contacts on the rung. The negative transition coil turns on for one scan if it senses a high-to-low transition of the rung. These transitional coils can be used as one-shots.

PLC APPLICATIONS

Programmable logic controllers are being used for a wide variety of applications (see Figures 1-43 to 1-45).

They are used to replace hardwired logic in older machines. This can reduce the downtime and maintenance of older equipment. More importantly PLCs can increase the speed and capability of older equipment. Retrofitting an older piece of equipment with a PLC for control is almost like getting a new machine.

PLCs are being used to control such processes as chemical production, paper production, steel production, food processing, and so on. In processes such as these they are used to control temperature, pressure, mixture, concentration, etc. They are used to control position and velocity in many kinds of production processes. For example, they can be used to control complex automated storage and retrieval systems. They can also be used to control equipment such as robots and production machining equipment.

GE Fanuc Coils		
Instruction	Function	Description
-()-	Normally Open	The associated reference is set on if the coil receives power.
-(SM)-	Retentive Set Coil	The associated discrete reference is set on if the coil receives power. The reference remains set until reset by a -(RM)- coil. Its state is retained through power failure and stop-to-run transitions.
-(RM)-	Retentive Reset Coil	The associated discrete reference is reset off if the coil receives power. The reference remains off until reset by a -(SM)- coil. Its state is retained through power failure and stop-to-run transitions.
---<+>	Continuation Coil	If the power to this coil is on, the continuation coil sets the next continuation contact on. If power is off, the continuation coil sets the next continuation contact off.
-(/M)-	Negated Retentive Coil	The associated discrete reference is set on if the function does not receive power. The state is retained through power failure and stop-to-run transitions.
-(/)-	Negated Coil	The associated discrete reference is set on if the coil does not receive power. This could also be called a normally closed coil.
-(M)-	Retentive Coil	The associated discrete reference is set on if the coil receives power. The state is retained through power failure and stop-to run transition.
-(S)-	Set Coil	The associated discrete reference is set on if the coil receives power. It remains on until it is reset by a reset coil.
-(R)-	Reset Coil	The associated discrete reference is set off if the coil receives power. It remains off until it is reset by a set coil.

Figure 1- 42. GE Fanuc coils.

$$-(\uparrow)- \quad -(\downarrow)-$$

Figure 1-43. GE Fanuc transition coils. The coil on the left is a positive transition coil and the one on the right is a negative transition coil.

Chapter 1: Overview of Programmable Logic Controllers

Figure 1-44. A PLC-controlled injection molding machine.
Photo courtesy of Rockwell Automation/Allen-Bradley Company
Inc.

Many small companies have started up recently to produce special-purpose equipment. This equipment is normally controlled by PLCs. It is very cost effective for these companies to use PLCs. Examples are conveyors and palletizing, packaging, processing, material handling, etc. Without PLC technology many small equipment design companies might not exist.

PLCs are being used extensively in position and velocity control. A PLC can control position and velocity much more quickly and accurately than can mechanical devices such as gears, cams, etc. An electronic system of control is not only faster, but does not wear out and lose accuracy as do mechanical devices.

PLCs are used for almost any process one can think of. There are companies that have PLC-equipped railroad cars that regrind and true the rail track as they travel. PLCs have been used to ring the perfect sequences of bells in church bell towers, at the exact times during the day and week. PLCs are used in lumber mills to grade, size, and cut lumber for optimal output.

The uses of PLCs are limited only by the imagination of the engineers and technicians who use them.

Figure 1-45. Large flexographic printing press. It is controlled by a PLC. Flexographic presses are used to produce packaging materials and other types of printed materials.

Questions

1. The PLC was developed to:
   a. make manufacturing more flexible.
   b. be easy for electricians to work with.
   c. make systems more reliable.
   d. all of the above.

2. The PLC was developed to replace:
   a. relay control panels.
   b. analog controllers.
   c. computers.
   d. programmers.

3. The basic difference between relay control and PLC control is:
   a. different types of input and output devices are used with PLC systems.
   b. one uses relay logic and one uses programmed instructions.
   c. different types of output devices are used with PLC systems.
   d. different types of input devices are used with PLC systems.

4. The PLC is programmed by technicians using:
   a. the C programming language.
   b. ladder logic.
   c. the choice of language used depends on the manufacturer.
   d. none of the above.

5. Changing relay control type circuits involves changing:
   a. the input circuit devices.
   b. the voltage levels of most I/O.
   c. the circuit wiring.
   d. the input and output devices.

6. The most common programming device for PLCs is the:
   a. dumb terminal.
   b. dedicated programming terminal.
   c. handheld programmer.
   d. personal computer.

7. CPU stands for central processing unit. True or false?

8. Opto-isolation:

      a. is used to protect the CPU from real-world inputs.

      b. opto-isolation is not used in PLCs. Isolation must be provided by the user.

      c. is used to protect the CPU from real-world outputs.

      d. both a and c.

9. EPROM is:

      a. electrically erasable memory.

      b. electrically programmable RAM.

      c. erased by exposing it to ultraviolet light.

      d. programmable.

      e. both c and d.

10. RAM typically holds the operating system. True or false?

11. Typical program storage devices for ladder diagrams include:

      a. computer disks.

      b. EEPROM.

      c. static RAM cards.

      d. all of the above.

      e. none of the above.

12. The IEC 1131 standard specifies characteristics for:

      a. PLC communications.

      b. EEPROM.

      c. memory.

      d. PLC programming languages.

      e. none of the above.

13. Input devices would include the following:

      a. switches.

      b. sensors.

      c. other smart devices.

      d. all of the above.

      e. none of the above.

14. Troubleshooting a PLC system:

      a. requires special PLC diagnostic equipment.

      b. is much more difficult than for relay type systems.

c. is easier because of indicators such as I/O indicators on I/O modules.

    d. all of the above.

    e. none of the above.

15. Output modules can be purchased with which of the following output devices:

    a. transistor outputs.

    b. triac outputs.

    c. relay outputs

    d. all of the above are correct.

    e. both a and c.

16. Field devices would include the following:

    a. switches.

    b. sensors.

    c. valves.

    d. all of the above.

    e. none of the above.

17. List at least four major advantages of the use of a microcomputer for programming PLCs.

18. What is a memory map? (Make sure that you describe the types of information that can be found on a memory map.)

19. If an output module's current rating is 1 A per output and there are eight outputs the current rating for the module is 8 A. True or false? Explain your answer.

20. List at least four reasons why PLCs have replaced hardwired logic in industry.

21. Draw a block diagram of a typical PLC that shows the major components.

22. Describe how the status of real-world inputs are stored in PLC memory.

23. Describe how the status of real-world outputs are stored in PLC memory.

24. Define the term *debounce*. Why is it so important?

25. What is the difference between on-line and off-line programming?

26. What does it mean to force I/O?

Additional Exercises (Allen-Bradley)

1. Explain and give an example of Allen-Bradley normally open and normally closed contacts.

2. Explain and give an example of Allen-Bradley coils.

3. Explain and give an example of Allen-Bradley immediate inputs.

4. Explain and give an example of Allen-Bradley immediate outputs.

5. Explain and give an example of Allen-Bradley latching and unlatching outputs.

6. Explain each of the following addresses.

   B3:14/1

   I:02/7

   C5:6.DN

   O:03/5

Additional Exercises (GE Fanuc)

1. Explain and give an example of GE Fanuc normally open and normally closed contacts.

2. Explain and give an example of GE Fanuc coils.

3. Explain and give an example of GE Fanuc retentive coils.

4. Explain and give an example of GE Fanuc set and reset outputs.

5. Explain each of the following addresses.

   %AQ0005

   %AI0007

   %I0003

   %Q0033

   %M0003

   %S0003

Chapter 2

Overview of Number Systems

A knowledge of various numbering systems is essential to the use of PLCs. In addition to decimal, binary, octal and hexadecimal are regularly used. An understanding of the systems will make the task of working with PLCs an easier one. In this chapter we examine each of these systems.

OBJECTIVES

Upon completion of this chapter, the student will be able to:

Explain each of the numbering systems.

Explain the benefits of typical number systems and why each is used.

Convert from one number system to another.

Explain how input or output modules might be numbered using the octal or hexadecimal number system.

Use each number system properly.

Explain such terminology as **most significant, least significant, nibble, byte,** *and* **word.**

DECIMAL

A short review of the basics of the decimal system will help in a thorough understanding of the other number systems. Although calculators will do the tedious work of number conversion between systems, it is vital that the technician be comfortable with the number systems. Binary, octal, and hexadecimal are regularly used to identify inputs/outputs, memory addresses, etc. The technician who is comfortable in understanding and using these systems will have an easier time with PLCs.

The decimal number system uses 10 digits: zero through nine. The highest digit is nine in the decimal system. Zero through nine are the only digits allowed.

Decimal System					
100,000s	10,000s	1000s	100s	10s	1s

Figure 2-1. Weights of the decimal system.

The first column in decimal can be used to count up to nine items. Another column must be added if the number is larger than nine. The second column can also use the digits zero through nine. This column is weighted, however (see Figure 2-1). The second column is used to tell how many 10s there are. For example, the number 23 represents two 10s and three 1s. By using one column we are able to count to nine. If we use two columns we can count to 99. The first column can hold up to nine. The second column can hold up to 90 (9 tens), exactly ten times as much as the first column. In fact, in the decimal system, each column is worth ten times as much as the preceding column. The third column represents the number of hundreds (10 times 10). For example, 227 would represent two 100s, two 10s, and seven 1s (see Figure 2-2).

$$2\ 2\ 7_{10} \longleftarrow \text{Decimal Number}$$

$$7 \times 10^0 = 7$$
$$2 \times 10^1 = 20$$
$$2 \times 10^2 = 200$$

$$\text{Decimal Number} \longrightarrow 227_{10}$$

Figure 2-2. Relationship between the weights of each column and the decimal number 227.

The decimal system is certainly the simplest for us because we have used it all of our lives. The other systems are based on the same principles as the decimal system. It would certainly be easier for us if there were only one system, but the computer cannot "think" in decimal. The computer can only work with binary numbers. In fact, the other number systems are very convenient for certain uses and actually simplify some tasks.

BINARY NUMBERING SYSTEM

The binary numbering system is based on only two digits: zero and one. A computer is a digital device. It works with voltages, on or off. Computer memory is a series of zeros and ones.

The binary system works just like decimal. The first column holds the number of 1s (see Figure 2-3). Since the only possible digits in the first column are zero or one, it should be clear that the first column can hold zero 1s or one 1. Thus we can only count up to 1 using the first column in binary.

The second column holds the number of 2s. There can be zero 2s or one 2. The binary number 10 would equal one 2 plus zero 1s. The number 10 in binary is two in the decimal system (one 2 + zero 1s). The binary number 11 would be 3 in decimal (one 2 + one 1) (see Figure 2-4).

The third column is the number of 4s. Thus binary 100 would be equal to decimal 4.

The fourth column is the number of 8s, the fifth column is the number of 16s, the sixth column is the number of 32s, the seventh is the number of 64s, and the eighth column is the number of 128s. As you can see, each column's value is twice as large as that of the previous column.

Binary System					
32s	16s	8s	4s	2s	1s

Figure 2-3. Weights of each column in the binary system. The first column is the number of 1s, the second is the number of 2s, and so on.

The value of the column in binary can be found by raising 2 to the power represented by that column. For example, the third column's weight could be found by raising 2 to the second power (2 X 2 = 4). Remember that the first column is column number 0, the second is column number 1, and so on. The weight of the fourth column is eight (2 X 2 X 2) (see Figure 2-5).

Binary				Decimal
8s	4s	2s	1s	
0	0	0	0	0
0	0	0	1	1
0	0	1	0	2
0	0	1	1	3
0	1	0	0	4
0	1	0	1	5
0	1	1	0	6
0	1	1	1	7
1	0	0	0	8
1	0	0	1	9
1	0	1	0	10
1	0	1	1	11
1	1	0	0	12
1	1	0	1	13
1	1	1	0	14
1	1	1	1	15

Figure 2-4. Comparison of the binary and decimal numbers from 0 through 15.

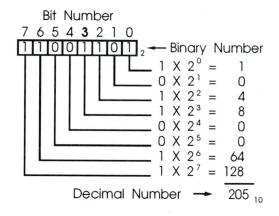

Figure 2-5. Relationship between binary and decimal. The binary number 11001101 is equal to 205 decimal.

Binary is used extensively because it is the only numbering system that a computer can deal with. It is also quite useful when considering digital logic, because a 1 can represent one state and a zero the opposite state. For example, a light is either on or off. See Figure 2-6 for the appearance of a binary word.

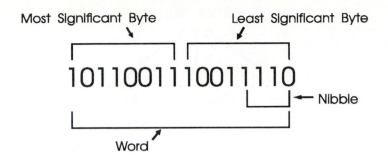

Figure 2-6. A 16-bit binary number. The bit on the right is the least significant bit. The bit on the left is the most significant bit. The next unit of grouping is the nibble. A nibble is 4 bits. The next grouping of a binary number is the byte. A byte is 8 bits. Note that the first 8 digits are called the least significant byte and the last 8 bits are called the most significant. The next grouping is called the word. The size of a word is dependent on the processor. A 16-bit processor has a 16-bit word. A 32-bit processor has a 32-bit word.

BINARY CODED DECIMAL SYSTEM

Binary coded decimal (BCD) involves the blending of the binary and decimal systems. In BCD 4 binary bits are used to represent one decimal digit. These 4 bits are used to represent the numbers zero through nine. Thus 0111 binary would be 7 decimal. (See Figure 2-7.)

The difference in BCD is in the way numbers above decimal nine are represented. For example, the decimal number 43 would be 0100 0011 in BCD. The first 4 bits (least significant bits) represent the decimal 3. The second 4 bits (most significant bits) represent the decimal 4. In BCD the first 4 bits represent the number of 1s in a decimal number, the second 4 represent the number of 10s, the third four represent the number of 100s, and so on.

The BCD format is very popular for output from instruments. Measuring devices will typically output BCD values. Some input devices use the BCD system to output their value. Thumbwheels are one example. A person dials in a decimal digit between zero and nine. The thumbwheel outputs 4 bits of data. The 4 bits are BCD. For example, if an operator were to dial in the number eight, the output from the BCD thumbwheel would be 1000. PLCs can easily accept BCD input.

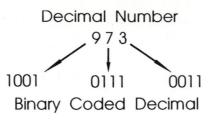

Decimal Number

973

1001   0111   0011

Binary Coded Decimal

Figure 2-7. How the decimal number 973 would be represented in the binary coded decimal (BCD) system. Each decimal digit is represented by its four-digit binary number. <u>Caution</u>: BCD is not the same as binary. The decimal number 973 is 1001 0111 0011 in BCD and is 0011 1100 1101 in binary.

OCTAL

The octal system is based on the same principles as binary and decimal except that it is base eight. There are eight possible digits in the octal system (0, 1, 2, 3, 4, 5, 6, and 7). The first column in an octal number is the number of 1s. The second column is the number of 8s, the third column is the number of 64s, the fourth column is the number of 512s, the fifth column is the number of 4096s, and so on.

The weights of the columns can be found by using the same method as that used for binary. The number eight is simply raised to the power represented by that column. The first column (column 0) represents eight to the zero power (1 by definition). Remember that the first column is column zero. The weight of the second column (column one) is found by raising eight to the first power (8X1). The weight of the third column (column two) is found by raising eight to the second power (8X8) (see Figure 2-8).

Octal System

32,768s	4,096s	512s	64s	8s	1s

Figure 2-8. Weights of the columns in the octal number system. The first column is the number of 1s, the second the number of 8s, and so on.

The actual digits in the octal number system are 1, 2, 3, 4, 5, 6, and 7. If we must count above seven we must use the next column. For example, let's count to ten in octal: 1, 2, 3, 4, 5, 6, 7, 10, 11, and 12. The 12 represents one 8 and two 1s (8+2=10).

The number 23 decimal would be 27 in octal. Two 8s and seven 1s is equal to 23. The number 3207 octal would be 1671 in decimal. Figure 2-9 shows how an octal number can be converted to a decimal number.

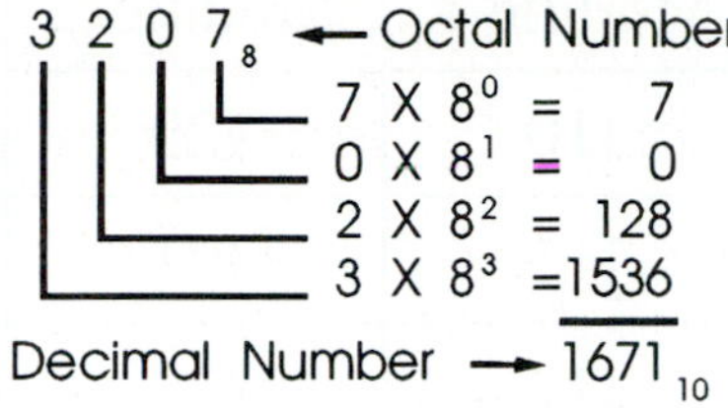

$$3\ 2\ 0\ 7_8 \leftarrow \text{Octal Number}$$

$$7 \times 8^0 = 7$$
$$0 \times 8^1 = 0$$
$$2 \times 8^2 = 128$$
$$3 \times 8^3 = 1536$$

$$\text{Decimal Number} \rightarrow 1671_{10}$$

Figure 2-9. How the octal number 3207 is converted to the decimal number 1671.

Some PLC manufacturers use octal to number input and output modules, and also to number memory addresses. For example (assume the use of input cards with eight inputs per card), the first eight inputs on the first card would be numbered 0, 1, 2, 3, 4, 5, 6, and 7. The next input card numbering would begin with octal 10, 11, 12, 13, 14, 15, 16, and 17 (see Figure 2-10).

This makes it very easy to find the location of inputs or outputs. The least significant digit is used to specify the actual input/output number, and the most significant digit is used to specify the particular card where the input/output is located.

Octal is also used by some manufacturers for numbering memory. For example, Siemens Industrial Automation, Inc. has 128 timers and counters available for the 405 series of PLCs. The timers are numbered 0 to 177 octal. This equates to 0 to 127 decimal.

HEXADECIMAL

Hexadecimal normally causes the most trouble for people. Hexadecimal is based on the same principles as the other numbering systems we examined. Hexadecimal has 16 possible digits. Hexadecimal has an unusual twist, however. Hexadecimal uses numbers and also the letters A to F. This is a little confusing at first.

In hexadecimal (hex for short) we count 0, 1, 2, 3, 4, 5, 6, 7, 8, and 9. After the number nine the counting changes: ten becomes A, eleven is B, twelve is C, thirteen is D, fourteen is E, and fifteen is F. (See Figure 2-11).

The first column (column 0) in hex is the number of 1s (see Figure 2-12). The second column is the number of 16s. The third column is the number of 256s, and so on.

Input Module 0	Input Module 1	Input Module 2	Input Module 3
I00	I10	I20	I30
I01	I11	I21	I31
I02	I12	I22	I32
I03	I13	I23	I33
I04	I14	I24	I34
I05	I15	I25	I35
I06	I16	I26	I36
I07	I17	I27	I37

Figure 2-10. I/O module addressing using the octal numbering system. The first input module is numbered 0. The first input on the module would be called input 00 (the first zero represents the first module, the second zero means that it is the first input). The eighth input on the first module would be called 07. The eighth input on the fourth module would be called input 37. (Remember that the first module is zero; the fourth module would be number three.) The first output module would begin numbering as output module 0 and each output would be numbered just as the input modules were. For example, the fifth output on the second output module would be numbered output 14.

The weights can be found in the same manner as in the binary system. In the hexadecimal system sixteen would be raised to the power of the column. For example, the weight of the third column (column 2) would be 16 to the second power (16X16=256).

Figure 2-13 shows how a hexadecimal number can be converted to a decimal number.

Hexadecimal	Decimal
0	0
1	1
2	2
3	3
4	4
5	5
6	6
7	7
8	8
9	9
A	10
B	11
C	12
D	13
E	14
F	15
10	16
11	17
12	18
13	19
14	20

Figure 2-11. Comparison of the hexadecimal and decimal systems.

Hexadecimal System					
1,048,576s	65,536s	4,096s	256s	16s	1s

Figure 2-12. Weights of the columns in the hexadecimal system. The first column is the number of 1s , the second column is the number of 16s, and so on.

Hexadecimal numbers are easier (less cumbersome) for us to work with than binary numbers. It is easy to convert between the two systems (see Figures 2-14 and 2-15). Each hex digit is simply converted to its four-digit binary equivalent. The result is the binary equivalent of the whole hex number.

The same process works in reverse. Any binary number can be converted to its hex equivalent by breaking the binary number into four-digit pieces and converting each four-bit piece to its hex equivalent. The resulting numbers are equal in value.

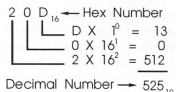

Figure 2-13. Weights of the hexadecimal number system converted to the decimal weighting system.

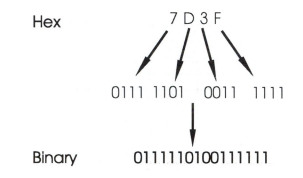

Figure 2-14. This figure shows the conversion of a hexadecimal number to its binary equivalent. Each hex digit is simply converted to its four-digit binary. The result is a binary equivalent. In this case hex 7D3F is equal to binary 0111110100111111. (Which number would you rather work with?)

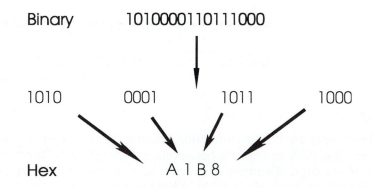

Figure 2-15. Conversion of a binary number to a hex number. The binary number is broken into four-bit pieces (4 bits are called a nibble) and each 4-bit nibble is converted to its hex equivalent.

The binary number 1010000110111000 is equal to hex A1B8.
(Which would you prefer to work with?)

Figure 2-16 shows an overview of the four number systems.

Hexadecimal	Decimal	Octal		Binary			
		8s	1s	8s	4s	2s	1s
0	0	0	0	0	0	0	0
1	1	0	1	0	0	0	1
2	2	0	2	0	0	1	0
3	3	0	3	0	0	1	1
4	4	0	4	0	1	0	0
5	5	0	5	0	1	0	1
6	6	0	6	0	1	1	0
7	7	0	7	0	1	1	1
8	8	1	0	1	0	0	0
9	9	1	1	1	0	0	1
A	10	1	2	1	0	1	0
B	11	1	3	1	0	1	1
C	12	1	4	1	1	0	0
D	13	1	5	1	1	0	1
E	14	1	6	1	1	1	0
F	15	1	7	1	1	1	1

Figure 2-16. A comparison of the four number systems for the
numbers 0 to 15.

Questions

1. Complete the following table.

	Binary	Octal	Decimal	Hexadecimal
a.	101			5
b.		11		
c.			15	
d.				D
e.		16		
f.	1001011			
g.		47		
h.			73	

2. Number the following input/output modules using the octal method.

	Input Module 0	Input Module 1	*Output Module 0*	Input Module 2
A.				
B.				
C.				
D.				
E.				
F.				
G.				
H.				

3. Define each of the following:
   - a. bit
   - b. nibble
   - c. byte
   - d. word

4. Why is the binary number system used in computer systems?

5. Complete the following table.

	Binary	Hexadecimal
a.	1011001011111101	
b.		1A07
c.	100100000111	
d.		C17F
e.	0010001111000010	
f.		D91C
g.	0011010111100110	
h.		ECA9
i.	0101001011000101	

6. True or false. A word is sixteen bits. Explain your answer.

7. True or false. One K of memory is exactly one thousand bytes.

8. How can the weight of the fifth column of a binary number be calculated?

9. How can the weight of the fourth column of a hex number be calculated?

10. What is the BCD system normally used for?

Chapter 3

Fundamentals of Programming

In this chapter we examine the basics of ladder logic programming. Terminology and common symbols are emphasized. The student will learn how to write basic ladder logic programs. Contacts, coils, timers, and counters are used.

OBJECTIVES

Upon completion of this chapter, the student will be able to:

> *Describe the basic process of writing ladder logic.*
>
> *Define such terms as* **contact, coil, rung, scan, normally open, normally closed,** *and* **timer.**
>
> *Write ladder logic for simple applications.*

LADDER LOGIC

Programmable controllers are primarily programmed in ladder logic. Ladder logic is really just a symbolic representation of an electrical circuit. Symbols were chosen that actually looked similar to schematic symbols of electrical devices. This made it easy for the plant electrician to learn to use the PLC. An electrician who has never seen a PLC can understand a ladder diagram.

The main function of the PLC program is to control outputs based on the condition of inputs. The symbols used in ladder logic programming can be divided into two broad categories: contacts (inputs) and coils (outputs).

Contacts

Most inputs to a PLC are simple devices that are either on or off. These inputs are sensors and switches that detect part presence, empty or full, and so on. The two common symbols for contacts are shown in Figure 3-1.

Normally Open Contact   **Normally Closed Contact**

Figure 3-1. A normally open and a normally closed contact.

Contacts can be thought of as switches. There are two basic kinds of switches, normally open and normally closed. A *normally open switch* will not pass current until pressed. A *normally closed switch* will allow current flow until it is pressed. Think of a doorbell switch. Would you use a normally open switch or a normally closed switch for a doorbell?

If you chose the normally closed switch, the bell would be on continuously until someone pushes the switch. Pushing the switch opens the contacts and stops current flow to the bell. The normally open switch is the necessary choice. If the normally open switch is used, the bell will not sound until someone pushes the button on the switch.

Sensors are used to sense for the presence of physical objects or quantities. For example, one type of sensor might be used to sense when a box moves down a conveyor, and a different type might be used to measure a quantity such as heat. Most sensors are switch-like. They are on or off depending on what the sensor is sensing. Like switches, sensors can be purchased that are either normally open or normally closed.

Imagine, for example, a sensor that is designed to sense a metal part as the part passes the sensor. We could buy a normally open or a normally closed sensor for the application. If it were desired to notify the PLC every time a part passed the sensor, a normally open sensor might be chosen. The sensor would turn on only if a metal part passed in front of the sensor. The sensor would turn off again when the part was gone.

The PLC could then count the number of times the normally open switch turned on (closed) and would know how many parts had passed the sensor. Normally closed sensors and switches are often used when safety is a concern. These are examined later.

ALLEN-BRADLEY CONTACTS

Examine-On

The examine-on instruction (XIC) is the normally open instruction. If a real-world input device is on, this type of instruction is true and passes power (see Figure 3-2).

If the input bit from the input image table associated with this instruction is a one, the instruction is true. If the bit in the input image table is a zero, the instruction associated with this particular input bit is false.

Figure 3-2. Examine-on instruction (XIC). If the CPU sees an on condition at bit I:012/07, this instruction is true. The numbering of the input instruction is as follows: This is rack 1, I/O group 2, real-world input 7 of the input module.

COILS

Contacts are input symbols, coils are output symbols. Outputs can take various forms: motors, lights, pumps, counters, timers, relays, and so on. A coil is simply an output. The PLC examines the contacts (inputs) in the ladder and turns the coils (outputs) on or off depending on the condition of the inputs. The basic coil is shown in Figure 3-3.

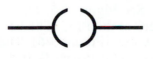

Coil

Figure 3-3. The ladder logic symbol for a coil.

ALLEN-BRADLEY COILS

Output Energize

The output energize instruction (OTE) is the normal output instruction. The OTE instruction sets a bit in memory. If the logic in its rung is true, the output bit will be set to a 1. If the logic of its rung is false, the output bit is reset to a zero. Figure 3-4 shows an output energize (OTE) instruction. This particular example is real-world output 1 of I/O group 3, of rack 1. If the logic of the rung leading to this output instruction is true, output bit O:013/01 will be set to a 1 (true). If the rung is false, the output bit would be set to a zero (false).

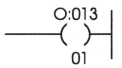

Figure 3-4. Output energize instruction (OTE).

LADDER DIAGRAMS

The basic ladder diagram looks similar to a ladder. There are two uprights and there are rungs that make up the PLC ladder. The left and right uprights represent power. If we connect the left and right uprights, power could flow through the rung from the left upright to the right upright.

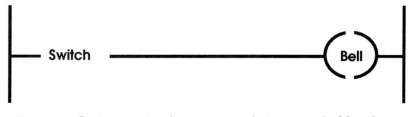

Figure 3-5. A very simple conceptual view of a ladder diagram.

Consider the doorbell example again, one input and one output. The ladder for the PLC would be only one rung. See Figure 3-5. The real-world switch would be connected to input number zero of the PLC. The bell would be connected to output number zero of the PLC (see Figure 3-6). The uprights represent a DC voltage that will be used to power the doorbell. If the real-world doorbell switch is pressed, power can flow through the switch to the doorbell.

The PLC would then run the ladder. The PLC will monitor the input continuously and control the output. This is called *scanning*. The amount of time it takes for the PLC to go through the ladder logic each time is called the *scan time*. Scan time varies from PLC to PLC. Most applications do not require extreme speed, so any PLC is fast enough; even a slow PLC scan time would be in milliseconds. The longer the ladder logic, the more time it takes to scan.

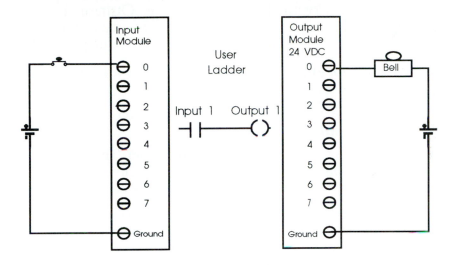

Figure 3-6. Conceptual view of a PLC system. The real-world inputs are attached to an input module (left side of the figure). Outputs are attached to an output module (right side of the figure). The center of the figure shows the logic that the CPU must evaluate. The CPU evaluates user logic by looking at the inputs and then turns on outputs based on the logic. In this case if input 0 (a normally open switch) is closed, output 0 (the doorbell) will turn on.

The scan cycle is illustrated by Figure 3-7. Note that this one rung of logic represents our entire ladder. Each time the PLC scans the doorbell ladder, it checks the state of the input switch *before* it enters the ladder (time period 1). While in the ladder, the PLC then decides if it needs to change the state of any outputs (evaluation during time period 2). *After* the PLC finishes evaluating the logic (time period 2), it turns on or off any outputs based on the evaluation (time period 3). The PLC then returns to the top of the ladder, checks the inputs again and repeats the entire process. It is the total of these three stages that makes up scan time. We discuss the scan cycle more completely later in this chapter.

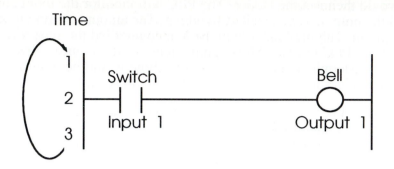

Figure 3-7. An example of how a user's ladder logic is continually scanned.

Normally Closed Contacts

The second type of contact is the normally closed contact. The normally closed contact will pass power until it is activated. A normally closed contact in a ladder diagram will pass power while the real-world input associated with it is off.

A home security system is an example of the use of normally closed logic. Assume that the security system was intended to monitor the two entrance doors to a house. One way to wire the house would be to wire one normally open switch from each door to the alarm, just like a doorbell switch (Figure 3-8). Then if a door opened, it would close the switch and the alarm would sound. This would work, but there are problems.

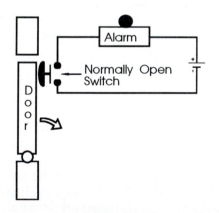

Figure 3-8. A conceptual diagram of a burglar alarm circuit. This is the wrong way to construct this type of application. In this case the homeowner would never know if the system failed. The correct method would be to use normally closed switches. The control system would then monitor the circuit continuously to see if the doors opened or a switch failed.

Assume that the switch fails. A wire might be cut accidentally, or a connection might become loose, or a switch breaks, etc. There are many ways in which the system could fail. The problem is that the homeowner would never know that the system was not working. An intruder could open the door, the switch would not work, and the alarm would not sound. Obviously, this is not a good way to design a system.

The system should be set up so that the alarm will sound for an intruder and will also sound if a component fails. The homeowner surely wants to know if the system fails. It is far better that the alarm sounds when the system fails and there is no intruder than not to sound if the system fails and there is an intruder.

Considerations such as these are even more important in an industrial setting where failure could cause an injury.

The procedure of programming to assure safety is called *fail-safe*. The programmer must carefully design the system and ladder logic so that if a failure occurs, people and processes are safe. Consider Figure 3-9. If the gate is opened, it opens the normally closed switch. The PLC would see that the switch had opened and would sound an alarm immediately to protect whomever had entered the work cell. (In reality, we would sound an alarm and stop the robot to protect the intruder.)

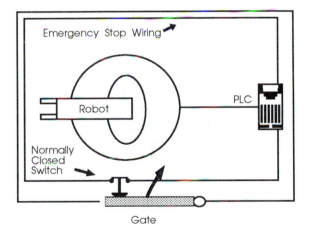

Figure 3-9. Cell application. This figure represents a robot cell. There is a fence around the cell with one gate. There is a PLC used as a cell controller. There is a safety switch to make sure no one enters the cell while the robot is running. If someone enters the cell the PLC will sense that the switch opened and sound an alarm. In this case a normally closed switch was used. If the wiring or switch fails the PLC will think someone entered the cell and sound an alarm. This system would be "fail-safe."

The normally closed switch as used in ladder logic can be confusing. The normally closed contact in our ladder passes electricity if the input switch is off. (The alarm would sound if the switch in the cell gate opened.) The switch in the gate of the cell is a normally closed switch. (The switch in the cell normally allows electricity to flow.) If someone opens the gate, the normally closed gate switch opens, stopping electrical flow (see Figure 3-10). The PLC sees that there is no flow, the normally closed contact in the ladder allows electricity to flow, and the alarm is turned on.

Assume that a tow motor drives too close to the cell and cuts the wire that connects the gate safety switch to the PLC. What will happen? The alarm will sound. Why? Because the wire being cut is similar to the gate opening the switch. Is it a good thing that the alarm sounds if the wire is cut? Yes. This warns the operator that something failed in the cell. The operator could then call maintenance and have the cell repaired. This is "fail-safe." Something in the cell failed and the system was shut down by the PLC so that no one would be hurt. The same would be true if the gate safety switch were to fail. The alarm would sound. If the switch were opened (someone opened the gate to the cell), the PLC would see that there is no power at the input (see Figure 3-11). The normally closed contact in the ladder logic is then closed, allowing electricity to flow. This causes the alarm to sound.

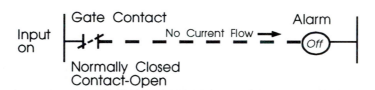

Figure 3-10. One rung of a ladder diagram. A normally closed contact is used in the ladder. If the switch associated with that contact is closed, it forces the normally closed contact open. No current flows to the output (the alarm). The alarm is off.

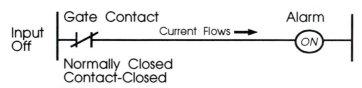

Figure 3-11. The input is off. (Someone opened the gate and opened the switch.) The normally closed contact is true when the input is false, so the alarm sounds. The same thing would happen if a tow motor cut the wire that led to the safety sensor. The input would go low and the alarm would sound.

Consider the rungs shown in Figure 3-12 and determine whether the output coils are on or off. (The answers follow.) Pay particular attention to the normally closed examples.

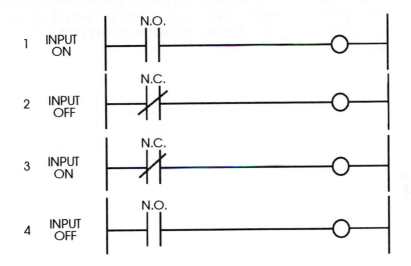

Figure 3-12. Ladder diagram exercise.

Answers to the ladder logic in Figure 3-10 are shown below.

1. The output in example 1 would be on. The input associated with the normally open contact (or examine-on) is one that would close the normally closed contact in the ladder and pass power to the output.

2. The output in example 2 would be on. The real-world input is off. The normally closed contact is closed because the input is off.

3. The output in example 3 is off. The real-world input is on, which forces the normally closed contact open (or the rung is false because the examine-off input is on).

4. The output in example 4 would be off. The real world input is off, so the normally open contact remains open and the output off.

ALLEN-BRADLEY SLC FILE ORGANIZATION AND I/O NUMBERING

Allen-Bradley data files contain the status information concerning all external I/O and all other instructions that the programmer uses in program files. These files also store status information concerning the PLC operation. The user can also use files to store information that may be used for data or recipes for system operation. Figure 3-13 shows the files that reside in the processor's memory. Figure 3-14 shows default file identifiers and numbers. Data file types are identified by a letter and a number. File numbers 0-7 are the default files that fixed, SLC 5/01, 5/02, and 5/03 OS300 processors create. File number 8 applies to only SLC 5/03, OS301 and 5/04 OS400 processors. If additional files are needed for storage the user can create files by specifying the appropriate identifier and a file number between 9 and 255 (see Figure 3-15).

0	Output Image
1	Input Image
2	Status
3	Bit
4	Timer
5	Counter
6	Control
7	Integer
8	Floating Point (SLC 5/03OS301 and SLC 5/04 only)
9	Can be used for network transfer on DH-485 network or ordinary data file if the processor is not on a network or is on a network with SLC 500 devices only.
10-255	Bit, Timer, Counter, Control, Integer, String, Floating Point, or ASCII assigned as needed

Figure 3-13. Files that reside in the processor's memory.

File Type	Identifier	File Number
Output	O	0
Input	I	1
Status	S	2
Bit	B	3
Timer	T	4
Counter	C	5
Control	R	6
Integer	N	7
Float	F	8

Figure 3-14. Default file identifiers and numbers.

Chapter 3: Fundamentals of Programming

User-Defined Files		
File Type	**Identifier**	**File Number**
Bit	B	
Timer	T	
Counter	C	
Control	R	9-255
Integer	N	
Float	F	
String	St	
ASCII	A	

Figure 3-15. User-defined file numbers.

Addresses are made up of several characters. The characters are separated by delimiters. Delimiters are used to separate the different portions of the address. The first portion specifies the file type (identifier). The identifier tells what type of file it will be. O would be an output, I would be an input, B would be a bit, and so on. Figure 3-14 and Figure 3-15 show the identifiers.

The file number is next. The file number is normally taken from the table shown in Figure 3-14. For an output the file type would be O and the file number would be 0. For an input the identifier would be I and the file would be 1. If we needed to use a status type the identifier would be S and the file number would be 2. The next character after the identifier and the file number would be a delimiter. A colon (:) is used for this delimiter. The next portion of the address is the element. The element is a number that specifies the slot and word number. Consider the controller shown in Figure 3-16. This is a fixed controller with an expansion chassis. Figure 3-17 shows that there are 24 inputs and 16 inputs available in slot 0. Slot 1 has 6 inputs and 6 outputs. Slot 2 has 0 inputs and 8 outputs available.

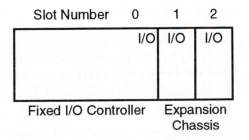

Figure 3-16. Drawing of PLC and slot numbers.

Slot	Inputs	Outputs
0	24	16
1	6	6
2	None	8

Figure 3-17. Table showing I/O allowed in slots 0, 1, and 2.

Figure 3-18 shows how input elements would be established. The first 16 inputs (slot 0) would be identified by element 0. Note that the element is shown after the I:. Remember that the I identifies this as an input type file. Slot 0 inputs 16-23 would have an element number of 0.1. Note that the element consists of the slot number plus the word number when needed. In this case it is needed because slot 0 has 24 inputs available. Each word can only hold 16 input bit states. In this case word 0 holds 16 and word 1(I:0.1) holds the remaining 8 (16-23). Slot 1 inputs (0-5) would have an element number of 1 (slot 1). Figure 3-19 shows an input element table for slots 0-2.

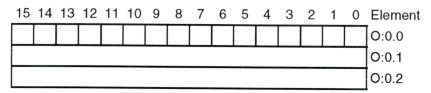

Figure 3-18. Input element addressing.

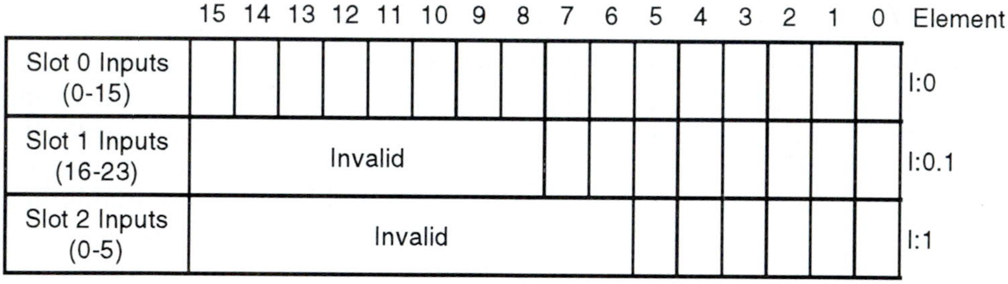

Figure 3-19. Input element table.

Figure 3-20 shows how output elements would be established. The first 16 outputs (slot 0) would be identified by element 0. Note that the element is shown after the O:. Remember that the O identifies this as an output type file. Slot 1 outputs 0-5 would have an element number of 1. Slot 2 outputs (0-7) would have an element number of 2 (slot 2).

So remember that the element is comprised of the slot number when needed and the word.

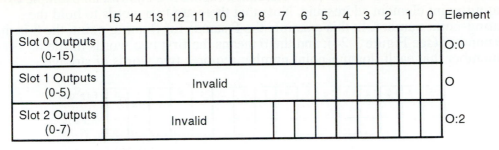

	15 14 13 12 11 10 9 8 7 6 5 4 3 2 1 0	Element
Slot 0 Outputs (0-15)		O:0
Slot 1 Outputs (0-5)	Invalid	O
Slot 2 Outputs (0-7)	Invalid	O:2

Figure 3-20. Output element table.

Figure 3-21 shows a few examples of element addresses. Pay particular attention to the numbering of the second input word element addressing.

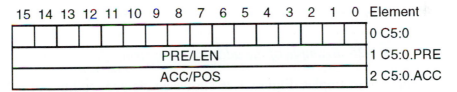

O:1	Output word 0, slot 1
I:0	Input word 0, slot 0
I:0.1	Input word 1, slot 0

Figure 3- 21. Examples of element addressing.

Elements for timers, counters, control and ASCII files consist of three words. Figure 3-22 shows an example for a counter. In this case the C stands for counter, the 5 is the file number, this is counter 0. C5:0.PRE word would hold the preset value and C5:0.ACC would hold the accumulated value.

15 14 13 12 11 10 9 8 7 6 5 4 3 2 1 0	Element
	0 C5:0
PRE/LEN	1 C5:0.PRE
ACC/POS	2 C5:0.ACC

Figure 3-22. Example of counter addressing.

Status, bit, and integer files have 1-word elements. Figure 3-23 shows an example of an integer element address. The N stands for integer and the 10 identifies the element.

15 14 13 12 11 10 9 8 7 6 5 4 3 2 1 0	Element
	N:10

Figure 3-23. Example of the addressing of integer elements.

Floating-point files have two-word elements. Figure 3-24 shows an example of floating-point element addressing. Note that there are two words to hold the floating-point number. In this example the F stands for floating point, the 8 is the file number (see Figure 3-24), and the 0 means the first two words in file 8. Remember that each floating-point number requires two words of storage.

15	14	13	12	11	10	9	8	7	6	5	4	3	2	1	0	Element
																F8:0

Figure 3-24. Example of the addressing of floating-point elements.

String files have 42-word elements. Figure 3-25 shows an example. Note the use of the decimal point delimiter to specify the particular word in the string file.

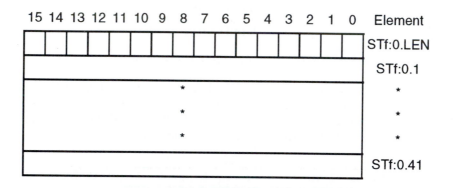

Figure 3-25. Example of the addressing of string file elements.

Consider a couple of examples. Figure 3-26 shows an element address of N7:15. The N identifier stands for an integer (see Figure 3-14). The file number is 7 (see Figure 3-14). The colon is the delimiter. The element in the file is 15.

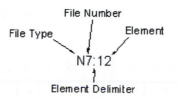

Figure 3-26. Integer element addressing example.

Figure 3-27 shows an element address of T4:7.ACC. The T identifier stands for a timer. The file number is 4. The colon is the delimiter. The element in the file is 7 and the ACC specifies the accumulated value word.

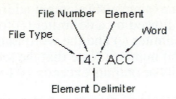

Figure 3-27. Timer element addressing example.

Figure 3-28 shows an example of bit element addressing. In this example the type identifier (B) specifies a bit type. The file number is 3. The element within the file is 64. The bit number is 12.

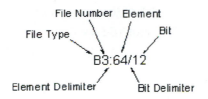

Figure 3-28. Bit element addressing example.

Format	Explanation		
O:e.s/b I:e.s/b	O	Output	
	I	Input	
	:	Element Delimiter	
	e	Slot Number (decimal)	Fixed I/O Controller: 0
			Left Side of Expansion Rack:1 Right Side of expansion Rack: 2
	.	Word Delimiter. Required only if a word number is necessary	
	s	Word Number	Required if the number of inputs or outputs exceeds 16 for the slot. Range 0-255 (range accomodates multiword "specialty cards")
	/	Bit Delimiter	
	b	Terminal Number	Inputs: 0-15 (or 0 to 23, slot 0). Outputs: 0-15

Figure 3-29. I/O element addressing.

Figure 3-29 shows a table that explains I/O element addressing. The general format is the type identifier followed by a delimiter (:) followed by the slot number (when required), a delimiter (.) and the word number and then the bit delimiter (/) followed by the bit number. Note that the word number is only required if the number of inputs or outputs exceeds 16 for the slot.

Figure 3-30 shows several example of element addressing. Pay particular attention to the last example which shows the use of the decimal because there are more than 16 inputs in this particular slot.

O:0/4	Controller Output 4 (slot 0)
O:2/7	Output 7, slot 2 of the expansion rack
I:1/4	Input 4, slot 1 of the expansion rack
I:0/15	Controller input 15 (slot 0)
I:0.1/7	Controller input 23 (bit 07, word 1 of slot 0)

Figure 3-30. Examples of I/O addressing.

ALLEN-BRADLEY NORMALLY CLOSED CONTACTS

Examine-Off

Allen-Bradley calls their normally closed contacts examine-off contacts. The examine-off instruction (XIO) can also be called a normally closed instruction. This instruction responds in the opposite fashion to the normally open instruction. If the bit associated with this instruction is a zero (off), the instruction is true and passes power. If the bit associated with the instruction is a one (true), the instruction is false and does not allow power flow.

Figure 3-31 shows an examine-off instruction (XIO). The input is number 7 of rack 1, I/O group 2. If the bit associated with I:012/07 is true (one) the instruction is false (open) and does not allow power flow. If bit I:012/07 is false (zero) the instruction is true (closed) and allows flow.

I:012

07

Figure 3-31. Examine-off instruction (normally closed).

MULTIPLE CONTACTS

More than one contact can be put on the same rung. For example, think of a drill machine. The engineer wants the drill press to turn on only if there is a part present and the operator has one hand on each of the start switches (see Figures 3-32 and 3-33). This would ensure that the operator's hands could not be in the press while it is running.

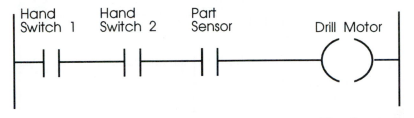

Figure 3-32. This figure shows a series circuit. Hand switches 1 AND 2 AND the part presence switch must be closed before the drill motor will be turned on. This will assure that there is a part in the machine and that both of the operator's hands are in a safe location.

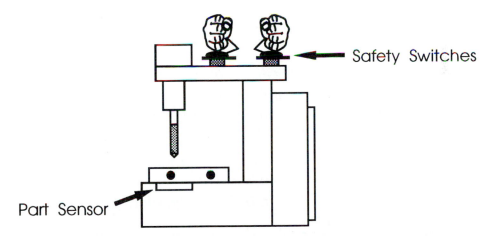

Figure 3-33. A simple drilling machine. There are two hand safety switches and one part sensor on the machine. Both hand switches and the part sensor must be true for the drill press to operate. This assures that the operator's hands are not in the way of the drill. This is an AND condition. Switch 1 and switch 2 and the part sensor must be activated to make the machine operate. The ladder for the PLC is shown in Figure 3-32.

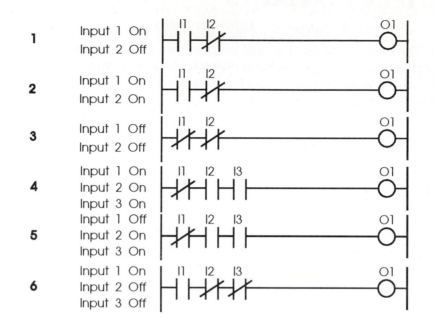

Figure 3-34.

Answers to the ladder logic shown in Figure 3-34.

1. The output for rung 1 will be on. Input 1 is on, which closes contact 1. Input 2 is off, so normally closed contact 2 is still closed. Both contacts are closed, so the output is on.

2. The output in rung 2 is off. Input 1 is on, which closes normally open contact 1. Input 2 is on, which forces normally closed contact 2 open. The output cannot be on, because normally closed contact 2 is forced open.

3. The input in rung 3 is on. Inputs 1 and 2 are off so that normally closed contacts 1 and 2 remain closed.

4. The output in rung 4 is off. Input 1 is on, which forces normally closed contact 1 open.

5. The output in rung 5 is on. Input 1 is off so normally closed contact 1 remains closed. Inputs 2 and 3 are on, which forces normally open contacts 2 and 3 closed.

6. The output in rung 6 is on. Input 1 is on forcing normally open contact 1 closed. Inputs 2 and 3 are off, which leaves normally closed contacts 2 and 3 closed.

Chapter 3: Fundamentals of Programming

Note that the switches were programmed as normally open contacts. They are all on the same rung (series). All will have to be on for the output to turn on. If there is a part present and the operator puts his/her hands on the start switches, the drill press will run.

If the operator removes one hand to wipe the sweat from his or her brow, the press will stop. Contacts in series such as this can be thought of as logical AND conditions. In this case, the part presence switch AND the left-hand switch AND the right-hand switch would have to be closed to run the drill press.

Study the examples in Figure 3-34 and determine the status of the outputs.

BRANCHING

There are often occasions when it is desired to turn on an output for more than one condition. For example, in a house, the doorbell should sound if either the front or rear door button is pushed (the two conditions under which the bell should sound). The ladder would look similar to Figure 3-35. This is called a *branch*. There are two paths (or conditions) that can turn on the doorbell. (This can also be called a logical OR condition.)

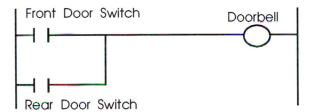

Figure 3-35. This figure shows a parallel condition. If the front door switch is closed the doorbell will sound, OR if the rear door switch is closed the doorbell will sound. These parallel conditions are also called OR conditions.

If the front door switch is closed, electricity can flow to the bell. Or if the rear door switch is closed, electricity can flow through the bottom branch to the bell. Branching can be thought of as an OR situation. One branch OR another can control the output. ORs allow multiple conditions to control an output. This is very important in industrial control of systems. Think of a motor that is used to move the table of a machine. There are usually two switches to control table movement: a jog switch and a feed switch (see Figure 3-36). Both switches are used to turn on the same motor. This is an OR condition. The jog switch OR the feed switch can turn on the table feed motor. Evaluate the ladder logic shown in Figure 3-37.

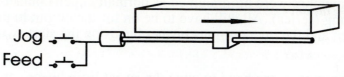

Jog

Feed

Figure 3-36. Conceptual drawing of a mill table. Note that there are two switches connected to the motor. These represent OR conditions. The jog switch or the feed switch can move the table.

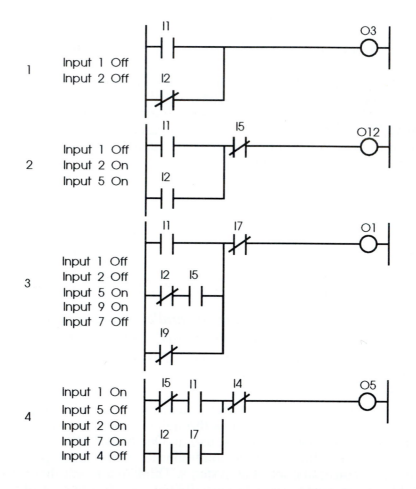

1   Input 1 Off
    Input 2 Off

2   Input 1 Off
    Input 2 On
    Input 5 On

3   Input 1 Off
    Input 2 Off
    Input 5 On
    Input 9 On
    Input 7 Off

4   Input 1 On
    Input 5 Off
    Input 2 On
    Input 7 On
    Input 4 Off

Figure 3-37. The answers are given on the next page.

These are the answers to the examples in Figure 3-37.

1. *The output in example 1 would be on. Input 2 is off, so that normally closed contact 2 is closed, allowing the output to be on.*

2. *The output in example 2 is off. Input 5 is on, which forces normally closed contact 5 open, so the output cannot be on whether or not input 1 OR input 2 is on. It should also be noted that in these branching examples we have combinations of ANDs and ORs. In English, this example would be: input 1 AND input 5 OR input 2 AND input 5 will turn on output 12.*

3. *The output in example 3 will be on. Input 2 is off, which leaves normally closed contact input 2 closed AND input 5 is on, which closes normally open input 5 AND normally closed input 7 is off, which leaves normally closed contact 7 closed, which turns on output 1. In this ladder there are 3 OR conditions and combinations of ANDs.*

4. *In example 4, the output will be on. Input 5 is off, which leaves normally closed contact 5 closed AND input 1 is on, which forces normally open contact 1 closed AND input 4 is off, which leaves normally closed contact 4 closed AND input 4 is off, which leaves normally closed contact 4 closed, turning on the output. Inputs 2 and 7 are also both on, closing normally open contacts 2 AND 7.*

Start/Stop Circuit

Start/stop circuits are extremely common in industry. Machines will have a start button to begin a process and a stop button to shut off the system. Several important concepts can be learned from the simple logic of a start/stop circuit.

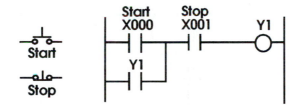

Figure 3-38. Start/stop circuit.

Examine Figure 3-38. Notice that the actual start switch is a normally open pushbutton. When it is pressed, it closes the switch. When the button is released, the switch opens. The stop switch is a normally closed switch. When pressed, it opens.

Now examine the ladder. When the start switch is momentarily pressed, power passes through X000. Power also passes through X001, because the real-world stop switch is a normally closed switch. The output (Y1) is turned on. Note that

Y1 is then also used as an input on the second line of logic. Output Y1 is on so contact Y1 also closes. This is called latching. The output latches itself on even if the start switch opens. Output Y1 will shut off only if the normally closed stop switch (X001) is pressed. If X001 opens, then Y1 is turned off. The system will require the start button to be pushed to restart the system. Note that the real-world stop switch is a normally closed switch, but that in the ladder, it is programmed normally open. This is done for safety.

There are as many ways to program start/stop circuits (or ladder diagrams in general) as there are programmers. Figures 3-39 and 3-40 show examples of the wiring of start/stop circuits. Safety is always the main consideration in start/stop circuits.

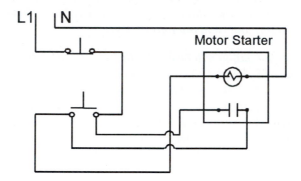

Figure 3-39. Start/stop circuit.

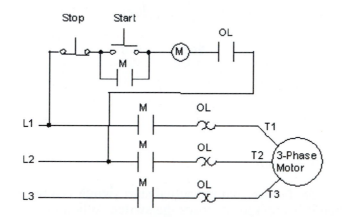

Figure 3-40. Start/stop circuit.

SPECIAL CONTACTS

There are many special-purpose contacts available to the programmer. The original PLCs did not have many available. Sharp programmers used normally open and normally closed contacts in ingenious ways to turn outputs on for one scan, to latch outputs on, and so on. PLC manufacturers added special contacts to their ladder programming languages to meet these needs. The programmer can now accomplish these special tasks with one contact instead of a few lines of logic.

Immediate Instructions

Immediate instructions are used when the input or output being controlled is very time dependent. For example, for safety reasons we may have to update the status of a particular input every few milliseconds. If our ladder diagram is 10 milliseconds long, the scan time would be too slow. This could be dangerous. The use of immediate instructions allows inputs to be updated immediately as they are encountered in the ladder. The same is true of output coils.

Allen-Bradley SLC 500 Immediate Input with Mask

The immediate input instruction (IIM) is an instruction that is used to acquire the present state of one word of inputs (see Figure 3-41). Normally, the CPU would have to finish all evaluation of the ladder logic and then update the output and input image tables. In this case when the CPU encounters this instruction during ladder evaluation, it interrupts the scan. Data from a specified I/O slot is transferred through a mask to the input data file. This makes the data available to instructions following the IIM instruction. This means that it gets the real-time states of the actual inputs at that time and puts them in that word of the input image table. The CPU then returns to evaluating the logic using the new states it acquired. This is used only when time is a crucial factor. Normally, the few milliseconds a scan takes is fast enough for anything we do. There are cases where the scan time takes too long for some I/O updates. Motion control is one case. Updates on speed and position may be required every few milliseconds or less. We cannot safely wait for the scan to finish to update the I/O. In these cases immediate instructions are used.

```
    ┌ IIM ──────────────────────┐
    │ IMMEDIATE INPUT w MASK    ├──
    │ Slot                      │
    │ Mask                      │
    └───────────────────────────┘
```

Figure 3-41. Instruction numbering. This is an immediate input instruction (IIM).

Figure 3-42 shows an immediate input instruction (IIM) addressing format. When the CPU encounters the instruction during evaluation it will immediately suspend what it is doing (evaluating) and will update the input image word associated with I/O slot 2, word 0.

I:2	Inputs of slot 2, word 0
I:2.1	Inputs of slot 2, word 1
I:1	Inputs of slot 1, word 0

Figure 3-42. Immediate input instruction (IIM).

Allen-Bradley Immediate Outputs

The immediate output instruction (IOM) is used to update output states immediately. In some applications the ladder scan time is longer than the needed update time for certain outputs. For example, it might cause a safety problem if an output were not turned on or off before an entire scan was complete. In these cases, or when performance requires immediate response, immediate outputs (IOMs) are used. When the CPU encounters an immediate instruction it exits the scan and immediately transfers data to a specified I/O slot through a mask. The CPU will then resume evaluating the ladder logic.

Figure 3-43. Format for an immediate output instruction (IOM).

Figures 3-43 and 3-44 show examples of the addressing format for an IOM.

O:2	Outputs of slot 2, word 0
O:2.1	Outputs of slot 2, word 1
O:1	Outputs of slot 1, word 0

Figure 3-44. Immediate output instruction I/O numbering.

Transitional Contacts

Transitional contacts are one type of special contact. They are also called *one-shot contacts*. The symbol for this type of contact is the normal contact symbol plus an arrow pointing either up or down (see Figure 3-45). A down arrow means that when this contact is energized it will transition from high to low for one scan. An arrow pointing up means that when this type of contact is energized it will transition from off to on for one scan. That is why they are called one-shots. They are only active for one scan when energized.

There are many reasons to use this type of contact. They are often used to provide a pulse for timing, counting, or sequencing.

They are also used when it is desired to perform an instruction only once (not every scan). For example, if an add instruction was used to add two numbers once, it would not be necessary to add them every scan. A transitional contact would assure that the instruction executes only on the desired transition.

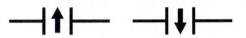

Figure 3-45. Two transitional contacts: a low-to-high transitional contact and a high-to-low transitional contact.

Allen-Bradley One-Shot Rising Instruction

The one-shot rising instruction (OSR) is a retentive input instruction that can trigger an event to happen one time (see Figure 3-46). When the rung conditions that precede the OSR go from false to true, the OSR instruction will be true for one scan. After the one scan, the OSR instruction becomes false, even if the preceding rung conditions that precede it remain true. The OSR will only become true again if the rung conditions preceding it make a transition from false to true. Only one OSR can be used per rung.

Figure 3-46. An Allen-Bradley OSR instruction.

Latching Instructions

Latches are used to lock in a condition. For example, if an input contact is on for only a short time, the output coil would be on for the same short time. If it were desired to keep the output on even if the input goes low, a latch could be used. This can be done by using the output coil to latch itself on (see Figure 3-47).

It can also be done with a special coil called a *latching coil* (see Figure 3-48). When a latching output is used, it will stay on until it is unlatched. Unlatching is done with a special coil called an *unlatching coil*. When it is activated the latched coil of the same number is unlatched.

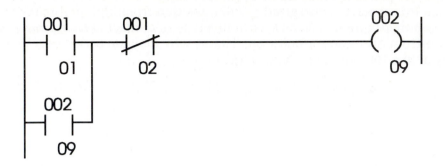

Figure 3-47. This figure shows an example of latching an output on. If input 00101 is true, coil 00209 will be energized. (Remember that contact 00102 is normally closed.) When coil 00209 energizes, it latches itself on by providing a parallel path around 00101. The only way to turn the latched coil off would be to energize normally closed contact 00102. This would open the rung and deenergize coil 00209.

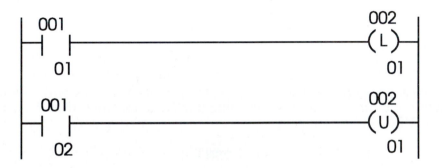

Figure 3-48. Use of a latching output. If input 00101 is true, output 00201 energizes. It will stay energized even if input 00101 becomes false. 00201 will remain energized until input 00102 becomes true and energizes the unlatch instruction (coil 00201). Note that the coil number of the latch is the same as that of the unlatch. This is an example of an Allen-Bradley latch instruction.

Allen-Bradley Output Latch Instruction

The output latch instruction (OTL) is a retentive instruction. If this input is turned on, it will stay on even if its input conditions become false. A retentive output can only be turned off by an unlatch instruction. Figure 3-49 shows an output latch instruction (OTL). In this case if the rung conditions for this output coil are true the output bit will be set to a 1. It will remain a 1 even if the rung becomes false.

Chapter 3: Fundamentals of Programming

The output will be latched on. Note that if the OTL is retentive, if the processor loses power the actual output turns off, but when power is restored the output is retentive and will turn on. This is also true in the case of switching from run to program mode. The actual output turns off, but the bit state of 1 is retained in memory. When the processor is switched to run again, retentive outputs will turn on again regardless of the rung conditions. Retentive instructions can help or hurt the programmer. Be very careful from a safety standpoint when using retentive instructions. You must use a unlatch instruction to turn a retentive output off.

$$-(\text{L})-$$

Figure 3-49. Output latch instruction (OTL).

Allen-Bradley Output Unlatch Instruction

The output unlatch instruction (OTU) is used to unlatch (change the state of) retentive output instructions. It is the only way to turn an output latch instruction (OTL) off.

$$-(\text{U})-$$

Figure 3-50. Output unlatch instruction (OTU).

Figure 3-50 shows an output unlatch instruction (OTU). If this instruction is true, it unlatches the retentive output coil of the same number.

PROGRAM FLOW INSTRUCTIONS

There are many types of flow control instructions available on PLCs. Flow control instructions can be used to control the sequence in which your program is executed. Flow instructions allow the programmer to change the order in which the CPU scans the ladder diagram. These instructions are typically used to minimize scan time and create a more efficient program. They can be used to help troubleshoot ladder logic as well. They should be used with great care by the programmer. Serious consequences can occur if they are improperly used because their use causes portions of the ladder logic to be skipped.

Master Control Relays

Master control relays are used to control blocks of a ladder diagram. They can control entire sections of a program or the whole program. The master control relay should not be confused with the hardwired master control relay that should be used to protect every application. Hardwired master control relays are used to provide for the immediate shutdown of power in the event of an emergency condition.

The master control relay instruction can be used to control one section or all of a ladder diagram (see Figure 3-51). The use of a master control relay normally requires the use of an MCR instruction to begin the controlled area and another MCR instruction to show the end of the area. The beginning MCR rung determines whether or not the MCR logic is active. If the inputs on the rung of the initial MCR instruction are true, the MCR is active. Each rung controls its outputs, just as in normal operation. If the initial MCR instruction rung is false, all nonretentive output instructions are deenergized regardless of their rung condition.

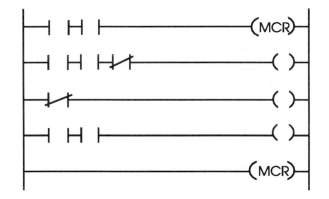

Figure 3-51. Use of a master control relay (MCR). Note that there are no input conditions for the second MCR. The second MCR marks the end of the controlled zone.

Allen-Bradley Master Control Reset Instructions

Allen-Bradley has master control reset (MCR) instructions available. The programmer must use a MCR to begin the controlled section of the program and one to end it. The end MCR must be unconditional. When the rung conditions for the MCL are true, the MCL is active. While the zone is active the actual rung conditions in the zone control their outputs (see Figure 3-52), just as if the MCR zone didn't exist.

In master control reset logic, if the MCR input conditions go false, all of the nonretentive outputs within the zone will be reset. These outputs may now be controlled by other sections of ladder. Master control reset logic can be used to break programs into sections of logic. Remember that most of the time used in writing a program is devoted to interlocking the logic (making sure that outputs are not on when they should not be). The use of zones can simplify this process by reducing the task of interlocking.

$$——(MCR)——$$

Figure 3-52. Allen-Bradley MCR instruction.

Allen-Bradley Jump Instructions

Allen-Bradley has jump (JMP) and label (LBL) instructions available (see Figure 3-53). These can be used to reduce program scan time by omitting a section of program until it is needed. It is possible to jump forward and backward in the ladder. The programmer must be careful not to jump backward an excessive amount of times. A counter, timer, logic, or the program scan register should be used to limit the amount of time spent looping inside a JMP/LBL instruction.

If the rung containing the JMP instruction is true, the CPU skips to the rung containing the specified label (LBL) and continues execution. You can jump to the same label from one or more JMP instructions.

$$——(JMP)——\qquad —|\ LBL\ |——$$

Figure 3-53. Allen-Bradley jump (JMP) and label (LBL) instructions.

Allen-Bradley Jump to Subroutine Instructions

Allen-Bradley also has subroutine instructions available. The jump-to-subroutine (JSR), subroutine (SBR), and return (RET) are used for this purpose (see Figures 3-54 and 3-55). Subroutines can be used to store recurring sections of logic that must be executed in several points in your program. A subroutine saves effort and memory because you only program it once. Subroutines can also be used for time critical logic by using immediate I/O in them.

The SBR instruction must be the first instruction on the first rung in the program file that contains the subroutine. Subroutines can be nested up to 8 deep. This allows the programmer to direct program flow from the main program to a subroutine and then on to another subroutine, and so on.

The desired subroutine is identified by the file number that you enter in the JSR instruction. This instruction serves as the label or identifier for a program file as a regular subroutine file. The SBR instruction is always evaluated as true. The RET instruction is used to show the end of the subroutine file. It causes the CPU to return to the instruction following the previous JSR instruction.

Figure 3-54. Allen-Bradley jump-to-subroutine (JSR) instruction.

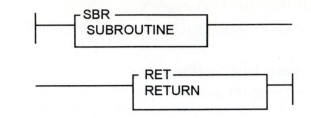

Figure 3-55. Allen-Bradley subroutine (SBR) and return (RET) instructions.

GE FANUC PROGRAM FLOW FUNCTIONS

GE Fanuc Master Control Relay (MCR) Function

When using the GE Fanuc (MCR) instruction, all rungs between an active MCR and the corresponding end master control relay (ENDMCR) function are executed without power flow to coils (see Figure 3-56). The ENDMCR function is used to resume normal program execution. MCRs can only be used in the forward direction.

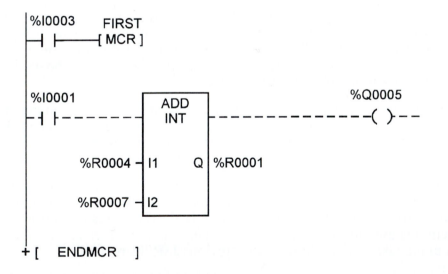

Figure 3-56. A GE Fanuc MCR.

GE Fanuc Jump Instructions

The jump instructions is used to bypass a section of program logic. There are two types of jumps: a normal jump (see Figure 3-57) and a nested jump (Figure 3-58). In the example shown in Figure 3-57, if input contact 1 is active, power flow will be transfered to the logic specified by the label LOOP1. A nested jump instruction can be placed anywhere in the program as long as it does not occur in the range of a non-nested jump or non-nested MCR. If input 1 is true in Figure 3-58 power flow will be transfered to the logic specified by the label LOOP2.

Figure 3-57. Example of a GE Fanuc jump instruction.

Figure 3-58. Example of a GE Fanuc nested jump instruction.

GE FANUC SUBROUTINE INSTRUCTIONS

GE Fanuc also has subroutine instructions available. Subroutines can be effectively used to control program flow. They can also be used to shorten and organize programs. The programmer can divide an application's logic into subroutines that can be called as needed by the program. Figure 3-59 shows an example of the use of a subroutine. In this case a section of program (subroutine 3) that is to be repeatedly used is called by the program.

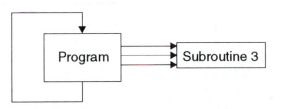

Figure 3-59. Example of a GE Fanuc subroutine instruction.

Figure 3-60 shows an example of a program used to call three separate subroutines. Figure 3-61 shows an example of subroutines calling other subroutines. The GE Fanuc PLC will only allow eight nested calls before an error occurs. The program is considered level 1.

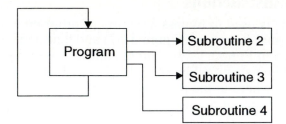

Figure 3-60. Example of a GE Fanuc subroutine instruction.

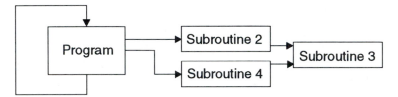

Figure 3-61. Example of a nested GE Fanuc subroutine instruction.

Figure 3-62 shows the use of a CALL statement to call a subroutine in a ladder diagram. Subroutines must be declared through the block declaration editor before a CALL instruction can be used for that subroutine. There is a limit of 64 subroutine declarations for a program; 64 CALL instructions are allowed for each logic block in a program.

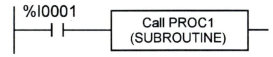

Figure 3-62. Example of the use of a GE Fanuc subroutine instruction.

PLC SCANNING AND SCAN TIME

Now that you are familiar with some basic PLC instructions and programming, it is important to understand the way a PLC executes a ladder diagram. Most people would like to believe that a ladder is a very sequential thing. We like to think of a ladder as first things first. We would like to believe that the first rung is evaluated and acted on before the next, and so on. We would like to believe that the CPU looks at the first rung, goes out and checks the actual inputs for their present state, comes back, immediately turns on or off the actual output for that rung, and then evaluates the next rung. This is not exactly true. Misunderstanding the way the PLC scans a ladder can cause programming bugs.

Chapter 3: Fundamentals of Programming

Scan time can be divided into two components: I/O scan and program scan. When the PLC enters run mode it first takes care of the I/O scan (see Figure 3-63). The I/O scan can be divided into the output step and the input step. During these two steps the CPU transfers data from the output image table to the output modules (output step) and then from the input modules to the input image table (input step).

The third step is logic evaluation. The CPU uses the conditions from the image table to evaluate the ladder logic. If a rung is true, the CPU writes a one into the corresponding bit in the output image table. If the rung is false, the CPU writes a zero into the corresponding bit in the output image table. Note that nothing concerning real-world I/O is occurring during the evaluation phase. This is often a point of confusion. The CPU is basing its decisions on the states of the inputs as they existed before it entered the evaluation phase. We would like to believe that if an input condition changes while the CPU is in the evaluation phase, it would use the new state. It cannot. (*Note*: It actually can if special instructions called immediate update contacts and coils are used. For the most part most ladders will not utilize immediate instructions. This is covered in more detail later.) The states of all inputs were frozen before it entered the evaluation phase. The CPU does not turn on/off outputs during this phase either. This phase is only for evaluation and updating the output image table status.

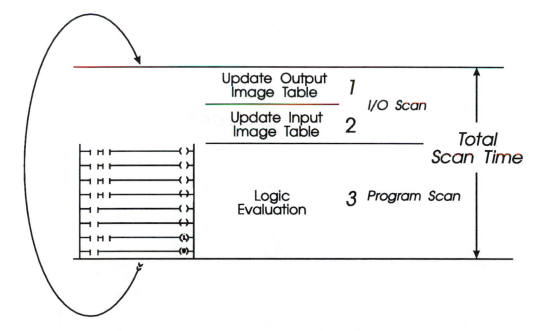

Figure 3-63. A generic example of PLC scanning.

Once the CPU has evaluated the entire ladder, it performs the I/O scan again. During the I/O scan the output states of real-world outputs are changed depending on the output image table. The real-world input states are then transferred again to the input image table.

All of this only took a few milliseconds (or less). PLCs are very fast. That is why troubleshooting can be so troublesome. Scan time is the sum of the times it takes to execute all of the individual instructions in the ladder. Simple contacts and coils take very little time. Complex math statements and other types of instructions take much more time. Even a long ladder diagram will normally execute in less than 50 milliseconds or so. There are considerable differences in the speeds of different brands and models of PLCs. Manufacturers normally will give scan time in terms of fractions of milliseconds per K of memory. This can help give a rough idea of the scan times of various brands.

Questions

1. What is a contact? A coil?

2. What is a transitional contact?

3. What are transitional contacts used for?

4. Explain the term *normally open* (examine-on).

5. Explain the term *normally closed* (examine-off).

6. What are some uses of normally open contacts?

7. Explain the terms *true* and *false* as they apply to contacts in ladder logic.

8. Design a ladder that shows series input (AND logic). Use X5, X6, AND NOT (normally closed contact) X9 for the inputs and use Y10 for the output.

9. Design a ladder that has parallel input (OR logic). Use X2 and X7 for the contacts.

10. Design a ladder that has three inputs and one output. The input logic should be: X1 AND NOT X2, OR X3. Use X1, X2, and X3 for the input numbers and Y1 for the output.

11. Design a three-input ladder that uses AND logic and OR logic. The input logic should be X1 OR X3, AND NOT X2. Use contacts X1, X2, and X3. Use Y12 for the output coil.

12. Design a ladder in which coil Y5 will latch itself in. The input contact should be X1. The unlatch contact should be X2.

13. Design a latching circuit using an Allen-Bradley latch and unlatch instruction. Use contact 00107 for the latch input, 00102 for the unlatch input, and 00513 for the coil.

14. What is the primary purpose of MCRs?

15. Is a ladder logic MCR sufficient to guarantee the safety of a system? Why or why not?

16. Explain the primary purpose of zone control logic.

17. Draw a diagram and thoroughly explain what occurs during a PLC scan.

18. Examine the rungs below and determine whether the output for each is on or off. The input conditions shown represent the states of real-world inputs.

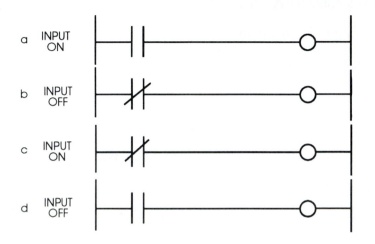

19. Examine the rungs below and determine whether the output for each is on or off. The input conditions shown represent the states of real-world inputs.

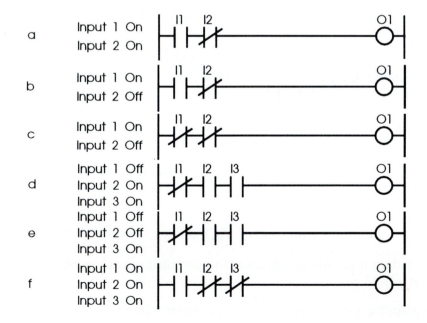

Chapter 3: Fundamentals of Programming

20. Examine the rungs below and determine whether the output for each is on or off. The input conditions shown represent the states of real-world inputs.

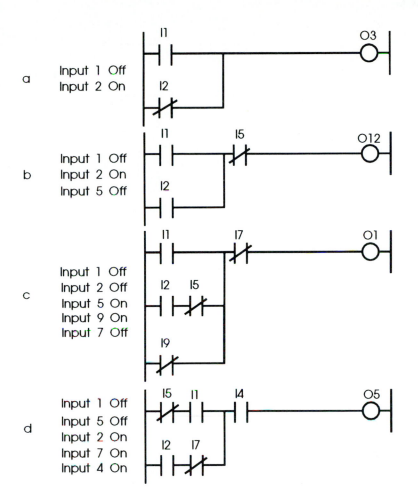

a
Input 1 Off
Input 2 On

b
Input 1 Off
Input 2 On
Input 5 Off

c
Input 1 Off
Input 2 Off
Input 5 On
Input 9 On
Input 7 Off

d
Input 1 Off
Input 5 Off
Input 2 On
Input 7 On
Input 4 On

Additional Exercises

Explain each of the following element addresses.

  1. N7:12

  2. B3:16/12

  3. O:1

  4. I:1/3

  5. O:2/5

6. I:1/2
7. T4:7.PRE
8. C5:0.ACC
9. F8:0
10. I:1/2

Chapter 4

Timers and Counters

Timers and counters are invaluable in PLC programming. Industry has a need to count product, time sequences, and so on. In this chapter we will examine the types and programming of timers and counters. Timers and counters are similar in all PLCs. In this chapter we will show examples of several of the leading brands.

OBJECTIVES

Upon completion of this chapter, the student will be able to:

Describe the use of timers and counters in ladder logic .

Define such terms as **retentive, cascade, delay-on, delay-off, flow diagram,** *and* **pseudocode.**

Develop flow diagrams and pseudocode for applications.

Utilize timers and counters to develop applications.

TIMERS

Timing functions are very important in PLC applications. Cycle times are critical in many processes. Timers are used to delay actions. They may be used to keep an output on for a specified time after an input turns off or to keep an output off for a specified time before it turns on.

Think of a garage light. It would be a nice feature if a person could touch the on switch and the light would immediately turn on, and then stay on for a given time (maybe 2 minutes). At the end of the time, the light would turn off. This would allow you to get into the house with the light on. In this example, the output (light) turned on instantly when the input (switch) turned on. The timer counted down the time (timed out) and turned the output (light) off. This is an example of a *delay-off timer*.

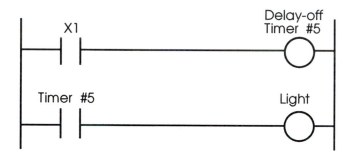

Figure 4-1. Delay-off timing circuit. If contact X1 closes, delay-off timer immediately turns on, which turns on the light. When the timer reaches the programmed time it will turn off, which turns the light off also.

Consider Figure 4-1. When switch X1 is activated, the timer turns on and starts counting. The timer is used in the second rung as a contact. If the timer is on, the contact in rung 2 (timer 5) closes and turns on the light. When the timer times out (in this case 2 minutes), the contact in rung 2 opens and the light turns off. As you see from this example, output coils can be used as contacts to control other outputs.

This type of timer is called a *delay-off timer*. The timer turns on instantly, counts down and then turns off (delay-off).

The other type of timer is called *delay-on*. When input X3 to the delay-on timer is activated, the timer starts counting, but remains off until the time has elapsed (see Figure 4-2). In this case, the switch is activated and the timer starts to count, but it remains off until the total time has elapsed. Then the timer turns on, which closes the contact in rung 2 and turns on the light.

Many PLCs use block-style timers and counters (see Figure 4-3). No matter what brand of PLC it is, there are many similarities in the way timers are programmed. Each timer will have a number to identify it. For some it will be as simple as T7 (timer 7). For others it may be the address of the storage register that holds the accumulated value of the timer, such as storage register 4001.

Each timer will have a time base. Timers can typically be programmed with several different time bases: 1 second, 0.1 second and 0.01 second are typical time bases. If a programmer entered .1 for the time base and 50 for the number of delay increments the timer would have a 5-second delay (50 X 0.1 second = 5 seconds).

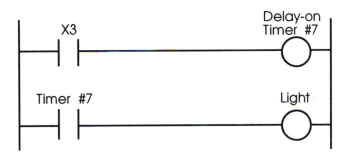

Figure 4-2. Delay-on timing circuit. In this example if contact X3 closes the timer will begin timing. When the time reaches the programmed time the timer will turn on which will turn the light on.

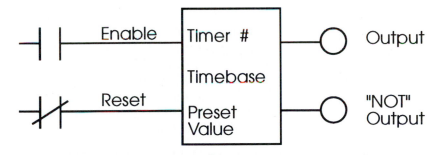

Figure 4-3. A typical block type timer.

Timers also must have a preset value. The preset value is the number of time increments the timer must count before changing the state of the output. The actual time delay would equal the preset value multiplied by the time base. Presets can be a constant value or a variable. If a variable is used, the timer would use the real-time value of the variable to calculate the delay. This allows delays to be changed depending on conditions during operation. An example is a system that produces two different products, each requiring a different time in the actual process. Product A requires a 10-second process time, so the ladder logic would assign ten to the variable. When product B comes along the ladder logic can change the value to that required by B. When a variable time is required, a variable number is entered into the timer block. Ladder logic can then be used to assign values to the variable.

Timers typically have two inputs. The first is the timer enable input. When this input is true (high) the timer will begin timing. The second input is the reset input. This input must be in the correct state for the timer to be active. Some brands of PLCs require that this input be low for the timer to be active; other manufacturers require a high state. They all function in essentially the same manner, however. If the reset line changes state, the timer clears the accumulated value. For example, the timer in Figure 4-3 requires a high for the timer to be active. If the reset line goes low, the timer clears the accumulated time to zero.

Timers can be retentive or nonretentive. *Retentive timers* do not lose the accumulated time when the enable input line goes low. They retain the accumulated time until the line goes high again. They then add to the count when the input goes high again. *Nonretentive timers* lose the accumulated time every time the enable input goes low. If the enable input to the timer goes low, the timer count goes to zero. Retentive timers are sometimes called accumulating timers. They function like a stopwatch. Stopwatches can be started and stopped and retain their timed value. There is also a reset button on a stopwatch to reset the time to zero.

Allen-Bradley PLC-2 Timers

Allen-Bradley timers require two values: an accumulated value (AC) and a preset value (PR). Each value requires one word of memory. The accumulated value holds the current number of elapsed time intervals. The upper four bits (most significant bits) are used as status bits. The preset value is the number of time intervals it is desired to time to. When the number of preset intervals equal the accumulated value of time intervals, a status bit is set. The status bit can turn devices on or off. Bit number 15 is the timed bit. Bit 15 is set to either a one or a zero, depending on which timer instruction is chosen when the timer has timed out. Bit number 17 is the enable bit. It is set to one when the rung conditions are true and reset to zero when the rung conditions are false. There are three time intervals that can be selected: 1.0 second, 0.1 second, or 0.01 second.

There are four types of timer instructions available: on-delay (TON), off-delay (TOF), retentive delay-on (RTO), and retentive timer reset (RTR).

On-Delay Timer (TON)

The Allen-Bradley on-delay timer (TON) is programmed like an output instruction. The TON timer does not accumulate the time if the timer loses its input signal. It will continue to count time intervals as long as the input condition is true. The timer will be "on" when the accumulated time intervals are equal or greater than the preset. As soon as the input to the timer goes low the accumulated value is set to zero and the timer is off.

Chapter 4: Timers and Counters

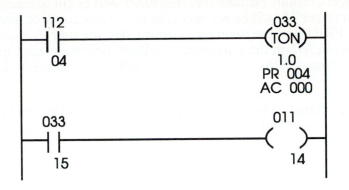

Figure 4-4. Use of an Allen-Bradley TON timer in a ladder diagram.

Figure 4-4 shows the use of a TON timer in a ladder diagram. The timer number is 33. If input 11204 becomes true, the timer will start to accumulate time.

Note that a 1.0-second time base is used. This means that the actual time delay will be the preset (4) multiplied by the time base (1.0 second). The delay will be 4 seconds. When the accumulated time equals the preset time, bit number 15 (03315 - timer 33 bit 15) will be set to one. The timer is then used as an input on the second rung to control output coil 01114. This means that when the timer times out, output 01114 will be energized.

Off-Delay Timer (TOF)

The off-delay timer (TOF) is programmed like an output instruction also. The main difference is that the timer increments the accumulated time when the input condition is false. The Allen-Bradley TOF timer does not accumulate the accumulated time if the input condition changes state. The accumulated value of the timer resets to zero if the input condition becomes true. The timer counts time intervals as long as the input condition remains false. In this manner the output will turn on instantly when the input condition becomes true and will remain on until the accumulated time equals the preset time.

Bit numbers 15 and 17 will be set as soon as the rung conditions become true. The timer will then begin to increment the accumulated (AC) value. When the accumulated value equals the preset value, bit number 15 will be reset. Bit number 17 is reset when the rung input condition becomes false.

Retentive Delay-On Timer (RTO)

The retentive delay-on timer (RTO) is very similar to the regular delay-on instruction. The only difference is that the accumulated time does not get reset if the input condition becomes false. The accumulated value remains at the present value until the input condition becomes true again. The timer then begins to add time interval counts to the accumulated value of the timer.

When the rung conditions become true, the timer will begin to count time base intervals. Bit number 15 will be set when the accumulated value equals the preset value (ACC = PRE), and the timer stops timing. Bit number 17 is set when the rung becomes true. When the rung input condition becomes false, the accumulated value is retained. Bit number 15's status will not be changed. Bit number 17 will be reset.

The timer accumulated value is reset to zero only if a special timer instruction is used. The instruction is called a retentive timer reset instruction.

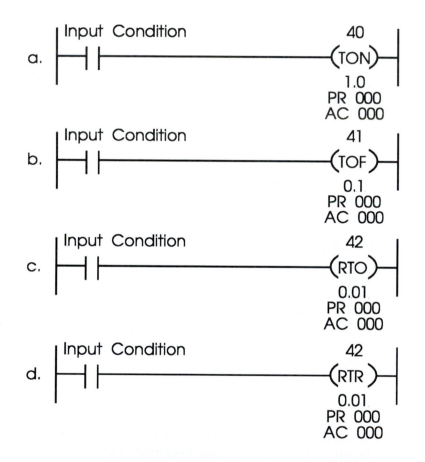

Figure 4-5. Use of Allen-Bradley timer instructions: (a) shows the use of a TON instruction; (b) shows the use of a TOF instruction; (c) shows the use of an RTO instruction; (d) shows the use of an RTR instruction.

Chapter 4: Timers and Counters

Retentive Timer Reset (RTR)

The retentive timer reset (RTR) must be given the same number as the timer that it is supposed to reset. If the input conditions to the RTR become true, the instruction will reset the accumulated value of its associated timer to zero. It will also reset the status bits for the associated timer. This means that bit numbers 15 and 17 will be reset to zero. The retentive timer reset is the only way to reset the accumulated value of a retentive timer instruction. See Figure 4-5 for examples of timer use.

ALLEN-BRADLEY PLC-5, SLC 500, AND MICROLOGIX 1000 TIMERS

Timer On-Delay

The timer on-delay instruction (TON) is used to turn an output on after a timer has been on for a preset time interval. The timer on-delay (TON) begins accumulating time when the rung becomes true and continues until one of the following conditions is met: The accumulated value is equal to the preset value; the rung goes false; a reset timer instruction resets the timer; or the associated SFC step becomes inactive. Figure 4-6 shows the numbering system for timers. The T stands for timer, the X would be the file number of the timer, and the Y is the actual timer number.

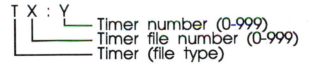

Figure 4-6. Format for a timer.

Status Bit Use

Timer status bits can be used in ladder logic. There are several bits available for use. The timer enable bit (.EN) is set when the rung goes true. It stays set until the rung goes false, or a reset instruction resets the timer, or the associated SFC step goes false. The .EN bit indicates that the timer is enabled. The .EN bit from any timer can be used for logic. For example, T4:0.EN could be used as the number for a contact in a ladder.

The timer timing bit (.TT) can also be used. The .TT bit is set when the rung goes true. It remains true until the rung goes false, or the .DN bit is set (accumulated value equals preset value), or the associated SFC step becomes inactive. For example, T4:0.TT could be used as a contact in a ladder.

Condition	Result
If the rung is true	.EN bit remains set (1)
	.TT bit remains set (1)
	.ACC value is cleared and begins counting up
If the rung is false	.EN bit is reset
	.TT bit is reset
	.DN bit is reset
	.ACC value is cleared and begins counting up

Figure 4-7. Use of special timer bits.

The timer done bit (.DN) is set until the accumulated value is equal to the preset value and the .DN remains set until the rung goes false, a reset instruction resets the timer, or the associated SFC step becomes inactive. When the .DN bit is set, it is an indication that the timing operation is complete. For example, T4:0.DN could be used as a contact in the ladder. The preset (.PRE) is also available to the ladder. For example, T4:0.PRE would access the preset value of T4:0.

Accumulated Value Use

The accumulated value can also be used by the programmer. The accumulated value (.ACC) is acquired in the same manner as the status bits and preset. For example, T4:0.ACC would access the accumulated value of timer T4:0.

Time Bases

Two time bases are available: 1-second intervals or 0.01-second intervals (see Figure 4-8). The potential time ranges are also shown. If a longer time is needed, timers can be cascaded (discussed later in this chapter). Figure 4-9 shows how timers are handled in memory. Three bits are used in the first storage location for this timer to store the present status of the timer bits (EN, .TT, and .DN). The preset value (PRE) is stored in the second 16 bits of this timer storage. The third 16 bits hold the accumulated value of the timer.

Time Base	Potential Time Range
1 Second	To 32,767 time-base intervals (up to 9.1 hours)
.01 Second (10 ms)	To 32,767 time-base intervals (up to 5.5 minutes)

Figure 4-8. Time bases available.

Chapter 4: Timers and Counters

15	14	13	12	11	10	9	8	7	6	5	4	3	2	1	0	Element
EN	TT	DN								Internal Use						0 T4:0
Preset Value PRE																1 T4:0.PRE
Accumulated Value ACC																2 T4:0.ACC

Figure 4-9. Use of control words for timers.

Figure 4-10 shows an example of the use of a TON timer in a ladder. When input I:012/10 is true, the timer begins to increment the accumulated value of TON timer 4:0 in 1-second intervals. The timer timing bit (.TT) for timer 4:0 is used in the second rung to turn on output O:013/01, while the timer is timing (.ACC < .PRE). The timer done bit (.DN) of timer 4:0 is used in rung 3 to turn on output O:013/02 when the timer is done timing (.ACC = .PRE). The preset for this timer is 180, which means that the timer will have to accumulate 180 1-second intervals to time out. Note that this is not a retentive timer. If input I:012/10 goes low before 180 is reached, the accumulated value is reset to zero.

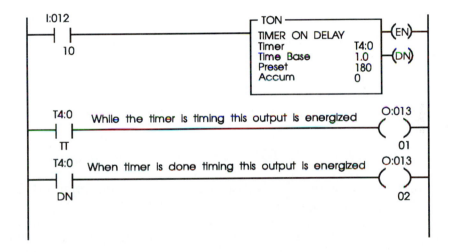

Figure 4-10. Use of a TON timer in a ladder logic program.

Timer Off-Delay

The timer off-delay instruction is used (TOF) to turn an output on or off after the rung has been off for a desired time. The TOF instruction starts to accumulate time when the rung becomes false. It will continue to accumulate time until the accumulated value equals the preset value, or the rung becomes true, or a reset timer instruction resets the timer, or the associated SFC becomes inactive. The timer enable bit (.EN bit 15) is set when the rung becomes true. It is reset when the rung become false, or a reset instruction resets the timer, or the associated SFC step becomes inactive. The timer timing bit (.TT bit 14) is set when the rung becomes false and .ACC < .PRE. The .TT bit is reset when the rung becomes false, or the .DN bit is reset (.ACC=.PRE), or a reset instruction resets the timer, or the associated SFC step becomes false.

The done bit (.DN bit 13) is reset when the accumulated value (.ACC) is equal to the preset (.PRE) value. The .DN bit is set when the rung becomes true, or a reset instruction resets the timer, or the associated SFC step becomes inactive.

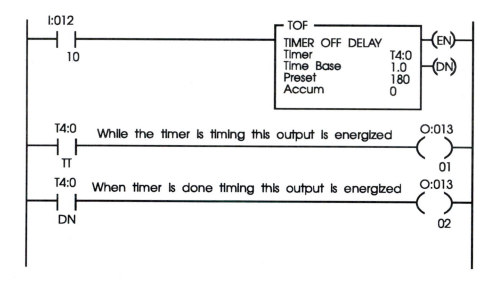

Figure 4-11. Use of a TOF timer in a ladder logic diagram.

Figure 4-11 shows the use of a TOF timer in a ladder diagram. Input I:012/10 is used to enable the timer. When input I:012/10 is false the accumulated value is incremented as long as the input stays false and .ACC < = .PRE. The timer enable bit for timer T4:0 (T:4.TT) is used to turn output O:013/01 on while the timer is timing. The done bit (.DN) for timer 4:0 (T4:0.DN) is used to turn on output O:013/02 when the timer has completed the timing (.ACC = .PRE).

Retentive Timer On

The retentive timer on instruction (RTO) is used to turn an output on after a set time period (see Figure 4-12). The RTO timer is an accumulating timer. It retains the accumulated value even if the rung goes false. The only way to zero the accumulated value is to use a reset instruction in another rung with the same number as the RTO you wish to reset. The RTO retains the accumulated count even if power is lost, or you switch modes, or the rung becomes false, or the associated SFC becomes inactive. Remember, the only way to zero the accumulated value is to use a reset instruction.

Chapter 4: Timers and Counters

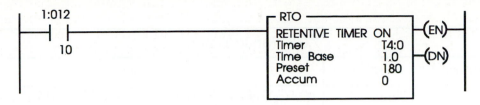

Figure 4-12. Use of an RTO timer.

The status bits can be used as contacts in a ladder diagram. The timer enable bit (.EN) is set when the rung becomes true. When the .EN bit is a 1 it indicates that the timer is timing. It remains set until the rung becomes false or a reset instruction zeros the accumulated value.

The timer timing bit (.TT) is set when the rung becomes true and remains set until the accumulated value equals the preset value or a reset instruction resets the timer. When the .TT bit is a 1 it indicates that the timer is timing.

The .TT bit is reset when the rung becomes false or when the done bit (.DN) is set. The timer done bit (.DN) is set when the timer's accumulated value is equal to the preset value. When the .DN bit is set it indicates that the timing is complete. The .DN is reset with the reset instruction.

GE FANUC TIMERS

Figure 4-13 shows a typical GE Fanuc timer block with a typical enable input and reset input. The preset value for the timer must be entered. The time base can also be changed. The actual delay can be calculated by multiplying the preset value by the time base. The programmer can choose a register where the present value of the timer will be stored or it can be chosen automatically. There is one output from each timer.

Figure 4-14 shows the use of a GE Fanuc TMR timer. Note that the preset value in this case is a constant (a variable time can be used also through the use of a register). The value of the constant is 50. The timebase is .1 second. This timer will time to 5 seconds (50 X .1 seconds). The current time in the timer will be held in register 1. If input 1 (%I0001) becomes energized, the timer will begin to accumulate time. When the current value reaches 50 in register 1 the timer will pass power to output coil 7 (%Q0007). Any time the current value (register 1) is larger than the preset value the timer will pass power to the coil. If the input coil (%I0001) becomes deenergized at any time, the timer current value is set to zero.

GE-Fanuc Timers		
Instruction	Function	Description
ondtr	On-Delay Stopwatch Timer	This type of timer accumulates time while receiving power. It passes power if the current value exceeds the preset value. The current value is reset to zero when the reset (R) input receives power.
oftd	Off-Delay Timer	This timer increments while power flow is off and resets to zero when power flow is on. Time bases available are tenths of seconds (.1 default), hundredths of seconds (.01), and thousandths of seconds (.001). The range for this timer is 0-32,767 time units. This timer is retentive on power failure. Note that no automatic initialization occurs at powerup.
tmr	On-Delay Timer	The current value of the timer function is set to zero when the function transitions on. The function accumulates time while receiving power and passes power if the current value is greater than or equal to a preset value.

Figure 4-13. The types of GE Fanuc timers.

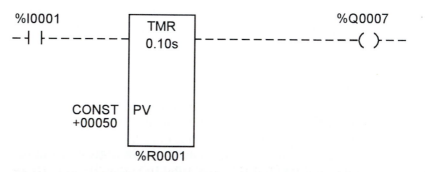

Figure 4-14. The use of a GE Fanuc timer.

Figure 4-15 shows the use of a GE Fanuc on-delay timer (ONDTR) timer. Note that the preset value in this case is a constant. The value of the constant is 50. The time base is .1 second. This timer will time to 5 seconds (50 X .1 seconds). The current time in the timer will be held in register 1. If input 1 (%I0001) becomes energized, the timer will begin to accumulate time.

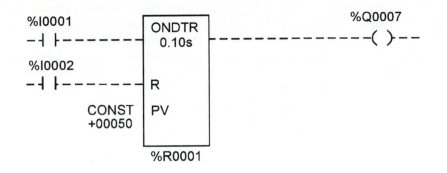

Figure 4-15. Use of an on-delay timer.

When the value reaches 50 in register 1 the timer will pass power to output coil 7 (%Q0007). Any time the current value (register 1) is larger than the preset value the timer will pass power to the coil. If the input coil (%I0001) becomes deenergized at any time, the timer current value is retained. When the input (%I0001) becomes energized again the timer begins to increment the time again. This is an accumulating timer. The timer's current value (accumulated time) is only reset to zero if the input to the reset input of the timer is energized.

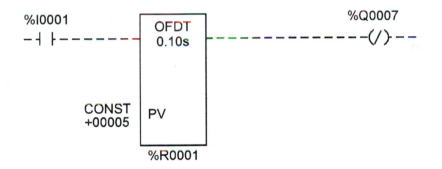

Figure 4-16. Example of the use of a GE Fanuc off-delay timer.

Figure 4-16 shows the use of a GE Fanuc off-delay timer (OFDT) timer. Note that the preset value in this case is a constant. The value of the constant is 50. The time base is .1 second. This timer will time to 5 seconds (50 X .1 seconds). This timer increments while power flow is off, and resets to zero when power flow is on. This means that while the present value exceeds the preset value the timer passes power. If the input (%I0001) is off the timer increments the time. It continues to increment the time until the input is energized. When the input is energized the timer is reset to zero.

GOULD MODICON TIMERS

Figure 4-17 shows a typical Gould Modicon timer block with a typical enable input and reset input. This is a TMR type timer. It is nonretentive. If the input goes low the timer's accumulated time is reset to zero. The preset value for the timer must be entered. The time base must be chosen. The actual delay can be calculated by multiplying the preset value by the time base. The timer number is actually a storage register where the accumulated value of the timer will be stored. There are two outputs possible from the timer. The first output is one that will turn on when the timer reaches the timer preset value. The second output is the opposite. It is "normally on" and it will turn off when the timer reaches the preset value. These two outputs make it very easy to program delay-on or delay-off logic.

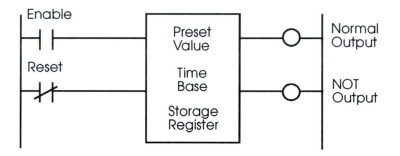

Figure 4-17. Gould Modicon timer block.

Figure 4-18 shows the actual use of a Gould Modicon timer. Note the first number of the inputs, outputs, and storage register. The 1 in the inputs denotes that the number will be an input. The first 0 in output 0005 shows that the number represents an output. The 4 of the number 4004 denotes shows that the number is a storage register. The time base is 1.0 second. The preset is 250, so the value of the delay is 250 seconds (250 X 1.0 second = 250 seconds). It will be used as a delay-on timer because the top output on the timer block was used. Note that input 7 is programmed as a normally closed contact. The reset line must be high for the timer to time. If this line goes low the timer is reset to a count of zero and is unable to time again until the reset line goes high. Input 2 is being used as the enable line. The current value of the timer will be stored in register 4004. If the timer value reaches 25 seconds, output 5 will turn on and stay on. If input 2 goes low or input 7 goes low, output 5 will turn off.

OMRON TIMERS

There are two types of timers: TIM and TIMH. The TIM timer measures in increments of 0.1 second. It is capable of timing from 0 to 999.9 seconds with an accuracy of plus or minus 0.1 second. The high-speed timer (TIMH) measures in increments of 0.01 second. Both timers are decrementing-style delay-on timers.

They require a timer number and a set value (SV). When the set value (SV) has elapsed, the timer output turns on. Timer counter numbers refer to an actual address in memory. Numbers must not be duplicated. You cannot use the same number for a timer and a counter. Timers/counters are numbered 0 to 511.

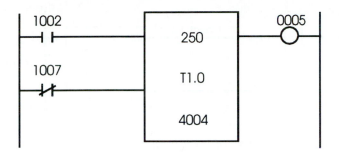

Figure 4-18. Gould Modicon timer example.

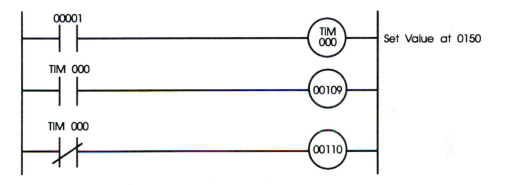

Figure 4-19. Omron TIM timer being used as a normally open contact and a normally closed contact.

Figure 4-19 shows the use of a TIM timer in a ladder diagram. When the timer decrements to zero the timer output turns on. Output 00110 will be on until the timer becomes true. Note that a normally closed timer 000 has been used as the contact for output coil 00110. This means that as long as the timer is false, output 001100 will be energized. When the timer becomes true output 00109 will turn on. This, of course, also means that output 00110 will deenergize. The timer will reset if input 00001 becomes false. A power failure would also cause the timer to reset. This can be solved if a retentive timer is used. A counter is used to make a retentive timer. Remember that there is very little difference between a timer and a counter. Both count increments or events. If a counter is used to count time increments, it becomes a timer.

Figure 4-20 shows the use of a counter to make a retentive timer. A special bit is being used to create the 1-second pulse. Bit 25502 is a special bit that Omron provides. It is a one-second clock pulse. If input 00001 is present and the one-second pulse bit makes 20 transitions, the counter will be true. Even if power is lost the present value of the count is retained. This means that we now have a retentive timer.

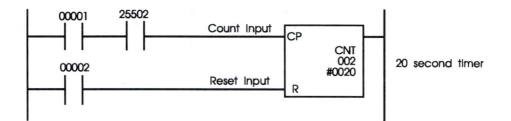

Figure 4-20. The use of a counter to create a retentive timer. This will preserve the present value of the timer in the event of a power interruption. Note that contact 25502 is a special internal bit. Bit 25502 is a 1-second timing pulse that can be used by the programmer.

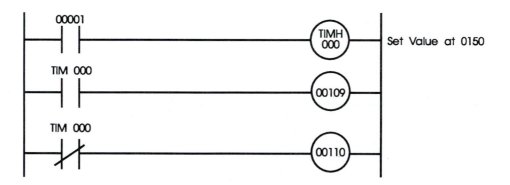

Figure 4-21. Omron high-speed timer (TIMH). Output 00110 will be on until the timer times out. This is the case because a normally closed contact was used. When the timer times out, output 00109 will be energized.

The high-speed timer (TIMH) times in 0.01-second increments. The TIMH timer has a range of 0.00 to 99.99 seconds. Scan time can affect TIMH timers numbered 48 to 511. Timers 48 to 511 may be inaccurate if the scan time exceeds 10 milliseconds. If the scan time is more than 10 milliseconds use timer numbers below 48. Figure 4-21 shows the use of a TIMH timer. It operates just like the TIM timer except that the time increment is smaller. Remember that the TIM and TIMH timers are not retentive; if the enable input is lost the present value is lost.

Chapter 4: Timers and Counters

SQUARE D TIMERS

Figure 4-22 shows a typical Square D timer. There are two inputs: a timing line (enable) and a clear (reset) line. There is a storage register (sometimes called data register) address. The storage register address is where the current value of the timer will be stored. There is a time base (0.1 second, 0.01 second, and 0.1 minute). There is a preset value, which is multiplied by the time base to calculate the time delay. When the number in the storage register equals the preset value the output is turned on. In this case when storage register 17 equals 50, the output will turn on. Square D calls the preset a decode value. There is also an output shown at the bottom of the block. This is the output number that will turn on when the timer reaches the required delay.

Figure 4-23 shows a Square D retentive timer used as a nonretentive timer. This same method can be used with most timers. The enable input to the timer has been programmed without a contact. This means that this input is always high. The counter is always enabled. This means that input 01-05 totally controls the timer. If input 01-05 is off, register S17 is reset to zero. If 01-05 is high, the timer begins to time. If register S17 equals 50 then output 04-10 will turn on. It will stay on until input 01-05 goes low. If input 01-05 goes low before S17 reaches 50 the output will stay off. As you can see, even though it is a retentive timer, it never retains the count if the input (01-05) goes low.

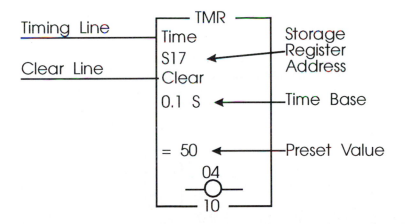

Figure 4-22. Example of a Square D timer.

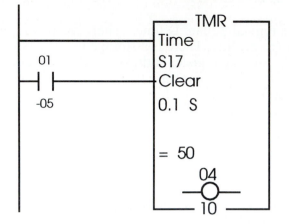

Figure 4-23. Square D timer. This timer is being used as a delay-on timer. When input 01-05 is true (on), the timer counts. If the timer reaches 5 seconds (50 X 0.1) output 04-10 will be turned on until input 01-05 turns off.

TEXAS INSTRUMENTS AND PLC DIRECT TIMERS

Figure 4-24 shows a PLC Direct/Texas Instruments nonretentive timer. There are two values required for the timer block. The first value is the timer number. The second is the preset value. If X0 becomes true (high), the timer will begin to time. If the time equals the preset, the timer contact turns on. The timer contact can then be used in another rung. In this case if the timer reaches its preset, output Y1 will turn on until X0 goes low again.

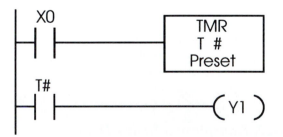

Figure 4-24. PLC Direct/Texas Instruments series 405 timer.

Figure 4-25 shows a PLC Direct/Texas Instruments accumulating timer. Accumulating timers can time up to a maximum of 99,999,999. Note that the TMRA in the block means "accumulating timer." The PLC Direct/Texas Instruments series 405 timer requires a low at the reset line to activate the timer. If the reset line goes high, the timer resets. In this case if X0 goes high, the timer will begin to time. If the timer count reaches the preset value, the timer contact will turn on. This would turn on output Y1. If X0 goes low before the count reaches the preset time, the timer will hold the present time and will not reset the time count to zero. When X0 goes high again the timing will continue. An accumulating timer will never reset to zero unless the reset line goes high.

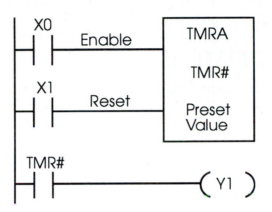

Figure 4-25. PLC Direct/Texas Instruments accumulating timer example.

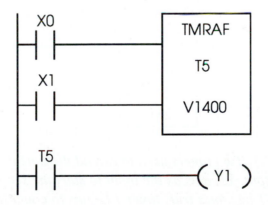

Figure 4-26. This figure shows a PLC Direct/Texas Instruments fast accumulating timer (TMRAF). The time increment for the fast timer is 0.01 second.

Figure 4-26 shows the use of another PLC Direct/Texas Instruments timer. Note that this is a fast accumulating timer (TMRAF). This is really a retentive timer with a time base of 0.01 second. The timer number is 5. The enable input is X0. The reset input is X1. T5 controls output Y1. (If the timer preset has been reached, the timer is on and output Y1 is on.) The preset in this case is a variable (V1400). The value for the preset value will equal whatever the value of V1400 is. Other ladder logic would establish the value of V1400.

The accumulated value of timers is kept in variables. There are 128 (0 to 177 octal) timers available. Their accumulated value is stored in variables V00000 to V00177. If we wanted to monitor the accumulated value of timer 5 we would examine variable V00005. These accumulated values can be used in ladder logic just as any other variable could be.

CASCADING TIMERS

Applications sometimes require longer time delays than one timer can accomplish. Multiple timers can than be used to achieve a longer delay than would otherwise be possible. One timer acts as the input to another. When the first timer times out it becomes the input to start the second timer timing. This is called *cascading* (see Figure 4-27).

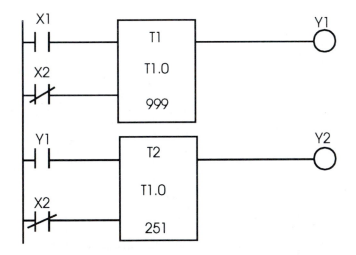

Figure 4-27. Two timers used to extend the time delay. The first timer output, Y1, acts as the input to start the second timer. When input X1 becomes true, timer 1 begins to count to 999 seconds. (The limit for these timers is 999 seconds. The limit for accumulating timers is 99,999,999.) When it reaches 999, output Y1 turns on. This activates input Y1 to timer 2. Timer 2 then counts to 251 seconds and then turns on output Y2. The delay was 1250 seconds.

Chapter 4: Timers and Counters

COUNTERS

Counting is very important in industrial applications. Often, the product must be counted so that another action will take place. For example, if 24 cans go into a case, the twenty-fourth can should be sensed by the PLC and the case should be sealed. Counters are required in almost all applications.

Several types of counters are available, including up counters, down counters, and up/down counters. The choice of which to use is dependent on the task to be done. For example, if we are counting the finished product leaving a machine, we might use an up counter. If we are tracking how many parts are left, we might use a down counter. If we are using a PLC to monitor an automated storage system, we might use an up/down counter to track how many are coming and how many are leaving to establish an actual, total number in stock.

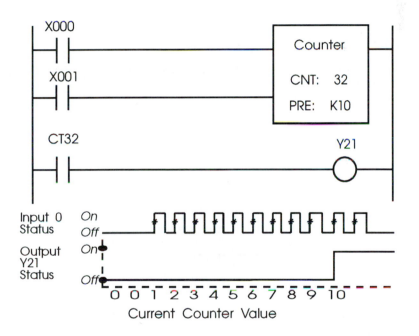

Figure 4-28. How a typical counter works. When 10 or more low-to-high transitions of input X000 have been made, counter CT32 is energized, which energizes output Y21.

Counters normally use a low-to-high transition from an input to trigger the counting action (see Figure 4-28). They also have a reset line to clear the accumulated count. There are actually a lot of similarities to timers. Timers count the number of time increments and counters count the number of low-to-high transitions on the input line.

Note that this timer in Figure 4-28 is edge-sensitive triggered. The rising, or leading edge, triggers the counter. Other devices may be level (magnitude) sensitive. X000 is used to count the pulses. Every time there is an off-to-on transition on X000, the counter adds one to its count. When the count equals the preset value, the counter turns on, which turns on output Y20. X001 is used as a reset/ enable. If contact X001 is closed, the counter is returned to zero. The counter is active (enabled or ready to count) only if X000 is off (open). The example just described is an up counter.

Down counters cause a count to decrease by one every time there is a pulse. There are also up/down counters. An up/down counter has one input that causes it to increment the count and another that causes it to decrement the count.

There are ladder diagram statements that can utilize these counts for comparing and/or decision making. Counts can also be compared to constants or variables to control outputs.

Allen-Bradley PLC-2 Counters

The Allen-Bradley counter is used to count events. Events are low-to-high transitions of the input conditions to the counter. The counter instructions are very similar to the AB timer instructions. They are programmed like output instructions. The upper 4 bits in the accumulated value (AC) are status bits. Bit number 14 is the overflow/underflow bit. It is set to a 1 when the accumulated value (ACC) of a CTU counter exceeds 999 or when the ACC value of a CTD instruction falls below zero. Bit number 15 is the count complete bit. This bit is set to a 1 when the accumulated value is greater than or equal to the preset value (PRE). Bit number 16 is the enable bit for the CTD instruction. It is set to a 1 when the rung's input condition is true. Bit number 17 is the enable bit for the CTU instruction. It is set to a 1 when the counter's rung conditions are true.

There are three types of counter instructions:

> ### Up-counter
>
> ### Down-counter
>
> ### Counter reset

Their use is shown in Figure 4-29.

Up-Counter Instruction (CTU)

The up-counter instruction (CTU) increments its count for every low-to-high transition of the counter input condition. There are two values for the instruction: accumulated value and preset value. The accumulated value is the present count. The preset value is the desired count. When the counter accumulated value (AC) is equal to or greater than the preset value (PR), the counter output status bit is set to 1.

Chapter 4: Timers and Counters

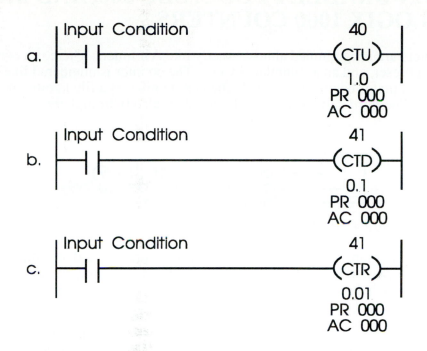

Figure 4-29. Use of three types of counters. (a) Count-up (CTU) counter. If the input condition makes a transition from low to high, the counter increments. (b) Count-down counter (CTD). When there is a high-to-low transition, the counter decrements one from the accumulated count. (c) Counter reset. If the input condition becomes true, the reset will set the accumulated count of counter 41 to zero.

Down-Counter Instruction (CTD)

The down-counter instruction (CTD) is similar but opposite. The counter decrements its value each time there is a low-to-high transition of the counter input conditions.

Counter Reset Instruction (CTR)

The counter reset instruction is used to reset the up- or down-counter accumulated value and status bits to zero.

ALLEN-BRADLEY PLC-5, SLC 500, AND MICROLOGIX 1000 COUNTERS

AB counters are programmed almost exactly like AB timers. There is a counter number, a preset, and an accumulated value. The counter is numbered like the timer except that it begins with a "C." The next number is a file number between 0 and 999. The third value is the counter number, also 0 through 999. (See Figure 4-30.)

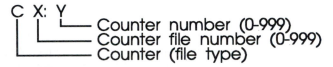

Figure 4-30. How counters are addressed.

Your ladder diagram can access counter status bits, presets, and accumulated values (see Figure 4-31). The .sb stands for status bit. You may use the .CU, .CD, .DN, .OV, or .UN bit for logic. These will each be covered a little later. You can also use the preset (.PRE) and the accumulated count (.ACC).

Counter values are stored in three 16-bit words of memory. The first eight bits of the first word are only for internal use of the CPU. The most significant bits of the first word are used to store the status of certain bits associated with the counter (see Figure 4-32). The count-up enable bit (.CU bit 15) is used to indicate that the counter is enabled. The .CU bit is reset when the rung becomes false or when it is reset by a RES instruction. The count-up done bit (.DN bit 13) when high indicates that the accumulated count has reached the preset value. It remains set even when the accumulated value (.ACC) exceeds the preset value (.PRE). The .DN bit is reset by a reset (RES) instruction.

Status Bit	Preset	Accumulated Value
CX:Y.sb	CX:Y.PRE	CX:Y.ACC

Figure 4-31. How counter bits and values can be accessed in a ladder diagram.

The count-up overflow bit (.OV bit 12) is set by the CPU to show that the count has exceeded the upper limit of +32,767. When this happens the counter accumulated value "wraps around" to -32,768 and begins to count up from there, back toward zero. (This has to do with the way computers store negative numbers. The preset value and accumulated value are stored as a two's complement number.) The .OV bit can be reset with a reset (RES) instruction.

15	14	13	12	11	10	9	8	7	6	5	4	3	2	1	0	Element
CU	CD	DN	OV	UN	UA					Internal Use						0 C5:0
Preset Value PRE																1 C5:0.PRE
Accumulated Value ACC																2 C5:0.ACC

Figure 4-32. How counter values and status bits are stored in memory.

Figure 4-33 shows the use of a count-up counter (CTU) in a ladder diagram. Each time input I:012/10 makes a low-to-high transition the counter accumulated value is incremented by one. The done bit of counter 5:0 (C5:0.DN) is being used to turn output I:013/01 on when the accumulated value is equal to the preset value (.ACC = .PRE). The overflow bit of counter 5:0 (C5:0.OV) is being used to set output I:013/02 on if the count ever reaches +32,767. The last rung uses input I:017/12 to reset counter 5's accumulated value to zero.

Figure 4-34 shows the use of a count-down counter (CTD) in a ladder diagram. Each time input I:012/10 makes a low-to-high transition the counter accumulated value is decremented by one. The done bit of counter 5:0 (C5:0.DN) is being used to turn output I:013/01 on when the accumulated value is equal to or exceeds the preset value (.ACC = .PRE). The accumulated value of counters is retentive. They are retained until a reset instruction is used. The underflow bit of counter

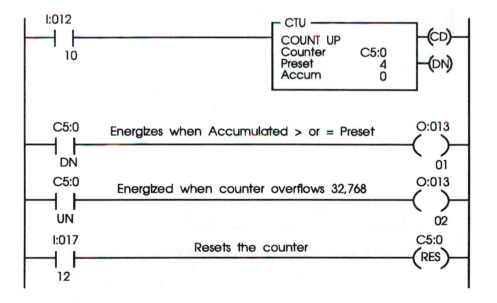

Figure 4-33. Use of a count-up counter (CTU) in a ladder diagram.

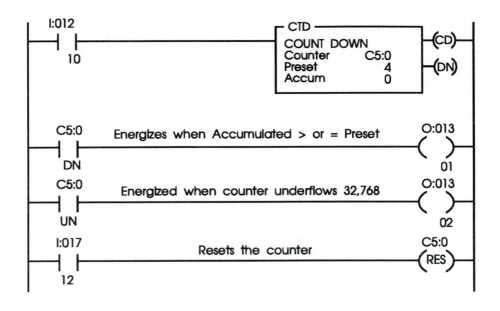

Figure 4-34. Use of a count-down counter (CTD).

5:0 (C5:0.OV) is being used to set output I:013/02 on if the count ever under-flows -32,768. Note the use of the reset instruction to reset the accumulated value of the counter to zero. The last rung uses input I:017/12 to reset counter 5's accumulated value to zero (see Figure 4-35).

Reset Instruction Use	The CPU Resets:
Timer (Do not use a reset instruction for a TOF timer.)	.ACC Value .EN Bit .TT Bit .DN Bit
Counter	.ACC Value .EN Bit .OV or .UN Bit .DN Bit

Figure 4-35. How the use of a reset (RES) instruction affects timer and counter values and bits.

GE FANUC COUNTERS

Figure 4-36 shows two types of GE Fanuc counters. The up-counter increments the count by 1 each time the counter receives transitional power. The down-counter decrements the count everytime the counter receives transitional power.

GE Fanuc Counters		
Instruction	Function	Description
upctr	Up-Counter	This type of counter increments by 1 each time the function receives transitional power. If the current value in the timer is greater than or equal to a preset value the function passes power. The R input is used to reset the counter to zero.
dnctr	Down-Counter	This counter counts down every time the function receives transitional power. If the current value of the counter is zero, the function passes power. The R input is used to set the current value to equal the preset value.

Figure 4-36. GE Fanuc counters.

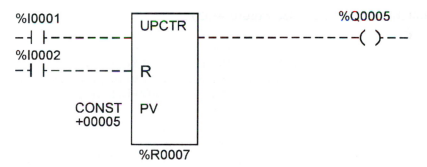

Figure 4-37. A GE Fanuc up-counter.

The GE Fanuc up-counter (UPCTR) is shown in Figure 4-37. This counter preset value is set to a constant value of 5. The current value of the count will be held in register 7 (%R0007). For every transition of the enable input (%I0001) the counter's current value will be incremented by one count. When the counter current value is equal to or greater than the preset value of 5 the counter will pass power and in this case turn coil 5 (%Q0005) on. The counter will continue to increment the current value with every power transition on the enable input. The counter's current value will only be reset to zero if the reset input (%I0002) is energized.

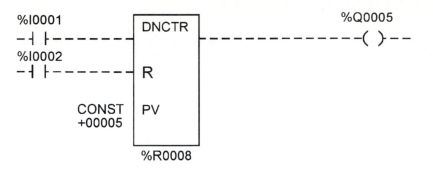

Figure 4-38. The use of a GE Fanuc down-counter.

The GE Fanuc down-counter (DNCTR) is shown in Figure 4-38. This counter preset value is set to a constant value of 5. The present value of the count will be held in register 8 (%R0008). For every transition of the enable input (%I0001) the counter's current value will be decremented from the preset value. When the current value is equal to zero the counter will pass power. In this case output 5 (%Q0005) will be turned on. The counter's current value will be reset to the preset value if the reset input (%I0002) is energized.

GOULD MODICON COUNTERS

Figure 4-39 shows an example of a Gould Modicon counter. The counter has two inputs, a enable and a reset. The reset line must be high for the counter to count. Every time the enable input makes a transition from low to high, the counter will increment. The preset value for this counter was set to 6. When the count equals 6, the outputs will change state. The current count is kept in data register 4005. The top output is off unless the actual count is equal to or greater than the preset. The bottom NOT output is on unless the count in 4005 is equal to or greater than the value of the preset.

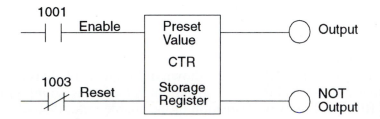

Figure 4-39. Gould Modicon counter. Note that two outputs are available.

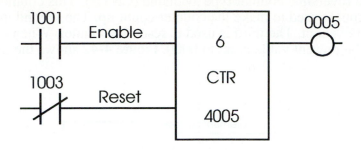

Figure 4-40. Gould Modicon counter.

Figure 4-40 shows the use of another Gould Modicon counter. Every time input 1001 makes a low-to-high transition the counter will increment the count in data register 4005. When the count reaches 6 or more, output 0005 will turn on. Output 0005 will stay on until the counter is reset by the reset line going low.

OMRON COUNTERS

Two types of Omron counters are available: CNT and CNTR. CNT is a decrementing counter and CNTR is a reversible counter instruction. Timers/counters can be numbered from 0 to 511. Numbers cannot be duplicated. If you use number 1 for a counter you may not use it for a timer. Both types of counters require a counter number, a set value (SV), inputs, and a reset input.

Figure 4-41 shows the use of an Omron counter (CNT). The set value is 20. For every low-to-high transition of the counter enable input (00001) the counter decrements the set value. When the count reaches zero the counter output turns on. A low-to-high transition of the reset input will reset the count to the set value.

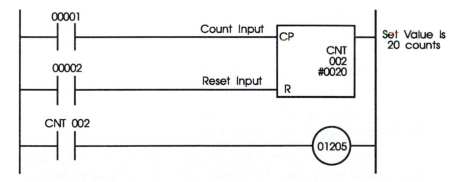

Figure 4-41. Use of an Omron CNT counter. Note that output 01205 will turn on when the counter becomes true. The output will remain on until the reset input to the counter goes high and resets the set value to 20.

There is also a reversible counter type available (CNTR). This counter has three inputs. One input is used to make the counter count up. The second input is used to decrement the count. The third is used to reset the counter. When the reset input is high the present value is set to 0000. Figure 4-42 shows the use of a CNTR counter.

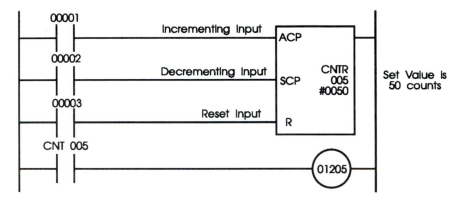

Figure 4-42. Use of a CNTR counter. The counter output will turn on whenever the present count becomes 0000.

Omron timers and counters can use external channels to receive their set values. Instead of giving a set value the programmer assigns a channel from which the timer/counter will "look up" its set value. In this way a timer/counter set value can be made a variable.

SQUARE D COUNTERS

Figures 4-43 and 4-44 illustrate Square D up/down counters. This type of counter has three inputs. The first is the up-count input. A low-to-high transition on this line will increment the count by one (provided the counter is active).

The second input is the down-count input. A low-to-high transition causes the counter to decrement the count by one.

The third input is the clear input. This line must be held high to activate the counter. If the line goes low, the count is cleared.

A storage register address is required. This is where the present value of the count is stored. A preset is required also. The preset is the desired count. In this case, when the counter reaches a count of 75, output 03-04 will turn on and remain on until the clear line goes low.

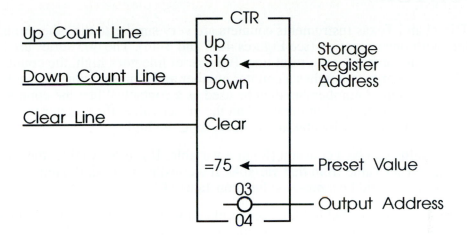

Figure 4-43. A Square D counter.

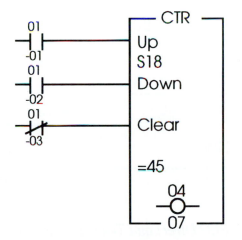

Figure 4-44. A Square D counter.

TEXAS INSTRUMENTS AND PLC DIRECT COUNTERS

PLC Direct and Texas Instruments counters are very similar to the other brands of counters with one exception (see Figures 4-45 and 4-46). The 405 series uses a normally open contact for the reset line. If the reset line goes high, the counter is reset. The programmer supplies a counter number and a preset value for each counter. The counter number can then be used as a contact. When the counter accumulated count is equal or greater than the preset value, the counter is true. The counter can be reset by making the reset input go high.

The preset value can be a constant (K) or a variable. If it is a variable, the CPU will get the preset value from that variable. If a constant is used, the letter K precedes it. K5 would be a preset value (constant) of 5.

The memory locations that hold the accumulated values can be monitored or used in comparison instructions. The accumulated values are kept in variable memory. For counters the variable numbers are V01000-V01177. Note that there are 128 counters available, or 0 to 177 octal. Note also that there are 128 (V01000 to V01177 octal) counter variables available.

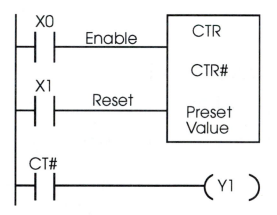

Figure 4-45. PLC Direct and Texas Instruments series 405 counter. The programmer must provide a counter number and a preset value. Note that the reset input is a normally open contact. If the reset line becomes high, the counter accumulated value will be set to zero.

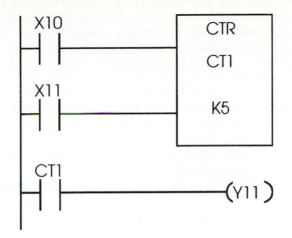

Figure 4-46. PLC Direct and Texas Instruments series 405 counter. For every low-to-high transition of input X10, CT1 will increment the count. When the accumulated count is equal to or greater than the preset value (5), counter CT1 will be true. The counter is used as contact CT1. When counter CT1 becomes true, it will energize output Y11.

PROGRAMMING HINTS

Timers and counters are indispensable to application development. There is much in common among PLC manufacturers' implementation of timers and counters. Once the technician becomes familiar with one, the rest become very easy to learn.

When you are asked to program a new brand or type of PLC there are some logical steps that can make the task less frustrating. First make sure that the programming device is really communicating with the PLC. There are several ways to be sure. Most programming software will check and warn you if communications are not established. Some PLC communications modules also have LEDs that flash when communications are taking place. If there is a problem, the most likely place to look is at the cable. Is it the right cable, and is it properly attached to the correct port? The other possible problem is that the PLC may have been assigned a station address different from the one that the software is trying to access. Another potential problem is that the communication parameters were set up incorrectly. Check the baud rates, number of data bits, stop bits, and parity of the software and PLC.

Assuming that the PLC is now communicating, the next step is to see if a simple ladder diagram can be entered and executed. Keep it simple. One contact and one coil are adequate. Make the contact a normally closed contact. Execute the ladder. If the output LED turns on, you have accomplished several important tasks. You have entered a ladder and executed it successfully. You have also figured out the correct I/O numbering.

The next step is probably to get a timer to work, followed by a counter, and so on. Then you can develop the real application quite quickly.

When writing any program there are some useful techniques that make the task less painful. One of these utilizes a *flow diagram*, which is a pictorial representation of a system or its logic. Figure 4-47 shows a few of the more common flow diagram symbols.

The decision block is used to answer a question. It could involve a quality check, a piece count, or another parameter. The question is generally a yes or no question. The process block is used to perform a process. It could be an arithmetic calculation, an assembly step, or any other logical processing step. The input/ output block is used to input information into the system and the output block is used to output system information. The input might be from an operator or from the system. The output might be a printed report or a display on a terminal. Arrows are used to connect the blocks and to show the direction of operation.

Flow diagramming can make an application program development much easier. Remember to break the application into small logical steps. Once the flow diagram is complete, it must be tested. This can be accomplished by applying some sample data and following it through the flow diagram. If the flow diagram works with sample data, the odds are good that your logic is good. Once the flow diagram is complete, it is a rather simple task to code the application into ladder logic.

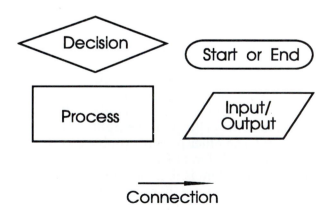

Figure 4-47. Common flow diagram symbols.

Figure 4-48 shows a heat treat application that was flow diagrammed. The sequence of operations is explained below.

The operator inputs the number of parts to be hardened.

The part is preheated to 1000 degrees Fahrenheit.

Next, the part is heated to 1650 degrees for hardening.

The part is then quenched in oil to harden it.

The hardness is then checked; if it is incorrect, the part is scrapped.

Then the piece count is checked to see if enough parts have been hardened; if not, more are hardened.

When enough have been made, the process is ended.

The second technique for developing programs is called *pseudocode*. Pseudocode is an English representation of the system steps. The use of pseudocode requires the programmer to write logical steps for the system in descriptive statements or blocks. Pseudocode would not look much different from the steps for the flow diagram that was just developed. In fact, they should be the same because the system should operate the same whether flow diagramming or pseudocode is used.

The purpose of either method is to help the programmer think logically before the actual programming is begun.

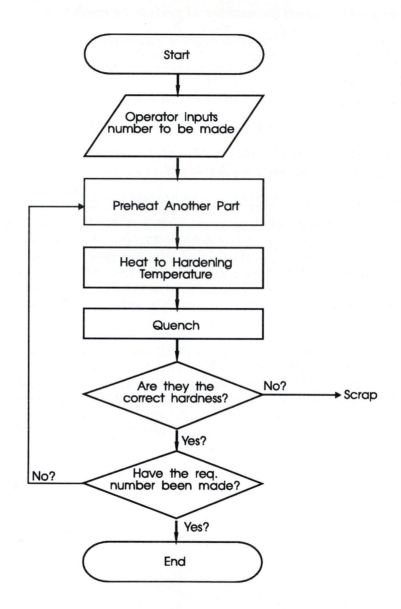

Figure 4-48. The flow diagram for a simple hardening process.

Questions

1. What are timers typically used for?

2. Explain the two types of timers and how each might be used.

3. What does the term *retentive* mean?

4. Draw a typical retentive timer and describe the purpose of the inputs.

5. What are counters typically used for?

6. In what way are counters and timers very similar?

7. Explain the two contacts that are usually required for a counter.

8. What is cascading?

9. You have been asked to program a system that requires that completed parts are counted. The largest counter available in the PLC's instruction set can only count to 999. We must be able to count up to 5000. Draw a ladder diagram that shows the method you would use to complete the task. *Hint*: Use two counters, one as the input to the other. The total count will involve looking at the total of the two counters.

10. Describe at least three reasons for using a flow diagram or pseudocode in program development.

11. Figure 4-49 is a partial drawing of a heat-treat system. You have been asked by your supervisor to troubleshoot the system. The engineer who originally developed the system no longer works for your company. He never fully documented the system. A short description of the system follows. You must study the drawings and data and complete this assignment.

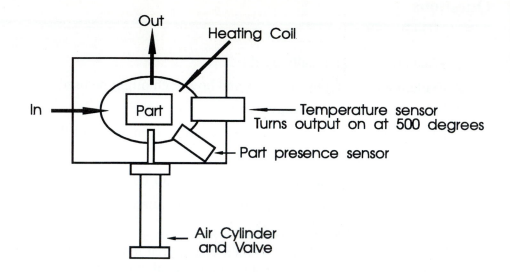

Figure 4-49. Drawing of a heat treat system.

A part enters this portion of the process. The temperature must then be raised from room temperature to 500 degrees Fahrenheit. There is also a part presence sensor. There is also a sensor that turns on when the temperature reaches 500 degrees. The part must then be pushed out of the machine. The cycle should take about 25 seconds. If it takes an excessive amount of time and the temperature has still not been reached, an operator must be informed and must reset the system. Study the system drawing, I/O chart, and ladder diagram, then complete the I/O table and answer the questions.

Complete the I/O chart (Figure 4-50) by writing short comments that describe the purpose of each input, output, timer, and counter. Make your comments very clear and descriptive so that the next person to troubleshoot the system will have an easier task. Then refer to Figure 4-51 and answer the following questions.

System I/O	
X3	Part Presence Sensor
X20	Operator Reset
Y5	
Y8	
Timer 1	
Timer 9	
Counter 1	
Counter 2	

Figure 4-50. I/O table.

Answer the following questions:

  a. What is the purpose of counter 1?

  b. What is the purpose of input X3?

  c. What is input X20 being used for?

  d. What is the purpose of counter 2?

  e. What is the purpose of contact Y15?

  f. Part of the logic is redundant. Identify that part and suggest a change.

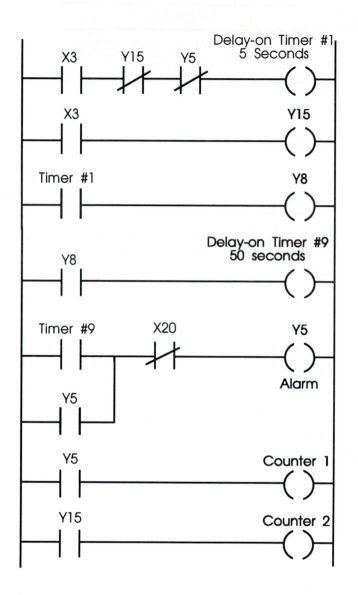

Figure 4-51.

12. Examine the ladder diagram in Figure 4-52. Assume that input 00001 is always true. What will this ladder logic do?

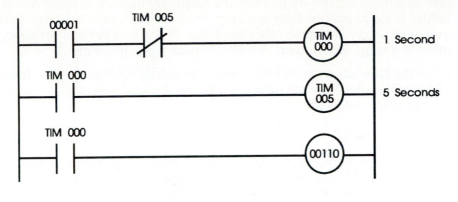

Figure 4-52.

13. You have been assigned the task of developing a stoplight application. Your company thinks there is a large market in intelligent street corner control. Your company is going to develop a PLC-based system in which lights will adapt their timing to compensate for the traffic volume. Your task is to program a normal stoplight sequence to be used as a comparison to the new system.

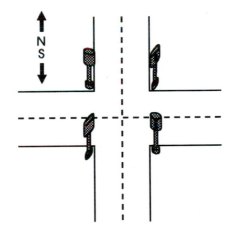

Figure 4-53.

Note in Figure 4-53 that there are really two sets of lights. The north and south lights must react exactly alike. The east-west set must be the complement of the north-south set. Write a program that will keep the green light on for 25 seconds and the yellow for 5 seconds. The red will then be on for 30 seconds. You must also add a counter because the bulbs are replaced at a certain count for preventive maintenance. The counter should count complete cycles. (*Hint*: To simplify your task, do one small

task at a time. Do not try to write the entire application at once. Write
ladder logic to get one light working, then the next, then the next, before
you even worry about the other stoplight. When you get one set done, the
other is a snap. Remember: A well-planned job is half done.)

Develop the flow diagram and then write the ladder diagram. (*Hint*: It might
be helpful to develop one flow diagram for the east-west lights and one
for the north-south set.)

Additional Exercises

1. Enter the stoplight program that you developed into programming software
for your PLC. Document the ladder with comments and names for all
contacts and coils. Print the complete ladder.

Chapter 5

Industrial Sensors

In this chapter we examine types and uses of industrial sensors. Digital and analog sensors are covered, and we also examine the wiring of sensors.

OBJECTIVES

Upon completion of this chapter, the student will be able to:

Describe at least two ways in which sensors can be classified.

Choose an appropriate sensor for a given application.

Describe the typical uses of digital sensors.

Describe the typical uses of analog sensors.

Explain common sensor terminology.

Explain the wiring of load and line-powered sensors.

Explain how field-effect sensors function.

Explain the principle of operation of thermocouples.

Explain such common thermocouple terms as **types** *and* **compensation.**

THE NEED FOR SENSORS

Sensors have become vital in industry. Manufacturers are moving toward integrating pieces of computer-controlled equipment. In the past, operators were the brains of the equipment. The operator was the source of all information about the operation of a process. The operator knew if there were parts available, which parts were ready, if they were good or bad, if the tooling was OK, if the fixture was open or closed, and so on. The operator could sense problems in the operation. He/she could see, hear, feel (vibration, etc.), and even smell problems.

Industry is now using computers (in many cases PLCs) to control the motions and sequences of machines. PLCs are much faster and more accurate than an operator at these tasks. PLCs cannot see, hear, feel, smell, or taste processes by themselves. Industrial sensors are used to give industrial controllers these capabilities.

Simple sensors can be used by the PLC to check if parts are present or absent, to size the parts, even to check if the product is empty or full. The use of sensors to track processes is vital for the success of the manufacturing process and also to assure the safety of the equipment and operator.

Sensors, in fact, perform simple tasks more efficiently and accurately than people do. Sensors are much faster and make far fewer mistakes.

Studies have been done to evaluate how effective human beings are in tedious, repetitive inspection tasks. One study examined people inspecting table tennis balls. A conveyor line was set up to bring table tennis balls past a person. White balls were considered good, black balls were considered scrap. It was found that people were only about 70 percent effective at finding the defective ping pong balls. Certainly people can find all of the black balls, but they do not perform mundane, tedious, repetitive tasks well. People become bored and make mistakes, whereas a simple sensor can perform simple tasks almost flawlessly.

SENSOR TYPES

Contact vs. Noncontact

Sensors are classified in a number of ways. One common way is to divide sensors into two categories: contact and noncontact. This is a simple way to identify a sensor. If the device must contact a part to sense the part, it would be called a *contact sensor*. A simple limit switch on a conveyor is an example. When the part moves the lever on the switch, the switch changes state. The contact of the part and the switch creates a change in state that the PLC can monitor.

Noncontact sensors are sensors that can detect the product without touching the product physically. *Noncontact sensors* do not operate mechanically (no moving parts). Mechanical devices are much less reliable than electronic devices. This means that noncontact sensors are less likely to fail.

Speed is another consideration. Electronic devices are much faster than mechanical devices. Noncontact devices can perform at very high production rates.

Another advantage of not touching the product is that you do not slow down or interfere with the process.

In the remainder of the chapter we examine noncontact sensors.

Digital vs. Analog

Another way in which sensors can be classified is digital or analog. Digital sensors are the easiest to use. Computers are digital devices. A computer actually works only with ones or zeros (on or off). A digital sensor has two states: on or off. Most applications involve presence/absence and counting. A digital sensor meets this need perfectly and inexpensively.

Digital output sensors are either on or off. They generally have transistor outputs. If the sensor senses an object the output will turn on. The transistor turns on and allows current to flow. The output from the sensor is usually connected to a PLC input module.

Sensors are available with either normally closed or normally open output contacts. Normally-open contact sensors are off until they sense an object. Normally closed contact sensors are on until they sense an object. When they sense an object the output turns off.

When photo sensors are involved the terms light-on and dark-on are often used. Dark-on means that the sensor output is on when there is no light returned to the sensor. This would usually be similar to a normally closed condition. A light-on sensor's output is on when light is returned to the receiver. This would normally be similar to a normally open sensor.

The current limit for most sensors' output is quite low. Usually output current must be limited to under 100 mA. Check the sensor before you turn on power. You must limit the output current or you will destroy the sensor! This is usually not a problem if the sensor is being connected to a PLC input. The PLC input will limit the current to a safe amount.

Analog sensors are more complex but can provide much more information about a process. Analog sensors are also called *linear output sensors*.

Think about a sensor used to measure the temperature. A temperature is analog information. The temperature in the Midwest is usually between 0 and 90 degrees. An analog sensor could sense the temperature and send a current to the PLC. The higher the temperature, the higher the output from the sensor. The sensor may, for example, output between 4 and 20 mA, depending on the actual temperature. There are an unlimited number of temperatures (and thus current outputs). Remember that the output from the digital sensor was on or off. The output from the analog sensor can be any value in the range from low to high.

Thus the PLC can monitor temperature very accurately and control a process closely. Pressure sensors are also available in analog style. They provide a range of output voltage (or current), depending on the pressure.

A 4-20 mA current loop system can be used for applications where the sensor needs to be mounted a long distance from the control device. A 4-20 mA loop is good to about 800 meters. A 4-20 mA sensor varies its output between 4 and 20 mA. There must be an adjustment on the sensor to adjust the range and sensitivity so that the sensor can measure the required values of the characteristic we are interested in, such as temperature.

There are needs for both digital and analog sensors in industrial applications. Digital sensors are more widely used because of their simplicity and ease of use. There are, however, applications that require information that only analog sensors can provide.

DIGITAL SENSORS

Digital sensors come in many types and styles. Types of digital sensors are examined next.

Optical Sensors

Optical sensors use light to sense objects. Optical sensors are increasing in use. In the past optical sensors were somewhat unreliable because they used common light and were affected by ambient lighting. This caused intermittent problems and made them somewhat unpopular. The optical sensors of today have solved these problems. Optical sensors today are very reliable because of the way they now operate.

All optical sensors function in approximately the same manner. There is a light source (emitter) and a photodetector to sense the presence or absence of light. Light-emitting diodes (LEDs) are typically used for the light source. An LED is a semiconductor diode that emits light. LEDs are a PN-type semiconductor. Forward-biased electrons from the N-type material enter the P-type material, where they combine with excess holes. When an electron and a hole combine, energy is released. These energy packets are called photons. Photons then escape as light energy. The type of material used for the semiconductor determines the wavelength of the emitted light.

LEDs are chosen because they are small, sturdy, very efficient, and can be turned on and off at extremely high speeds. They operate in a narrow wavelength and are very reliable. LEDs are not sensitive to temperature, shock, or vibration. They also have an almost endless life.

The LEDs in sensors are used in a pulsed mode. The emitter is pulsed (turned off and on repeatedly). The "on time" is extremely small compared to the "off time." LEDs are pulsed for two reasons: so that the sensor is unaffected by ambient light, and to increase the life of the LED. This can also be called modulation.

The pulsed light is sensed by the photodetector. The photo emitter and photo receiver are both "tuned" to the frequency of the modulation. The photodetector essentially sorts out all ambient light and looks for the pulsed light. The light sources chosen are typically invisible to the human eye. The wavelengths are chosen so that the sensors are not affected by other lighting in the plant. The use of different wavelengths allows some sensors, called color mark sensors, to differentiate between colors. Visible sensors are usually used for this purpose. The pulse method and the wavelength chosen make optical sensors very reliable.

There are some applications that utilize ambient light. Red-hot materials such as glass or metal emit infrared light. Photo receivers that are sensitive to infrared light can be used in these applications.

All of the various types of optical sensors function in the same basic manner. The differences are in the way in which the light source (emitter) and receiver are packaged.

Light/Dark Sensing
Optical sensors are available in either light or dark sensing. This is also called light-on or dark-on. In fact, many sensors can be switched between light and dark modes. Light/dark sensing refers to the normal state of the sensor, whether its output is on or off in its normal state.

Light Sensing (Light-on)
The output is energized (on) when the sensor receives the modulated beam. In other words, the sensor is on when the beam is unobstructed.

Dark Sensing (Dark-on)
The output is energized (on) when the sensor does not receive the modulated beam. In other words, the sensor is on when the beam is blocked.

Timing Functions
Timing functions are available on some optical sensors. They are available with on-delay and off-delay. *On-delay* delays the turning on of the output by a user-selectable amount. *Off-delay* holds the output on for a user-specified time after the object has moved away from the sensor.

Types of Optical Sensors

Reflective Sensors
One of the more common types of optical sensors is the reflective or diffuse reflective type. The emitter and receiver are packaged in the same unit. The emitter sends out light which bounces off the product to be sensed. The reflected light returns to the receiver where it is sensed (Figure 5-1). Reflective sensors have less sensing distance (range) than other types of optical sensors because they rely on light reflected off the product.

Figure 5-1. Reflective-type sensor. The light emitter and receiver are in the same package. When the light from the emitter bounces off an object it is sensed by the receiver and the output of the sensor changes state. The broken-line style of the arrows represents the pulsed mode of lighting, which is used to assure that ambient lighting does not interfere with the application. The sensing distance (range) of this style is limited by how well the object can reflect the light back to the receiver.

Polarizing Photo-Sensors

There is a special kind of retroreflective sensor for sensing shiny objects. This sensor uses a special reflector. The reflector actually consists of small prisms that polarize the light from the sensor. The sensor emits horizontally polarized light.

The reflector vertically polarizes the light and reflects it back to the sensor receiver. Thus if a very shiny object moves between the sensor and reflector and reflects light back to the sensor it will be ignored because it is not vertically polarized (see Figure 5-2).

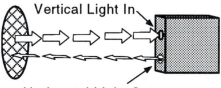

Figure 5-2. A polarizing photo sensor. Note the use of the special reflector.

Retroreflective

The retroreflective is similar to the reflective sensor (see Figure 5-3). The emitter and receiver are both mounted in the same package. The difference is that the retroreflective sensor bounces the light off a reflector instead of the product. The reflector is similar to the reflectors used on bicycles.

Chapter 5: Industrial Sensors

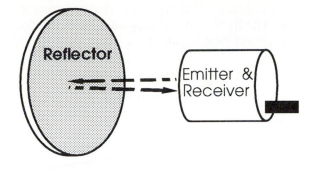

Figure 5-3. Retroreflective sensor. The light emitter and receiver are in the same package. The light bounces off a reflector (similar to the reflector on a bicycle) and is sensed by the receiver. If an object obstructs the beam, the output of the sensor changes state. The excellent reflective characteristics of a reflector give this sensor more sensing range than a typical diffuse style, where the light bounces off the object. The broken-line arrows represent the pulsed method of lighting that is used.

Retroreflective sensors have more sensing distance (range) than do reflective (diffuse) sensors but less sensing distance than that of a thru-beam sensor. They are a good choice when scanning can be done from only one side of the application. This usually occurs when space is the limitation.

Thru-Beam

Another common sensor is the thru-beam (Figure 5-4). This mode can also be called the "beam break" mode. In this configuration the emitter and receiver are packaged separately. The emitter sends out light across a space and the light is sensed by the receiver. If the product passes between the emitter and receiver, it stops the light from hitting the receiver and the sensor knows there is product present. This is probably the most reliable sensing mode for opaque (nontransparent) objects.

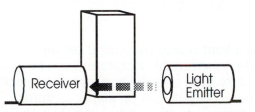

Figure 5-4. Thru-beam sensor. The emitter and receiver are in separate packages. The broken-line arrow symbolizes the pulsed mode of the light that is used in optical sensors.

Convergent Photo-Sensors

Convergent photo-sensors are a special type of reflective sensor. They emit light to a specific focal point. The light must be reflected from the focal point to be sensed by the sensor's receiver (see Figure 5-5). They are also called focal length sensors.

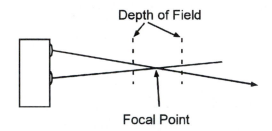

Figure 5-5. A convergent-style photo-sensor.

Fiber-Optic Sensors

Fiber-optic sensors are simply mixes of the other types. The actual emitter and receiver are the same. Fiber-optic cables are just attached to both. One cable is attached to the emitter, one to the receiver. Fiber-optic cables are very small and flexible. Fiber-optic cables are transparent strands of plastic or glass fibers that are used as a "light pipe" to carry light.

The light from the emitter passes through the cable and exits from the other end. The light enters the end of the cable attached to the receiver, passes through the cable, and is sensed by the receiver. The cables are available in both thru-beam and reflective configuration. (See Figure 5-6.)

Color Mark Sensors

Color mark sensors are a special type of diffuse reflective optical sensor. This type of sensor can differentiate between colors. These are typically used to check labels and for sorting packages by color mark. They are chosen according to the color to be sensed. A sensitivity adjust is provided to make fine adjustments. The background color (behind the object) is an important consideration. Color mark sensors actually detect contrast between two colors. Sensor manufacturers have charts available for the proper selection of color mark sensors for various colors.

Laser Sensors

Lasers are also used as a light source for optical sensors. These sensors can be used for precision-quality inspections. Resolution can be as small as a few microns. Laser sensors can be used to make very accurate measurements. A laser LED is used as the source. Outputs can be analog or digital. The digital outputs can be used to signal pass/fail, or other indication. The analog output can be used to monitor and record actual measurements.

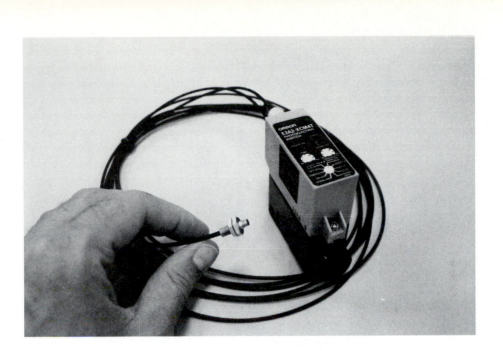

Figure 5-6. Fiber-optic sensor. Courtesy of Omron Electronics.

Encoders

Encoders are used for position feedback and in some cases for velocity feedback also. An exploded view of an encoder is shown in Figure 5-7. The most common type of encoder is incremental. The resolution of an encoder is determined by the number of lines on the encoder disk. The more lines the higher the resolution. Encoders with 500, 1000, or even more lines are common. LEDs are used as light sources. The light shines through the lines on the encoder disk and a mask and is then sensed by light receivers (photo transistors).

Absolute Encoders

The absolute encoder provides a whole word of output with a unique pattern that represents each position (see Figure 5-8). The LEDs and receivers are aligned to read the disk pattern (see Figure 5-9). There are many types of coding schemes that could be used for the disk pattern. The most commonly used patterns are gray, natural, binary, and binary-coded decimal (BCD). Gray and natural binary are available up to about 256 counts (8 bits). Gray code is very popular because it is a nonambiguous code. Only one track changes at a time. This allows any indecision that may occur during any edge transition to be limited to plus or minus one count. Natural binary code can be converted from gray code through digital logic. If the output changes while it is being read a latch option locks the code to prevent ambiguities.

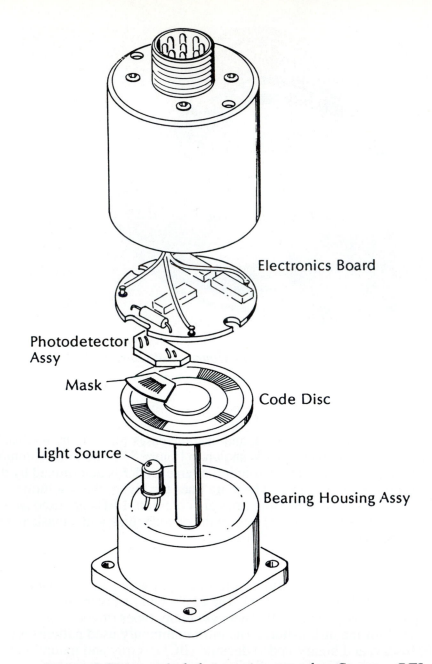

Electronics Board

Photodetector
Assy

Mask

Light Source

Code Disc

Bearing Housing Assy

*Figure 5-7. An exploded view of an encoder. Courtesy BEI
Sensors & Systems Company.*

8 Bit Absolute Disc

Figure 5-8. A disk from an absolute encoder. Courtesy BEI Sensors & Systems Company.

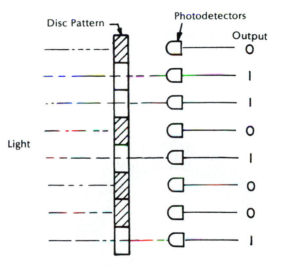

Figure 5-9. How photodetectors read the position of an absolute encoder. Courtesy BEI Sensors & Systems Company.

Incremental Encoders

An incremental encoder creates a series of square waves. Figure 5-10 shows a disk from an incremental encoder. Incremental encoders are available in various resolutions. The number of slits for light to pass through determines the resolution of the encoder. For example, a 500-count encoder would produce 500 square waves in one revolution; 250 pulses would be produced with a 180-degree turn.

Incremental encoders provide more resolution at a lower cost than absolute encoders. They also have fewer interface problems because they have fewer output lines. There are two main types of incremental encoders, tachometer (single-track) and quadrature (multitrack).

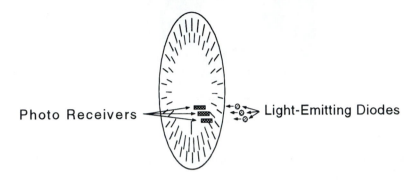

Photo Receivers ← ··· → Light-Emitting Diodes

Figure 5-10. An incremental encoder.

Tachometer Encoders
The disk in Figure 5-11 shows a tachometer-style encoder disk. They are sometimes called single-track encoders because they only have one output and cannot detect direction of travel. The output is a square wave. Velocity can be determined by looking at the frequency of pulses.

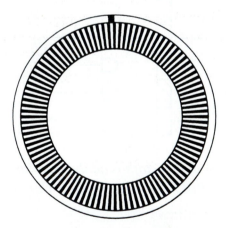

Incremental Disc

Figure 5-11. A disk from a tachometer-style incremental encoder. Courtesy BEI Sensors & Systems Company.

Quadrature Encoders

Most encoders use two quadrature output channels for position sensing. Figure 5-12 shows what the A and B pulses would look like on a oscilloscope. Note that the A and B pulses are a square wave. Every slit in the encoder disk is represented by a high pulse. Note the relationship between the A and B pulse train.

The transitions from high to low or low to high on these channels can be sensed by a computer, PLC, or special controller like a CNC, or robot controller. There are three tracks on most incremental quadrature encoders (see Figure 5-13). The A track and the B track have the same number of slits. The index track has only one pulse per revolution. This is usually called a zero or index pulse. This is used for establishing the home position for position controllers.

The A and B channels can be compared to determine direction of rotation. By using one channel as a reference we can tell whether the transition on B leads or lags the A transition. This tells the direction of rotation. Study Figure 5-14. This figure shows how direction can be determined by comparing the two channels.

A quadrature encoder might have a total of 6 lines: 3 for the A, B, and home pulse signals and 2 to supply power to the encoder and 1 for case ground. A 13-bit absolute encoder would have 13 output wires plus 2 power lines. If the absolute encoder is provided with complementary outputs the encoder would require 28 wires.

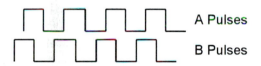

Figure 5-12. What the A and B ring output from the encoder look like on an oscilloscope.

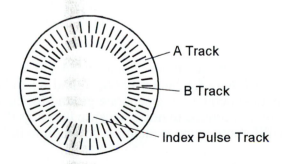

Figure 5-13. An incremental encoder.

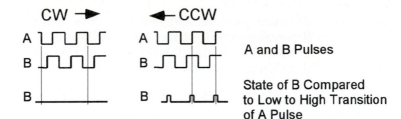

Figure 5-14. Encoder pulses for clockwise and counterclockwise rotation.

If the number of transitions are counted the amount of rotation will be known. The quadrature encoder has a distinct advantage over the tachometer encoder for position sensing. Imagine what would happen if a tachometer encoder was to stop right on a transition and the machine was to vibrate in such a way that the encoder continued to output transitions due to the vibration forcing the unit back and forth across a transition edge, even though there was no rotation. This cannot happen with a quadrature encoder. Even if the encoder stopped on a transition and vibration caused the output to show transitions on one channel there would be no transition on the other channel. By using quadrature detection on a two-channel encoder and viewing the transition in its relationship to the condition of the other channel, we can generate accurate and reliable position and direction information.

Direction Sensing and Pulse Multiplication
Pulses from the A and B channels can be fed to an up-down counter or programmable controller input card. Many PLC high-speed input modules have the capability of accepting encoder input. With quadrature detection we can derive 1, 2, or 4 times the basic resolution of the encoder. For example, if a 1000-count encoder is used its basic resolution would be 1000 pulses per revolution, or 360 degrees / 1000. If the detection circuit looks at the high to low transition and the high to low transition of the B channel we would have 2000 pulses per revolution. If the detection circuit looks at the high to low transition and the low to high transition for the A and the B channel we would have 4000 pulses per revolution. With a quality encoder 4X signal will be accurate to better than 1/5 pulse.

Encoder Wiring
Encoders are available in two wiring types: single ended and differential. In single-ended encoders there is one wire for each ring (A, B, and X) and a common ground wire. A differential encoder has a separate ground for each ring. The differential encoder is more immune to noise because the rings do not share a common ground. The differential type is the most common in industrial applications.

Ultrasonic Sensors
Ultrasonic sensors use a narrow ultrasonic beam to detect and even measure (see Figure 5-15). The ultrasonic sensor is much like radar. A narrow ultrasonic beam (about 5 mm) is bounced off the object back to the sensor.

Ultrasonic sensors use sound to measure distance. They send out sound waves and measure how long it takes for them to return. The distance to the object is proportional to the time it takes for the high-frequency sound to travel to the object and to be reflected back to the sensor. The sensor is then able to determine the distance to the object.

They can make very accurate measurements. Objects as small as 1 mm can be inspected to an accuracy of plus or minus 0.2 mm. They utilize ultrasonic waves. They use frequencies well in excess of the audible range. Some cameras use ultrasonic sensing to determine the distance to the object to be photographed.

Another type of ultrasonic sensor is the interferometer sensor (see Figure 5-16). An interferometer sensor sends out a sound wave (or light wave) which is reflected back to the sensor. The transmitted wave interferes with the reflected wave. If the peaks of the two waves coincide the amplitude will be twice that of the transmitted wave.

If the transmitted wave is 180 degrees out of phase with the reflected wave the amplitude of the result will be zero. Between these two extremes the amplitude will be between zero and twice the amplitude and the phase will be shifted between 0 and 180 degrees.

This type of sensor can detect distances to within a fraction of a wavelength. This is very fine resolution indeed because some light has wavelengths in the size range of .0005 mm. Laser light can be used for longer distances because the light is coherent and doesn't scatter much.

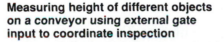

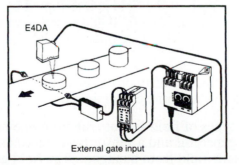

Figure 5-15. Use of an ultrasonic sensor. Note the use of the "gate" sensor to notify the PLC when a part is present. Courtesy of Omron Electronics.

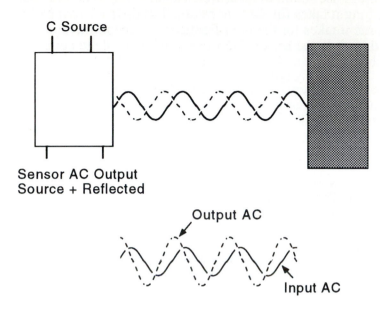

Figure 5-16. How an ultrasonic sensor works.

ELECTRONIC FIELD SENSORS (FIELD SENSORS)

Field sensors are used to sense objects. There are two types: capacitive and inductive. Both types produce a field. If the field is interrupted by an object the sensor turns on. An inductive sensor creates a field that is sensitive to metal. A capacitive sensor is sensitive to any object. There are many special types of field sensors available for special uses.

Field sensors are a good choice in dirty or wet environments. A photo sensor can be affected by dirt, liquids or airborne contamination.

The two most common types of field sensors function in essentially the same way. They have a field generator and a sensor to sense when the field is interfered with. Imagine the magnetic field from a magnet. The field generator puts out a similar field. The two types of field sensors are inductive and capacitive.

Inductive Sensors

The inductive sensor, used to sense metallic (ferrous) objects, is commonly used in the machine tool industry. The inductive sensor works by the principle of electromagnetic induction (see Figure 5-17). Inductive sensors function in a manner similar to the primary and secondary windings of a transformer. There is an oscillator and a coil in the sensor. Together they produce a weak magnetic field. As an object enters the sensing field, small eddy currents are induced on the surface of the object. Because of the interference with the magnetic field, energy is drawn away from the oscillator circuit of the sensor. The amplitude of the oscillation decreases, causing a voltage drop. The detector circuit of the sensor senses the voltage drop of the oscillator circuit and responds by changing the state of the sensor.

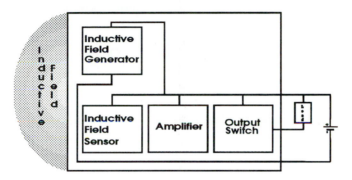

Figure 5-17. Block diagram of an inductive sensor. The inductive field generator creates an inductive field in front of the sensor. This field is monitored by the field sensor. When a ferrous object (an object containing the element iron) enters the field, the field is disrupted. The disruption in the field is sensed by the field sensor and the output of the sensor changes state. The sensing distance of these sensors is determined by the size of the field. This means that the larger the required sensing range, the larger the diameter of the sensor will be.

Inductive sensors are available in very small sizes. If the area in which the sensor will be mounted is restricted, or if the object to be sensed is small, this style of sensor works very well. (See Figure 5-18.)

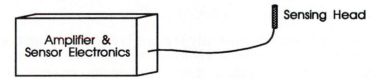

Figure 5-18. Use of a remote inductive sensing head. The electronics are mounted away from the application. The small sensing head is mounted in the application.

Electromagnetic Induction

Refer to Figure 5-19. The output is initially deenergized (off). As the target (object) moves into the field of the sensor, eddy currents are produced along the target surface. As the voltage drops in the oscillator circuit, the detector senses the drop and changes the state of the sensor. The output is then energized.

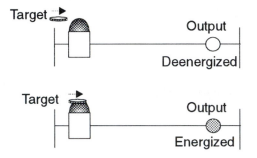

Figure 5-19. How an object (target) is detected as it enters a field.

Sensing Distance

Sensing distance (range) is related to the size of the inductor coil and whether the sensor coil is shielded or nonshielded (see Figure 5-20). This sensor coil is shielded. In this case a copper band is placed around the coil. This prevents the field from extending beyond the sensor diameter. Note that this reduces the sensing distance.

The shielded sensor has about half the sensing range of an unshielded sensor. Sensing distance is affected by temperature. Sensing distance will generally vary by about 5 percent due to changes in ambient temperature.

Hysteresis

Hysteresis means that the object must be closer to turn a sensor on than to turn it off (see Figure 5-21). Direction and distance are important. If the object is moving toward the sensor, it will have to move to the closer point to turn on. Once the sensor turns on (operation point or on-point), it will remain on until the object moves away to the release point (off-point). This differential gap, or differential travel, is caused by hysteresis. The principle is used to eliminate the possibility of "teasing" the sensor. The sensor is either on or off.

Hysteresis is a built-in feature in proximity sensors. Hysteresis benefits us because it helps stabilize part sensing. Imagine a bottle moving down a conveyor line. Vibration causes the bottle to wiggle as it moves along the conveyor. If the on-point was the same as the off-point and the bottle was wiggling as it went by the sensor it could be sensed many times as it wiggles in and out past the on point. When hysteresis is involved, however, the on-point and off-point are at different distances from the sensor.

To turn the sensor on the object must be closer than the on-point. The sensor output will remain on until the object moves farther away than the off-point. This prevents multiple unwanted reads.

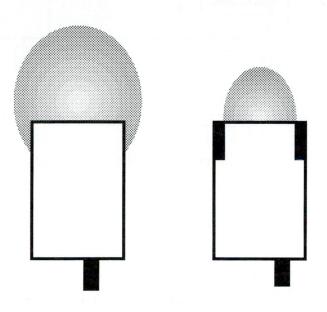

Figure 5-20. The use of a copper band in a shielded field sensor. Note that the sensing distance is reduced. It does allow the sensor to be mounted flush, however. If the unshielded sensor were mounted flush, it would detect the object in which it was mounted.

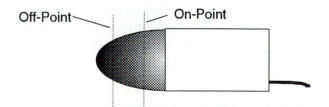

Figure 5-21. Example of hysteresis. Note that the on-point and off-point are different.

Capacitive Sensors

Capacitive sensors (Figures 5-22 and 5-23) can be used to sense both metallic and nonmetallic objects. They are commonly used in the food industry. Capacitive sensors can also be used to sense for product inside nonmetallic containers (see Figure 5-24).

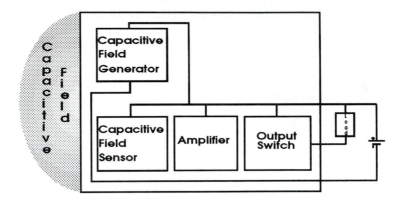

Figure 5-22. Block diagram of a capacitive sensor.

Capacitive sensors operate on the principle of electrostatic capacitance. They function in a manner similar to the plates of a capacitor. The oscillator and electrode produce an electrostatic field (remember that the inductive sensor produced an electromagnetic field). The target (object to be sensed) acts as the second plate of a capacitor. An electric field is produced between the target and the sensor. As the amplitude of the oscillation increases, the oscillator circuit voltage increases and the detector circuit responds by changing the state of the sensor.

Almost any object can be sensed by a capacitive sensor. The object acts like a capacitor to ground. When the target (object) enters the electrostatic field, the DC balance of the sensor circuit is disturbed. This starts the electrode circuit oscillation and maintains the oscillation as long as the target is within the field.

Sensing Distance

Capacitive sensors are unshielded, nonembeddable devices (see Figure 5-23). This means that they cannot be installed flush in a mount because they would then sense the mount. Conducting materials can be sensed farther away than nonconductors because the electrons in conductors are freer to move. The target mass affects the sensing distance: The larger the mass, the larger the sensing distance.

Capacitive sensors are more sensitive than inductive sensors to temperature and humidity fluctuation. Sensing distance can fluctuate as much as plus or minus 15 to 20 percent. Capacitive sensors are not as accurate as inductive sensors. Repeat accuracy can vary by 10 to 15 percent in capacitive sensors.

Some capacitive sensors are available with a sensitivity adjustment. This can be used to sense product inside a container (see Figure 5-24). The sensitivity can be reduced so that the container is not sensed but the product inside is.

Figure 5-23. Capacitive sensors. Courtesy of Omron Electronics.

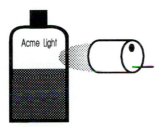

Acme Light

Figure 5-24. Use of a capacitive sensor to check inside a container. They are also used to check fluid and solid levels inside tanks. There is an adjustment screw on some capacitive sensors. The sensor would be adjusted so that the container is not sensed but the material inside is.

SENSOR WIRING

There are basically two wiring schemes for sensors: load-powered and line-powered. This applies to AC- and DC-powered sensors.

Sensors are available in two- and three-wire types. The most important consideration must be to limit the sensor's output current to an acceptable level. Output current refers to the sensor's output. The output device in the sensor is normally a transistor. If the output current exceeds the output current limit the sensor will fail.

Check the specifications for the sensor. A sensor with a transistor output can generally handle up to about 100 mA. Sensors with a relay output can handle 1 amp or more.

Load-Powered Sensors

Two-wire sensors are called load-powered sensors. One wire is connected to power, the other wire is connected to one of the load's wires (see Figure 5-25). The other load wire is then connected to power. Wiring diagrams can usually be found on the sensor.

The current required for the sensor to operate must pass through the load. A load is anything that will limit the current of the sensor output. Think of the load as being an input to a PLC. The small current flow that must flow to allow the sensor to operate is called leakage current, or operating current.

This current is typically under 2.0 mA (see Figure 5-26). This is enough current for the sensor to operate, but not enough to turn on the input of the PLC. (The leakage current is usually not enough current to activate a PLC input. If it is enough current to turn on the PLC input it will be necessary to connect a bleeder resistor as shown in Figure 5-27.) When the sensor turns on, it allows enough current to flow to turn on the PLC input.

Response time is the lapsed time between the target being sensed and the output changing state. Response time can be crucial in high production applications. Sensor specification sheets will give response times.

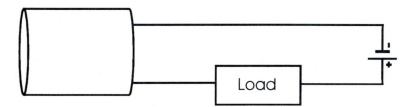

Figure 5-25. Two-wire sensor (load-powered). The load represents whatever the technician will be monitoring the sensor output with. It is normally a PLC input. The load must limit the current to an acceptable level for the sensor, or the sensor output will be blown.

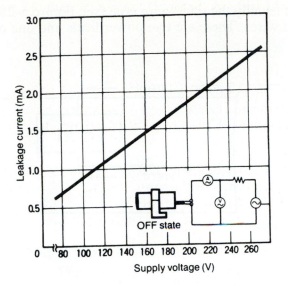

Figure 5-26. Leakage current vs. supply voltage.

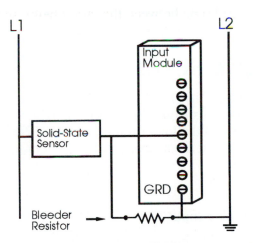

Figure 5-27. Use of a two-wire sensor (load-powered). In this case the leakage current was enough to cause the input module to sense an input when there was none. A resistor was added to bleed the leakage current to ground so that the input could not sense it.

Line-Powered Sensors

Line-powered sensors are usually the three-wire type (see Figure 5-28), but there can be either three or four wires. There are two power leads and one output lead in the three-wire variety.

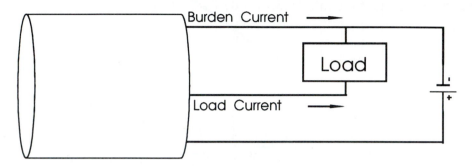

Figure 5-28. Three-wire (line-powered) sensor. Note the load. The load must limit the current to an acceptable level.

The sensor needs a small current, called burden current or operating current, to operate. This current flows whether or not the sensor output is on or off. The load current is the output from the sensor. If the sensor is on, there is load current. This load current turns the load (PLC input) on. The maximum load current is typically between 50 to 200 mA for most sensors. Make sure that you limit the load (output) current or the sensor will be ruined.

SOURCING AND SINKING SENSORS

PNP Sensor (Sourcing Type)

Conventional current flow goes from plus to minus. When the sensor is off, the current does not flow through the load. In the case of a PNP, when there is an output current from the sensor, the sensor sources current to the load (sourcing type) (see Figure 5-29).

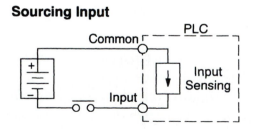

Figure 5-29. A sourcing switch connected to a PLC input. Courtesy PLC Direct by Koyo.

NPN (Sinking Type)

When the sensor is off (nonconducting), there is no current flow through the load. When the sensor is conducting, there is a load current flowing from the load to the sensor (see Figure 5-30). The choice of whether to use an NPN or a PNP sensor is dependent on the type of load. In other words, choose a sensor that matches the PLC input module requirements for sinking or sourcing.

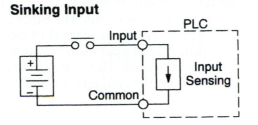

Sinking Input

Figure 5-30. A sinking switch connected to a PLC input. Courtesy PLC Direct by Koyo.

Sinking = Path to supply ground (-)

Sourcing = Path to supply source (+)

ANALOG SENSORS

There are many types of analog sensors available. Many of the types already discussed are also available with analog output. Photo sensors and field sensors are available with analog output.

Analog Considerations

Analog sensors provide much more information about a process than digital sensors do. They provide an output that varies depending on the conditions being measured.

Accuracy, Precision, and Repeatability

Accuracy can be defined as how closely a sensor indicates the true quantity being measured. In a temperature measurement accuracy would be defined as how closely the sensor indicates the actual temperature being measured.

Precision refers to a group of sensors. Precision is a measure of how closely each measures the same variable. For example, if we measured a given temperature the output of each should be the same. This is very important in many applications. It is important when we need to replace a sensor.

Repeatability is the ability of a sensor to repeat its previous readings. For example, in temperature measurement the sensor's output should be the same every time it senses a given temperature.

Thermocouples

The thermocouple is one of the most common devices for temperature measurement in industrial applications. A thermocouple is a very simple device: two pieces of dissimilar metal wire joined at one or both ends (see Figure 5-31). The typical industrial thermocouple is joined at one end. The other ends of the wire are connected via compensating wire to the analog inputs of a control device such as a PLC (see Figure 5-32). The principle of operation is that when dissimilar metals are joined, a small voltage is produced. The voltage output is proportional to the difference in temperature between the cold and hot junctions (see Figures 5-33 and 5-34).

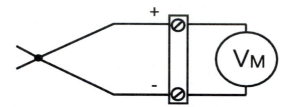

Figure 5-31. Common thermocouple connection. Note that the actual thermocouple wires are only connected to each other on one end. The other end of each is connected to copper wire, which in this case is connected to a meter. They are normally connected at a terminal strip. The terminal strip assures that both ends remain at the same temperature (ambient temperature). The net loop voltage remains the same as the double-ended loop.

The cold junction is assumed to be at ambient temperature (room temperature). Industrial thermocouple tables use 75 degrees Fahrenheit for the reference temperature (see Figure 5-35).

In reality, temperatures vary considerably in an industrial environment. If the cold junction varies with the ambient temperature, the readings will be inaccurate. This would be unacceptable in most industrial applications. It is too complicated to try to maintain the cold junction at 75 degrees. Industrial thermocouples must therefore be compensated. This is normally accomplished with the use of resistor networks that are temperature sensitive. The resistors that are used have a negative coefficient of resistance. Resistance decreases as the temperature increases. This adjusts the voltage automatically so that readings remain accurate. PLC thermocouple modules automatically compensate for temperature variation.

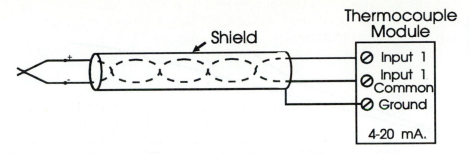

Figure 5-32. Thermocouple. Note that the wire connecting the thermocouple to the PLC module is twisted-pair (two wires twisted around each other) shielded cable. The shield around the twisted pair is to help eliminate electrical noise as a problem. Twisted-pair wiring also helps reduce the effects of electrical noise. Note also that the shielding is grounded only at the control device. Typical PLC modules would allow four or more analog inputs.

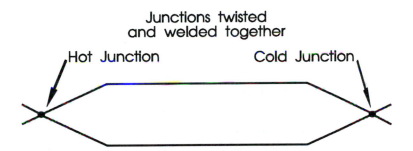

Figure 5-33. Simple thermocouple. A thermocouple can be made by twisting the desired type of wire together and silver soldering the end. To measure the temperature change the wire would be cut in the middle and a meter would be inserted. The voltage would be proportional to the difference in temperature between the hot and cold junctions.

The thermocouple is a very accurate device. The resolution is determined by the device that takes the output from the thermocouple. The device is normally a PLC analog module. The typical resolution of an industrial analog module is 12 bits; 2 to the twelfth power is 4096. This means that if the range of temperature to be measured was 1200 degrees the resolution would be 0.29296875 degrees/bit (1200/4096 = 0.29296875), which would mean that our PLC could tell the temperature to about one-fourth of one degree. This is reasonably good resolution, close enough for the vast majority of applications. Higher-resolution analog modules are available.

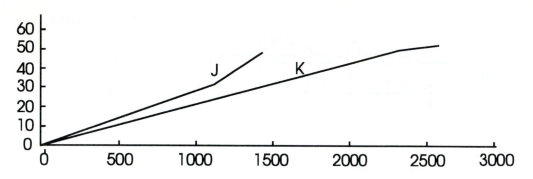

Figure 5-34. Voltage output vs. temperature for J- and K-type thermocouples. Note that the relationship is approximately linear. For example, if the output voltage was 20 mV with a J-type thermocouple the temperature would be approximately 525 degrees (600 - 75). Note that the chart shows the voltage produced by the difference in temperature between the cold and hot junctions. The chart assumes a 75 degree ambient temperature. The temperatures shown are Fahrenheit.

Thermocouple Temperature Ranges

Type	Temperature Fahrenheit	Temperature Celsius
J	-50 to +1400	-40 to +760
K	-100 to +2250	-80 to +1240
T	-200 to +660	-130 to +350
E	-200 to +1250	-130 to +680
R	0 to +3200	-20 to +1760
S	0 to +3200	-20 to +1760
B	+500 to +3200	+260 to +1760
C	0 to +4200	-20 to +2320

Figure 5-35. Temperature ranges for various types of thermocouples. Note the output for the difference in temperature between the hot and cold junctions. The table is based on a cold junction temperature of 75 degrees Fahrenheit.

Thermocouple Types	
E	Nickel-Chromium vs. Copper-Nickel (Chromel-Constantan)
J	Iron vs. Copper-Nickel (Iron-Constantan)
K	Nickel-Chromium vs. Nickel-Aluminum (Chromel-Alumel)
T	Copper vs. Copper-Nickel (Copper-Constantan)
C	Tungsten 5% Rhenium vs. Tungsten 26% Rhenium
D	Tungsten vs. Tungsten 26% Rhenium
G	Tungsten 3% Rhenium vs. Tungsten 25% Rhenium
R	Platinum vs. Platinum 13% Rhodium
S	Platinum vs. Platinum 10% Rhodium
B	Platinum 6% Rhodium vs. Platinum 30% Rhodium

Figure 5-36. Composition of various thermocouple types.

Thermocouples are the most widely used temperature sensors. There are a wide variety of thermocouples available (see Figure 5-36). They are not generally as accurate as thermistors or RTDs.

RTDs

RTD stands for resistance temperature device. An RTD is a device that changes resistance with temperature. One of a metal's basic properties is that its electrical resistivity changes with temperature. Some metals have a very predictable change in resistance for a given change in temperature. An RTD is really just a precision resistor. The chosen metal is made into a resistor with a nominal resistance at a given temperature. The change in temperature can then be determined by comparing the resistance at the unknown temperature to the known nominal resistance at the known temperature.

RTDs are made from a pure wire-wound metal that has a positive temperature coefficient. These are naturally occurring metals. This means that as the temperature increases the resistance to current flow increases (see Figure 5-37). There is a very small change in resistance.

RTDs come in two configurations. One is a wire-wound construction. This coiled construction allows for a greater change in resistance for a given change in temperature. This increases the sensor's sensitivity and helps improve the resolution. The coil is wrapped around a nonconductive material such as ceramic. This helps the wire respond more quickly to temperature change. The coil is then covered by a sheath to protect it from abuse and the environment. The sheath is also filled with a dry thermal conducting gas to increase the temperature conductivity. The sheath is constructed from materials that are very temperature conductive. Stainless steel is usually used.

There are tables available that show the temperature-resistance relationships for various metals used in RTDs.

Platinum is the most popular material for RTDs. Other materials include copper, nickel, balco, tungsten, and iridium. Platinum sensors are now being made with very thin film resistance elements that use a very small amount of platinum. This makes platinum RTDs more competitive in price. Platinum has a very linear change in resistance vs. temperature and has a wide operating range. Platinum is also a very stable element, which assures long-term stability.

RTDs are connected like resistors. The most common resistance for an RTD is 100 ohms at zero degrees C. Other RTDs are available in the 50 to 1000 ohm range.

Wiring

An RTD is really a 2-wire device just like a resistor. The lead wires from the RTD can affect the accuracy of the RTD because it is additional uncompensated resistance. A third wire can be added to compensate for lead wire resistance.

Four-wire RTDs are also available but are generally only used in laboratory applications. A four-wire RTD can be used with a three-wire device but it doesn't improve the accuracy over a three-wire RTD. A three-wire RTD cannot be used with a four-wire device. It is best to use the appropriate RTD for the measurement device.

The other RTD configuration is in the form of a thin film metal layer that has been deposited on a nonthermally conductive substrate. Ceramic is often used for the substrate. These can be made very small. This type doesn't need an external sheath. The ceramic acts as a protective sheath. Simple circuitry can be used to linearize RTDs.

| 2-Wire | 3-Wire | 4-Wire |

Figure 5-37. Three RTD wiring configurations.

Thermistors

A thermistor is a temperature-measuring sensor. Thermistors are made from man-made materials. They are semiconductors. Thermistors have a negative temperature coefficient (see Figure 5-38). The resistance decreases as the temperature increases. They can be very precise and stable. One major advantage of a thermistor is that it produces a large change in resistance for a small change in temperature. One of the thermistor's main problems is that the output is not very linear. The resistance does not vary proportionately with a change in temperature. It will only be linear within a small temperature range. If the range of temperature to be measured is relatively small, the thermistor is a good device.

Thermistors are semiconductors so they cannot operate above about 300 degrees C. Thermistors are more sensitive to temperature change than RTDs. There are various package styles available for various applications. Thermistor networks are available that have very linear voltage change with temperature change. One of the uses for thermistors is in motor applications. Thermistors can be used to monitor the temperature of electric motors. The thermistor is fastened to the housing of the motor and connected to a bridge circuit. The output of the bridge circuit is compared to a reference voltage. The reference voltage is chosen to be a safe value for the maximum operating temperature of the motor. This can then be used to shut off the motor circuit.

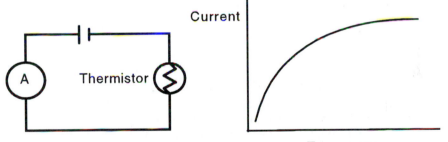

Figure 5-38. The graph of current versus temperature change for a thermistor.

Integrated Circuit (Semiconductor) Temperature Sensors

Some semiconductors can be used to sense temperature change. They respond to temperature increases by increasing their reverse bias current across PN junctions. This generates a small, detectable current or voltage that is proportional to the change in temperature (see Figure 5-39).

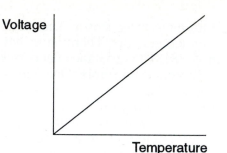

Figure 5-39. The graph of voltage versus temperature change for an integrated circuit (semiconductor).

Magnetic Reed Sensors

Magnetic reed sensors usually have two sets of contacts. They have a normally closed set and a normally open set of contacts. When a small magnet is brought close to the reed sensor the common contact is drawn toward the normally open contact and moves away from the normally closed contact. These are commonly used for applications such as sensing the location of the piston in a cylinder. We could use them to sense if the piston is extended or retracted for feedback in a system.

Hall Effect Sensor

The hall effect sensor is based upon a principle that was discovered shortly after the transistor was developed in the 1950s. A hall effect sensor is biased with a current. If a magnetic field is introduced to the sensor material, it allows a small current flow through the sensor that can then be sensed (see Figure 5-40). Hall effect sensors are widely used for many applications. They are available as digital or analog sensors.

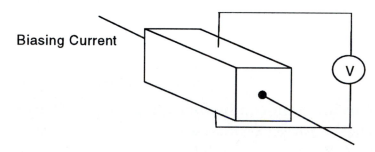

Figure 5-40. The basic principle of the hall effect sensor.

Hall effect sensors are used in the medical field as blood flow sensors. Hall effect transducers can be used as linear displacement devices. Hall effect sensors can be used to sense rotary motion for speed. They are often used to detect the velocity of a rotating shaft. They can be used to determine displacement of a shaft in the same manner. They are often used in cruise (speed) control on cars.

Strain Gages

Strain gages are used to measure force. They are based on the principle that the thinner the wire, the higher the resistance. A smaller diameter wire would have higher resistance than a larger diameter wire (see Figure 5-41).

Figure 5-41. Two pieces of wire. Which one will have lower resistance? The larger the wire diameter, the lower the resistance.

If we had an elastic wire we could stretch it and measure the change in resistance (see Figure 5-42). The change in resistance compared to the change in voltage would be quite linear. If we put a constant current through the strain gage and the resistance of the strain gage varies with the force we can measure the change in voltage.

Strain gages are used for pressure measurement by bonding them to a membrane, which is then exposed to the pressure. They are also used in accelerometers. Most strain gages are zig-zagged wire mounted on a paper or membrane backing (see Figure 5-43). Since the resistance of wire is low a long piece is used to maximize the change. It is important to mount the strain gage correctly because the strain gage is only sensitive to change in one direction. Strain gages normally have an arrow which indicates which direction they should be mounted in. The gages are normally mounted with adhesives.

Figure 5-42. What would happen if we had an elastic wire. If the wire is stretched, the resistance increases.

Strain gages are often set up in a bridge configuration to increase the change in resistance and to compensate for temperature fluctuations. Since strain gages are made from long metal strands (or metallic paths) and metal responds to temperature change, strain gages are susceptible to variation in temperature. These temperature fluctuations can cause a variance in the output.

Temperature fluctuations can be compensated for (see Figure 5-44). A dummy strain gage is inserted and is not subjected to the strain. It is, however, subjected to the same temperature change. This eliminates the effect of temperature fluctuation in the stress measurement.

Figure 5-45 shows a second way that strain gages can be set up to overcome fluctuation due to variation in temperature. In this case 4 strain gages are used. Two balancing resistors are also added (R5 and R6). These are used to compensate for thermal drift and to balance the bridge.

Any temperature fluctuation will be felt by all of the strain gages and will thus cancel any temperature-induced fluctuation.

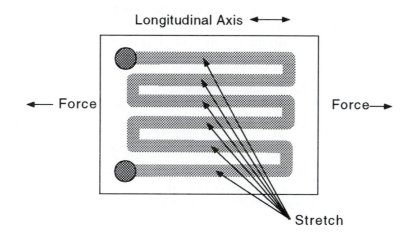

Figure 5-43. One style of strain gage. There are many different styles available.

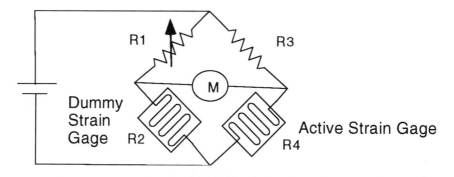

Figure 5-44. One way a strain gage can be set up to compensate for temperature fluctuation.

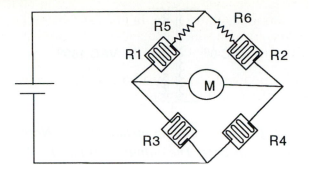

Figure 5-45. A second way that strain gages can be set up to compensate for temperature fluctuation.

Linear Variable Displacement Transformer (LVDT)

An LVDT uses the principle of a transformer to sense position. An LVDT is a cylindrical sensor. It has three fixed windings. The center of the cylinder is hollow, which allows a core to slide in the cylinder (see Figure 5-46). One of the coils, the primary, is connected to a low-voltage AC source (a sine wave). As the core moves through the LVDT a voltage is induced in the other two windings. The voltage is proportional to the core position.

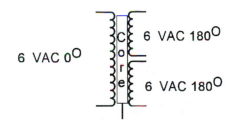

Figure 5-46. This figure shows a typical LVDT. Note that the core is aligned with both windings. Note the output of each of the output windings is 6 VAC with a phase of 180 degrees.

There are two methods of wiring LVDTs. They can be wired in the series adding mode or the series opposition mode. In the series adding mode, the secondary coils are wired in series, thus adding their output. In the series opposition the leads from one secondary are reversed and then the secondaries are wired in series. The output voltages are then in opposition. In the series adding mode the voltage increases smoothly as the core moves into and though the core (see Figures 5-47 and 5-48). In the series opposition mode, the voltage is zero when the core is centered and the voltage increases as it is moved from the centered position (see Figure 5-49). The signal also changes phase on each side of the centered position.

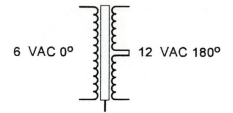

6 VAC 0°   12 VAC 180°

Figure 5-47. The series adding method of use. Note that the output is added in this case. The output would be 12 VAC 180 degrees.

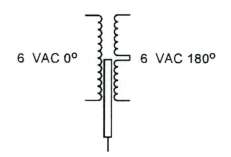

6 VAC 0°   6 VAC 180°

Figure 5-48. In this example the core is located halfway into the windings. The total output in the series adding mode is 6 VAC 180 degrees.

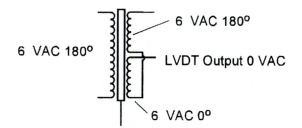

6 VAC 180°

6 VAC 180°

LVDT Output 0 VAC

6 VAC 0°

Figure 5-49. The series opposition method. Note that the core is centered. Note also that the output of the top winding is 6 VAC 180 degrees. The bottom winding is 6 VAC 0 degrees because, in effect, the leads have been reversed. The output when the core is centered is 0 volts.

Figures 5-50, 5-51, and 5-52 show examples of the output for an LVDT that is wired in the series opposition mode.

Chapter 5: Industrial Sensors

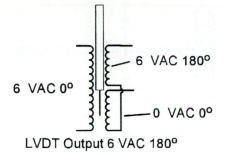

6 VAC 180°

6 VAC 0°

0 VAC 0°

LVDT Output 6 VAC 180°

Figure 5-50. In this example the core is aligned with the top output winding. The output in this example is 6 volts 180 degrees.

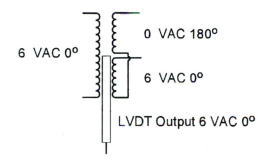

0 VAC 180°

6 VAC 0°

6 VAC 0°

LVDT Output 6 VAC 0°

Figure 5-51. The core is aligned with the lower winding. The output is 6 VAC 0 degrees.

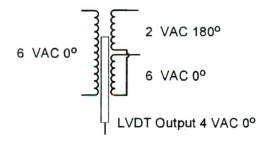

2 VAC 180°

6 VAC 0°

6 VAC 0°

LVDT Output 4 VAC 0°

Figure 5-52. The core is aligned with the lower winding and part of the upper winding. The output of the LVDT in the series opposition mode is 4 VAC 0 degrees.

The output from an LVDT can be rectified to provide a DC signal (see Figure 5-53). It is easier for PLCs to work with a DC signal than an AC signal. This is called signal conditioning. LVDTs can be purchased with DC output.

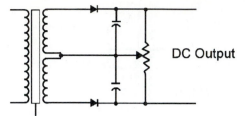

Figure 5-53. The output from an LVDT can be rectified to provide a DC signal that can be read by an analog input card.

Resolvers

A resolver is an analog position transducer. The resolver uses the principle of the transformer to operate.

A 0-degree sine wave (AC voltage) is applied to one stationary coil (see Figure 5-54). A 90-degree sine (cosine) wave is applied to a second stationary winding. The frequency of these signals is typically about 2 kHz. The rotor of the resolver has a third winding. This winding rotates with the rotor. Because of the transformer action this winding produces an output. There are usually half as many windings on the rotor as there are on the stator windings so the output voltage will be one half the input voltage.

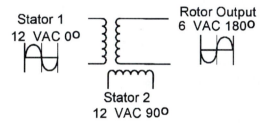

Figure 5-54. In this example the rotor is aligned with stator winding 1. The rotor output would be 6 VAC (1/2 the number of windings on the rotor) 180 degrees.

In Figure 5-54 we see that if the rotor is aligned with the 0-degree sine winding the output will be 6 volts 180 degrees out of phase (remember the transformer). So if we put the resolver rotor winding output on a scope, it would be 6 volts 180 degrees out of phase.

If we rotate the rotor 1/2 turn (180 degrees) it will align with the 0-degree sine winding again, but the output will be 6 volts 0 degrees (see Figure 5-55).

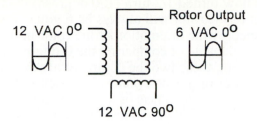

12 VAC 0°   Rotor Output
6 VAC 0°

12 VAC 90°

Figure 5-55. The rotor is aligned with the 0-degree stator but note that the rotor has been rotated 180 degrees.

In Figure 5-56 we see the rotor aligned with the 90-degree stator. In this case the output is 6 volts 270 degrees. Note that the output is 180 degrees out of phase with the input 90-degree sine wave.

If we rotate the rotor 1/2 turn (180 degrees) the output will be 6 volts 90 degrees (see Figure 5-57).

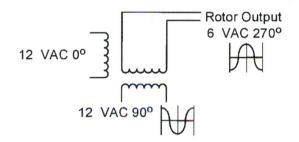

12 VAC 0°   Rotor Output
6 VAC 270°

12 VAC 90°

Figure 5-56. The rotor lined up with the 90-degree stator.

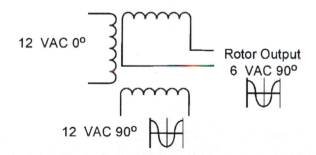

12 VAC 0°

Rotor Output
6 VAC 90°

12 VAC 90°

Figure 5-57. The output with the rotor aligned with the 90-degree stator but revolved 180 degrees.

What if the rotor is not aligned with either stator but is located at some other angle? Examine Figure 5-58. In this case the rotor is at a 45-degree angle to the 0- and 90-degree stator. In this case the rotor winding output is a combination of the field created by the 0- and 90-degree stator. The output would be 6 volts 225 degrees. If the rotor is now turned 1/2 turn (180 degrees) the output will be 6 volts 45 degrees. The output of the rotor will always be 6 volts. The phase of the rotor output will be determined by the orientation of the rotor in relation to the stator windings. A computer is capable of comparing the 0-degree sine wave to the output phase. By comparing these two the computer always knows the position of the rotor. One manufacturer of position controllers uses a clock that pulses 4000 times per cycle. They compare the phase of the input and output and can break the cycle into 4000 pieces. This would give a resolution of 4000 for 360 degrees of rotation. This is very fine resolution.

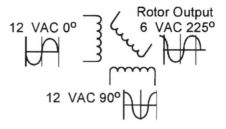

Figure 5-58. The output if the rotor were at a 45-degree angle to the 0- and 90-degree stators.

There is another way that resolvers are used. It is the reverse of what we just studied. It is often called ratiometric tracking. In the ratiometric tracking method, a sine wave input is applied to the rotor. This causes an output voltage and phase on the stators. The voltage and phase of the stators vary as the rotor turns. The computer can compare the two stator outputs and determine position (see Figure 5-59).

Resolvers are very reliable devices. There is very little that can go wrong with a resolver. Typically the bearings are really the only thing that can fail.

Pressure Sensors

Pressure sensors are typically used to measure and control fluids such as gases and liquids. There are a number of different types of pressure sensors on the market. Some operate through a change in resistance, some through a change in capacitance, and some through changes in inductance. A strain gage type pressure sensor has a strain gage attached to a membrane that stretches in proportion to the pressure applied to it (see Figure 5-60). If a constant current is applied to the strain gage the output voltage will vary. This voltage change will correspond to the change in pressure.

A second type of pressure sensor utilizes an LVDT (see Figure 5-61). In this style sensor, pressure applied to the bellows causes it to expand and move the core in the LVDT. This change in core position can then be measured by the output of the LVDT.

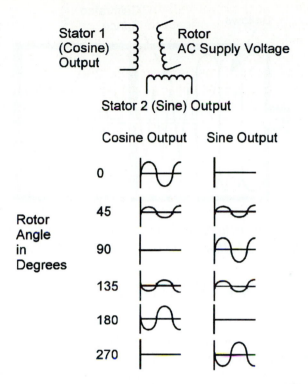

Figure 5-59. The output for a resolver using the ratiometric method.

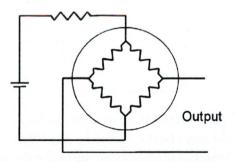

Figure 5-60. A strain gage pressure sensor. The circle represents a membrane that the pressure is applied to.

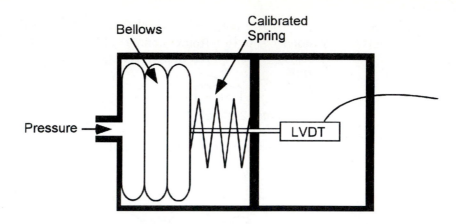

Figure 5-61. A bellows-style pressure sensor. Any change in pressure affects the bellows and moves the core in the LVDT.

INSTALLATION CONSIDERATIONS

Electrical

The main consideration for sensors is to limit the load current. The output (load) current must be limited on most sensors to a very small output current. The output limit is typically between 50 and 200 mA. If the load draws more than the sensor current limit, the sensor fails and you buy a new one. More sensors probably fail because of improper wiring than actually fail from use. It is crucial that you carefully limit current to a level that the sensor can handle (see Figure 5-62). PLC input modules limit the current to acceptable levels. Some sensors are available with relay outputs. These can handle higher load currents (typically 3A).

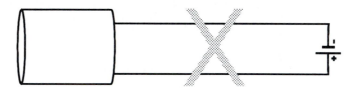

Figure 5-62. How __not__ to hook up a two-wire sensor. Remember, do not connect any sensor without a load that will limit the current to an acceptable level, or you will ruin the sensor.

If high-voltage wiring is run in close proximity to sensor cabling, the cabling should be run in a metal conduit to prevent the sensor from false sensing, malfunction, or damage. The other main consideration is to buy the proper polarity sensors. If the PLC module requires sinking devices, make sure that sinking devices are purchased.

Mechanical

Sensors should be mounted horizontally whenever possible (see Figure 5-63). This prevents the buildup of chips and debris on the sensor that could cause misreads.

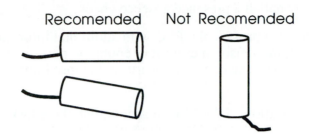

Recomended   Not Recomended

Figure 5-63. Sensor installation. A horizontal mount helps minimize chips and debris from collecting on the sensor and possibly causing misreads. Courtesy of Omron Electronics.

In a vertical position, chips, dirt, oil, and so on can gather on the sensing surface and cause the sensor to malfunction. In a horizontal position, the chips fall away. If the sensor must be mounted vertically, provision must be made to remove chips and dirt periodically. Air blasts or oil baths can be used.

Care should also be exercised that the sensor does not detect its own mount. For example, an inductive sensor mounted improperly in a steel fixture might sense the fixture. You must also be sure that two sensors do not mutually interfere. If two sensors are mounted too close together, they can interfere with each other and may cause erratic sensing.

You must also be careful not to use too much force when installing a sensor. Many sensor cases are plastic and can be damaged if deformed or impacted by a mounting or holding screw (see Figure 5-64). This deformation can easily ruin a sensor.

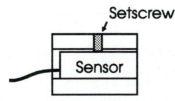

Setscrew

Sensor

Figure 5-64. How a setscrew is used to hold a sensor in a mount. Care must be exercised that the setscrew is not overtightened. This could easily damage the sensor.

TYPICAL APPLICATIONS

One of the most common uses of a sensor is in a product feeding situation. It may involve parts moving along a conveyor or in some type of parts feeder (see Figure 5-65). The sensor is used to notify the PLC when a part is in position, ready to be used. This is typically called a presence/absence check.

The same sensor can also provide the PLC with additional information. The PLC can take the data from the sensor and use it to count the parts as they are sensed. The PLC can also compare the completed parts and elapsed time to compute cycle times.

This one simple sensor allowed the PLC to accomplish three tasks:

> *Are there parts present?*
>
> *How many parts have been used?*
>
> *What is the cycle time for each part?*

Simple sensors can be used to decide which product is present. Imagine a manufacturer that produces three different size packages on the same line. Now imagine the product sizes moving along a conveyor at random. When each package arrives at the end of the line, the PLC must know what size product is present.

This can be done very easily with three simple sensors. If only one sensor is on, the small product is present. If two sensors are on, it must be the middle-sized product. If three sensors are on, the product must be the large size. The same information could then be used to track production for all product sizes and cycle times for each size.

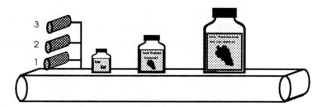

Figure 5-65. Use of sensors to check the size of containers as they move along a conveyor. This information could be used to divert the product to the appropriate processes.

Sensors can even be used to check whether or not containers have been filled. Imagine aspirin bottles moving along a conveyor with the protective foil and the cover on. There are simple sensors that can sense right through the cap and seal and make sure that the bottle was filled. One sensor would be set up to sense when there is a bottle present. This is often called a gate sensor. A gate sensor is used to show when a product is in place. The PLC then knows that a product is present and can perform further checks.

A second sensor would be set up to sense the aspirin in the bottle. If there is a bottle present, but the sensor does not detect the aspirin inside, the PLC knows that the aspirin bottle was not filled. The PLC can then make sure that no empty bottles leave the plant. The PLC can also track scrap rates, production rates, and cycle times.

Sensors can be used to monitor temperature. Imagine an oven used in a bakery. The sensor can monitor the temperature. The PLC can then control the heater element in the oven to maintain the ideal temperature.

Pressure is vital in many processes. Imagine a plastic injection machine. Heated plastic is forced into a mold under a given pressure. The pressure must be accurately maintained or the parts will be defective. Sensors can be used to monitor these pressures. The PLC can then monitor the sensor and control the pressure.

Flow rates are important in process industries such as papermaking. Sensors can be used to monitor the flow rates of raw material. The PLC can use these data to adjust and control the flow rate of the system. Think about your water supply at home. The water department monitors the flow of water to calculate your bill.

The applications noted above are but a few of the simple uses to which sensors can be put. The innovative engineer or technician will invent many other uses. You should now be aware that the data from one sensor can be used to provide many different types of information (i.e., presence/absence, part count, cycle time, etc.).

When choosing the sensor to use for a particular application there are several important considerations. The characteristics of the object to be sensed are crucial. Is the material plastic? Is it metallic? Is it a ferrous metal? Is the object clear, transparent, or reflective? Is it large or very small?

The specifics of the physical application are very important. Is there a large area available to mount the sensor? Are there problems with contaminants? What speed of response is required? What sensing distance is required? Is excessive electrical noise present? What accuracy is required?

Answering these questions will help narrow down the available choices. A sensor must then be chosen based on such criteria as the cost of the sensor, the cost of failure, and reliability.

Physical Arrangement

The physical arrangement of sensors can be used to differentiate between products (see Figure 5-66). Normally, one sensor is used as a gate sensor. A gate sensor is used as a trigger to the PLC to tell the PLC that a product is in position and further actions should take place. In the case of Figure 5-66, when the gate sensor triggers the PLC, the PLC looks at the state of the two other sensors. If both top sensors are on, it means that the product must be a square. If only the right sensor is on, it means that the product must be a circle. If the gate sensor is on and neither top sensor is on, the product must be a triangle.

There are two important concepts here. The first is that the physical characteristics of an object can be sensed by the innovative use of sensors. Shape can be checked. Sensors can check for holes, protruding surfaces, and to see if parts have been attached to the product.

The second important concept is the use of a sensor as a gate. I have seen a manufacturer scrap a very sophisticated PLC control system, complete with radio-frequency scanning, because it exhibited erratic performance. The erratic performance was due to the fact that the product was being scanned for characteristics without a gate sensor to choose the correct time to scan. The lack of a gate sensor meant that the PLC sometimes scanned the product ahead or behind the desired product. This company scrapped a multithousand-dollar system when a $30 sensor could have solved the problem. The gate sensor could have assured that the proper product was scanned.

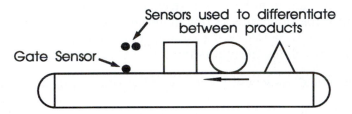

Figure 5-66. How the position of sensors can be used to determine which product is present.

The physical characteristics of the sensor are also important. Special shapes and configurations are available. Horseshoe-shaped optical sensors are available to sense objects passing through the horseshoe. Ring sensors are available to sense objects passing through the ring. Miniature sensing heads are available when space is at a premium or when the object to be sensed is small. Flat, very thin sensor packages are available when space is at a minimum. The point is that there are sensors available to meet any need. A technician should take time to glance through sensor catalogs to see what is available.

As systems become more automated they reduce the opportunity for human observation and intervention. When an operator drills holes in an engine block he/she knows immediately if the drill breaks. The operator then stops the process and replaces the drill. When the process is automated, there is no operator to observe problems. The PLC could care less if the drill breaks. Sensors must be used to sense problems in an automated system. The cost of failure is usually the guide to how much sensing must be done. If the cost is high, sensors should be used to notify the PLC.

Sensors can be used to check whether product has been correctly assembled. Figure 5-67 shows a small inductive sensor being used to sense when the drill has driller completely through the part. Sensors can be used to check whether or not machining operations have been performed. Sensors can also be used to check for the presence/absence of electrical components. In Figure 5-69 sensors are being used to sort packages.

Chapter 5: Industrial Sensors

Sensor

Figure 5-67. Use of a sensor to make sure that the drill actually drills through the gear. Note that the sensor is mounted sideways to allow the chips to fall through.

Two typical applications for ultrasonic sensors are shown in Figure 5-68. These applications utilize ultrasonic sensors to measure distances. The application on the left is using the analog output of the sensor for precise control of the web. The application on the right is measuring the height of objects. The fiber optic is used as a gate to indicate part presence. The ultrasonic sensor then takes a reading and the height of the part is known.

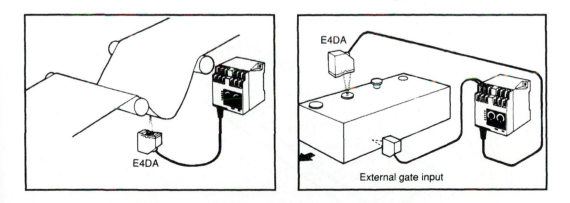

Figure 5-68. Two typical applications, in which ultrasonic sensors are used to measure distances. Application concept courtesy of Omron Electronics.

Figure 5-69 shows a package sorting application. Color mark sensors were used for this application.

Sorting packages by color mark and size

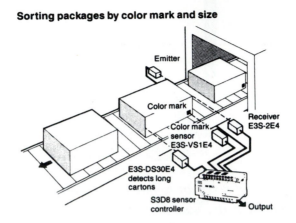

Figure 5-69. Sorting packages with color mark sensors. Application concept courtesy of Omron Electronics.

Figure 5-70 shows how inductive sensors could be used in a bottle capping application. The output from this sensor could be used to distinguish between those bottles that are capped and those that are not. Note that a field sensor can work in hostile environments.

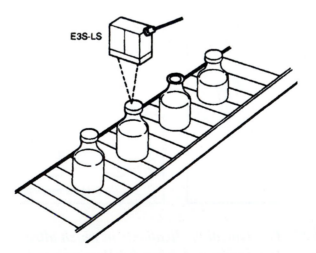

Figure 5-70. An inductive sensor can be used to sense for the proper capping of a container. Application concept courtesy Omron Electronics.

Chapter 5: Industrial Sensors

Figure 5-71 shows the use of a laser sensor. This sensor has 5-micron resolution. The sensor uses a laser LED to emit a 10-mm-wide beam with a maximum sensing distance of 300 mm (11.8 inches). Linear and discrete outputs are available. In this case the analog output is used to detect wire breakage.

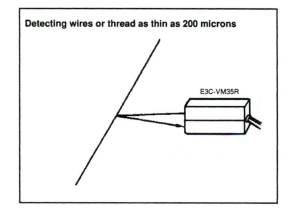

Figure 5-71. Wire can be sensed by a laser sensor. Application concept courtesy Omron Electronics.

Figure 5-72 shows three fiber-optic applications. In the first, the sensor checks for part presence. The second application uses a special fiber-optic head that speads the beam. The third is used to check diameter in a tape winding application.

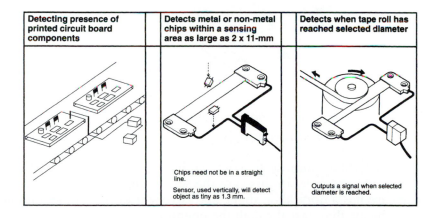

Figure 5-72. Small objects can be sensed by a laser sensor. Application concept courtesy Omron Electronics.

Figure 5-73 shows how transparent film can be detected. In this case the sensor is mounted perpendicular to the film to maximize the reflected light to the receiver.

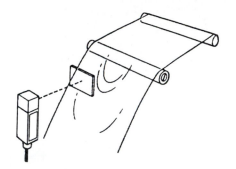

Figure 5-73. Transparent film can be detected. Application concept courtesy Omron Electronics.

Figure 5-74 shows an example of an inductive sensor being used to check to see that screws have been correctly assembled. This inductive sensor provides digital outputs for high/pass/low detection. This could be used to pass or fail the parts into pass bins, too tall bins, and too short bins.

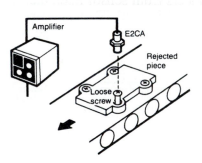

Figure 5-74. Small metal objects can be detected. Application concept courtesy Omron Electronics.

There are also many special-purpose sensors that have unique characteristics. Figure 5-75 shows one example. In this case the sensor and receiver are packaged in one case. The case has a horseshoe shape and is very useful for detecting holes in disks or small objects that pass through the opening.

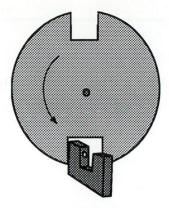

Figure 5-75. A horseshoe-style optical sensor being used to sense cutouts in a rotating disk.

Choosing a Sensor for a Special Application

There are always applications that present special problems for the technician or engineer. The sensor may fail to sense every part or fail randomly. In these cases a different type or different model of sensor may need to be chosen. Sales representatives and applications engineers from sensor manufacturers can be very helpful in choosing a sensor to meet a specific need. They have usually seen the particular problem before and know how to solve it.

The types of sensors and the complexity of their use in solving application problems grows daily. New sensors are constantly being introduced to solve needs. There are almost as many sensor types available as there are applications. There are even magazines devoted to the topic of sensors.

The innovative use of sensors can help increase the safety, reliability, productivity, and quality of processes. Sensors are crucial to the future of American manufacturing. The technician must be able to choose, install, and troubleshoot sensors properly.

Questions

1. Describe at least four uses of digital sensors.

2. Describe at least three analog sensors and their uses.

3. List and explain at least four types of optical sensors.

4. Explain how capacitive sensors work.

5. Explain how inductive sensors work.

6. Explain the term *hysteresis*.

7. Draw and explain the wiring of a load-powered sensor.

8. Draw and explain the wiring of a line-powered sensor.

9. What is burden current? Load current?

10. Why must load current be limited?

11. Explain the basic principle on which a thermocouple is based.

12. How are changes in ambient temperature compensated for?

13. What is the temperature range for a type J thermocouple?

14. Explain at least three electrical precautions as they relate to sensor installation.

15. Explain at least three mechanical precautions as they relate to sensor installation.

Chapter 6

Input/Output Modules and Wiring

Originally, PLCs were designed strictly for simple digital (on/off) control. Over the years, PLC manufacturers have added to the capabilities of the PLC. Today, there are I/O cards available for almost any application imaginable.

OBJECTIVES

Upon completion of this chapter, the student will be able to:

List at least two I/O cards that can be used for communication.

Describe at least five special-purpose I/O cards.

Define such terms as **resolution, high density, discrete, TTL,** *and* **RF.**

Choose an appropriate I/O module for a given application.

I/O MODULES

Originally, PLCs were used to control one simple machine or process. The changes in American manufacturing have required much more capability. The increasing speed of production and the demand for higher quality require closer control of industrial processes. PLC manufacturers have added modules to meet these new requirements.

Special modules have been developed to meet almost any imaginable need. Modules have been developed to control processes closely. Temperature control is one example.

Industry is beginning to integrate its equipment so that data can be shared. Modules have been developed to allow the PLC to communicate with other devices, such as computers, robots, and machines.

Velocity and position control modules have been developed to meet the needs of accurate high-speed machining. These modules also make it possible for entrepreneurs to start new businesses that design and produce special-purpose manufacturing devices, such as packaging equipment, palletizing equipment, and various other production machinery.

These modules are also designed to be easy to use. They are intended to make it easier for the engineer to build an application. In the balance of this chapter we examine many of the modules that are available.

DIGITAL (DISCRETE) MODULES

Digital modules are also called discrete modules because they are either on or off. A large percentage of manufacturing control can be accomplished through on/off control. Discrete control is easy and inexpensive to implement.

Digital Input Modules

These modules accept an "on" or "off" state from the real world. The module inputs are attached to devices such as switches or digital sensors. The modules must be able to buffer the CPU from the real world. Assume that the input is 250 V AC. The input module must change the 250 V AC level to a low-level DC logic level for the CPU. The modules must also optically isolate the realworld from the CPU. Input modules usually have fuses for module protection.

Input modules typically have light-emitting diodes (LEDs) for monitoring the inputs. There is one LED for every input. If the input is on, the LED is on. Some modules also have fault indicators. The fault LED turns on if there is a problem with the module. The LEDs on the modules are very useful for troubleshooting.

Most modules also have plug-on wiring terminal strips. All wiring is connected to the terminal strip. The terminal strip is plugged onto the actual module. If there is a problem with a module the entire strip is removed, a new module inserted, and the terminal strip plugged into the new module. Note that there is no rewiring. A module can be changed in minutes or less. This is vital considering the huge cost of a system being down. (The term *down* means "unable to produce product.")

Input modules usually need to be supplied with power. (Some of the small PLCs are the exception.) The power must be supplied to a common terminal on the module, through an input device, and back to a specific input on the module. Current must enter at one terminal of the I/O module and exit at another terminal (see Figure 6-1). The figure shows a power supply, field input device, the main path, I/O circuit, and the return path. Unfortunately this would require two terminals for every I/O point. Most I/O modules provide groups of I/O that share return paths (commons). Figure 6-2 shows the use of a common return path.

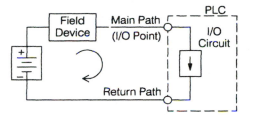

Figure 6-1. Typical I/O current path. Courtesy PLC Direct by Koyo.

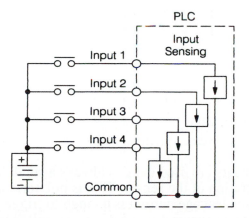

Figure 6-2. Typical shared return path (common). Courtesy PLC Direct by Koyo.

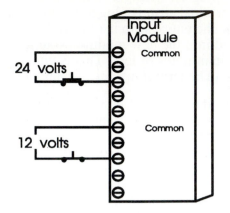

Figure 6-3. Input module with dual commons. This allows the user to mix input voltages.

Some modules provide multiple commons. This allows the user to mix voltages on the same module (see Figure 6-3). These commons can be jumpered together if desired (see Figure 6-4).

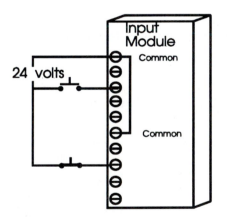

Figure 6-4. How dual commons can be wired together. All inputs would use the same voltage.

When load-powered sensors are used there is always a small leakage current that is necessary for the operation of the sensor. This is not normally a problem. In some cases, however, this leakage current is enough to trigger the input of the PLC module. In this case a resistor can be added that will "bleed" the leakage current to ground (see Figure 6-5). When a bleeder resistor is added, most of the current goes through it to common. This assures that the PLC input turns on only when the sensor is really on.

Chapter 6: Input/Output Modules and Wiring

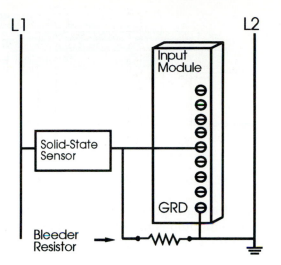

Figure 6-5. Use of a "bleeder" resistor.

High-Speed Counter Modules

High speed counter modules are used to count pulses from sensors, encoders, switches, and so on, at very high speeds. Figure 6-6 shows how an encoder would be wired to a high-speed input module. Figure 6-7 shows an example of a spinning disk with a slot in it. In this case only one input is needed. Modules to count pulses up to 75 kHz are common. These units are designed to make the engineer's job simpler. The counter module counts and accumulates the high-speed pulses. The PLC can do its other tasks and check on the count as it needs to. These are really high-speed special-purpose, digital input modules.

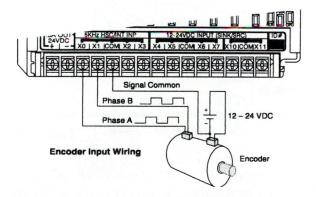

Figure 6-6. Encoder wiring. Courtesy PLC Direct by Koyo.

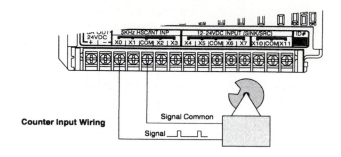

Figure 6-7. Pulse counting. Courtesy PLC Direct by Koyo.

Figure 6-8 shows the use of a high-speed counter module to control a cutting application. The high-speed module is used to count the encoder pulses. The encoder pulses are used to track position. In this case the motor is run until the encoder pulses are equal to the required number of pulses. The number of pulses relates to the required length of product. When the number of pulses equals the desired number of pulses the motor is stopped and the brake is reengaged. The cutters then cut the material to the correct length. This module is capable of bidirectional counting of high-speed inputs from quadrature encoders and other high-speed sources.

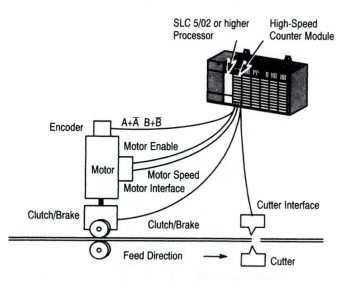

Figure 6-8. Example of the use of a high-speed counter module to control a cutting application. Courtesy of Allen-Bradley Company, Inc., a unit of Rockwell International.

Digital Output Modules

Discrete output modules are used to turn real-world output devices either on or off. Discrete output modules can be used to control any two-state device. Output modules are available in AC and DC versions. They are available in various voltage ranges and current capabilities.

The current specifications for a module are normally given as an overall module current and as individual output current. The specification may rate each output at 1 A. If it is an eight-output module, one might assume that the overall current limit would be 8 A (8X1). This is normally not the case. The overall current limit will probably be less than the sum of the individuals. For example, the overall current limit for the module might be 5 A. The user must be careful that the total current to be demanded does not exceed the total that the module can handle. Normally the user's outputs will not each draw the maximum current, nor will they normally all be on at the same time. The user should consider the worst case when choosing an appropriate output module.

Output modules are normally fused to protect each output. There will be a fuse for each output. Many modules also have lights to indicate when fuses are burned out. The overall module is normally fused to prevent damage in case the overall current limit is exceeded. The fuses are normally very easily replaced. The wiring harness is removed, the module is removed from the rack, and the fuse is replaced. Check the technical manual for the PLC to find the exact procedure.

Some output modules provide more than one common terminal. This allows the user to use different voltage ranges on the same card (see Figure 6-9). These multiple commons can be tied together if the user desires. All outputs would be required to use the same voltage, however (see Figure 6-10).

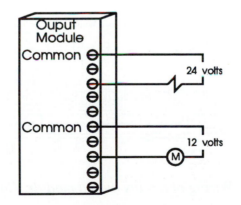

Figure 6-9. Use of a dual-common output module.

Output modules can be purchased with transistor output, triac output, or relay output. The transistor output would be used for DC outputs. There are various voltage ranges and current ranges available, as well as TTL (transistor-transistor logic) output. Triac outputs are used for AC devices. They are also available in various voltage ranges and current ranges. The relay outputs are found quite often on small PLCs. Relay outputs can be used with AC or DC voltages. The voltages can even be mixed (see Figure 6-11). Figure 6-12 shows the use of a bleeder resistor to "bleed off" unwanted leakage through an output module. This leakage can occur with the solid-state devices that are used in output modules. Relay output modules are also available for larger PLCs. Figure 6-13 shows what the inside of the relay module looks like. Note the relays. Relay outputs should be protected with a diode to prolong contact life (see Figure 6-14).

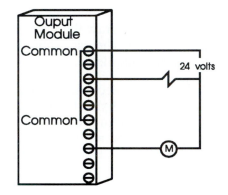

Figure 6-10. Use of a dual-common output module with the commons tied together. Note that the voltages must be the same if the commons are tied together.

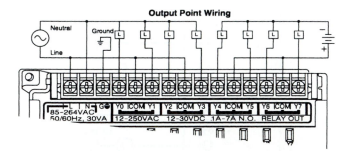

Figure 6-11. Wiring of a relay output module. Courtesy of PLC Direct by Koyo.

Chapter 6: Input/Output Modules and Wiring

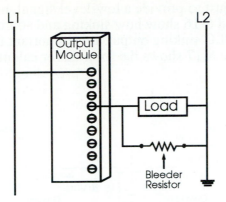

Figure 6-12. Use of a bleeder resistor to "bleed off" unwanted leakage through an output module.

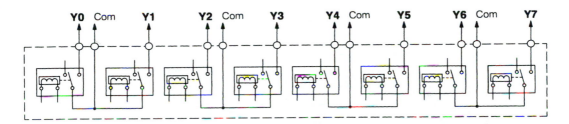

Figure 6-13. The wiring of relay outputs. Courtesy PLC Direct by Koyo.

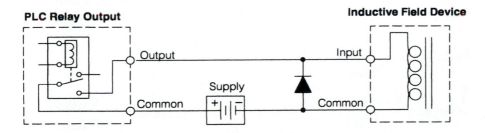

Figure 6-14. The use of a diode to help prolong contact life. Courtesy PLC Direct by Koyo.

Some applications require connecting a PLC output to the solid state input of a device. This is usually just to provide a low-level signal, not to power an actuator or coil. Figures 6-15 and 6-16 show how sinking and sourcing, solid state devices can be connected to a PLC sinking output. It is important to size the pull-up resistor properly. Figure 6-17 shows the formula for calculating the proper size resistor.

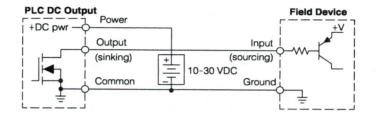

Figure 6-15. A PLC sinking output connected to a solid state sourcing input on an output device. Courtesy PLC Direct by Koyo.

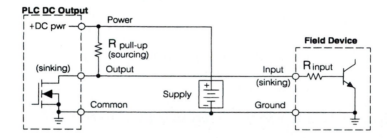

Figure 6-16. A PLC sinking output connected to a solid state sinking input on an output device. Courtesy PLC Direct by Koyo.

$$I_{input} = \frac{V_{input \, (turn-on)}}{R_{input}}$$

$$R_{pull-up} = \frac{V_{supply} - 0.7}{I_{input}} - R_{input} \qquad P_{pull-up} = \frac{V_{supply}^2}{R_{pullup}}$$

Figure 6-17. The formula for calculating the correct size of pull-up resistor. Courtesy PLC Direct by Koyo.

Chapter 6: Input/Output Modules and Wiring

High-Density I/O Modules

High-density modules are digital I/O modules. A normal I/O module has eight inputs or outputs. A high-density module may have up to 32 inputs or outputs. The advantage is that there are a limited number of slots in a PLC rack. Each module uses a slot. With the high-density module, it is possible to install 32 inputs or outputs in one slot. The only disadvantage is that the high-density output modules cannot handle as much current per output.

ANALOG MODULES

Computers (PLCs) are digital devices. They do not work with analog information. Analog data such as temperature must be converted to digital information before the computer can work with it.

Analog Input Modules

Cards have been developed to take analog information and convert it to digital information. These are called analog-to-digital (A/D) input cards. There are two basic types available: current sensing and voltage sensing. These cards will take the output from analog sensors (such as thermocouples) and change it to digital data for the PLC.

Voltage input modules are available in two types: unipolar and bipolar. *Unipolar modules* can take only one polarity for input. For example, if the application requires the card to measure only 0 to +10 V, a unidirectional card will work. The *bipolar* card will take input of positive and negative polarity. For example, if the application produces a voltage between -10 V and +10 V, a bidirectional input card is required; that is because the measured voltage could be negative or positive. Analog input modules are commonly available in 0 to 10 V models for the unipolar and -10 to +10 V for the bipolar model.

Analog models are also available to measure current. These typically measure 4 to 20 mA. Four milliamps represents the smallest input value and 20 milliamps represents the largest input value.

Many analog modules can be configured by the user. Dip switches or jumpers are used to configure the module to accommodate different voltages or current. Some manufacturers make modules that will accept voltage or current for input. The user simply wires to either the voltage or the current terminals, depending on the application.

Resolution in Analog Modules

Resolution can be thought of as how closely a quantity can be measured. Imagine a 1-ft. ruler. If the only graduations on the ruler were inches, the resolution would be 1 in. If the graduations were every 1/4 in., the resolution would be a 1/4 in.

The closest we would be able to measure any object would be 1/4 inch.

That is the basis for the measure of an analog signal. The computer can only work with digital information. The analog-to-digital (A/D) card changes the analog source into discrete steps.

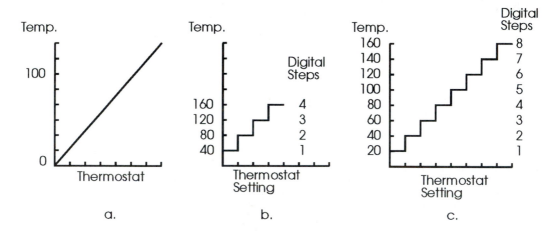

Figure 6-18. Graphs of temperature vs. thermostat setting. Note that graph (a) represents a linear relationship of temperature vs. setting. In reality, when analog control is used, the analog is really a series of steps (resolution). Graph (b) shows what a four-step system would look like. The resolution would be 40 degrees per step. Graph (c) shows an eight-step system. The resolution of graph (c) is 20 degrees.

Examine Figure 6-18a. Ideally, the PLC would be able to read an exact temperature for every setting of the thermostat. Unfortunately, PLCs work only digitally. Consider Figure 6-18b. The analog input card changes the analog voltage (temperature) into digital steps. In this example, the analog card changed the temperature from 40 degrees to 160 degrees in four steps. The PLC would read a number between one and four from the A/D card. A simple math statement in the ladder could change the number into a temperature. For example, assume that the temperature was 120 degrees. The A/D card would output the number 3. The math statement in the PLC would take the number and multiply by 40 to get the temperature. In this case, it would be 3 X 40, or 120 degrees. If the PLC read 4 from the A/D card, the temperature would be 4 X 40 or 160. Assume now that the temperature is 97. The A/D card would output the number 2. The PLC would read 2 and multiply by 40. The PLC would believe the temperature to be 80 degrees. The closest the PLC can read the temperature is about 20 degrees if four steps are used. (The temperature that the PLC calculates will always be in a range from 20 degrees below the actual temperature to 20 degrees above the measured temperature.) Each step is 40 degrees. This is called *resolution*. The smallest

Chapter 6: Input/Output Modules and Wiring

temperature increment is 40 degrees. The resolution would be 40.

Consider Figure 6-18c. There are eight steps on this A/D card. The resolution would be twice as fine or 20 degrees. For a temperature of 67 degrees, the A/D would output 3. The PLC would multiply 3 X 20 and assume the temperature to be 60. The largest possible error would be approximately 10 degrees. (The PLC calculated temperature would be within 10 degrees below the actual temperature to 10 degrees above the actual temperature.)

Industry requires much finer resolution. Typically, an industrial A/D card for a PLC would have 12-bit binary resolution, which means that there would be 4,096 steps. In other words, the analog quantity to be measured would be broken into 4,096 steps. Very fine resolution! There are cards available with even finer resolution than this. The typical A/D card is 12 bit binary (4,096 steps).

Analog modules are available that can take between one and eight individual analog inputs. Special-purpose A/D modules are also available. One example would be thermocouple modules. These are just A/D modules that have been adapted to meet the needs of thermocouple input. Thermocouples output very small voltages. To be able to use the entire range of the module resolution a thermocouple module amplifies the small output from the thermocouple so that the entire 12-bit resolution is used. The modules also provide cold junction compensation.

These modules are available to make it easy to accept input from thermocouples. The modules are available for various types of thermocouples.

Figure 6-19 shows a tank filling application.

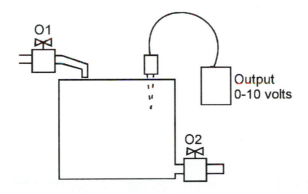

Figure 6-19. A tank fill system.

Analog Output Modules

Analog output modules are also available. The PLC works in digital, so the PLC outputs a digital number (STEP) to the digital-to-analog (D/A) converter module. The D/A converts the digital number from the PLC to an analog output. Analog output modules are available with voltage or current output. Typical outputs are 0 to 10 volts, -10 to +10 volts and 4 to 20 milliamps.

Imagine a bakery. A temperature sensor (analog) in the oven could be connected to an A/D input module in the PLC. The PLC could read the voltage (steps) from the A/D card. The PLC would then know the temperature. The PLC can then send digital data to the D/A output module, which would control the heating element in the oven. This would create a fully integrated, closed-loop system to control the temperature in the oven.

REMOTE I/O MODULES

Special modules are available for some PLCs that allow the I/O module to be positioned separately from the PLC. In some processes it is desirable (or necessary) to position the I/O at a different location. In some cases the machine or application is spread over a wide physical area. In these cases it may be desirable to position the I/O modules away from the PLC. Figure 6-20 shows an example of the use of an Allen-Bradley remote I/O adapter module.

There are many ways to link the I/O module to the PLC. Two common methods are twisted pair (wire) and fiber optics. Twisted pair is the cheaper method. Two wires are twisted around each other and connected between the PLC and the remote I/O. Twisting reduces the possibility of electrical interference (noise). Twisted-pair connections can transmit data thousands of feet.

Fiber-optic links are also available. Fiber-optic cables are noise immune because the data is transmitted as light. Much higher speeds are possible with fiber-optics. The cost of fiber-optics is rapidly decreasing. Fiber-optics can be used to transmit over distances of miles. Remote I/O modules may be purchased for either type of transmission medium.

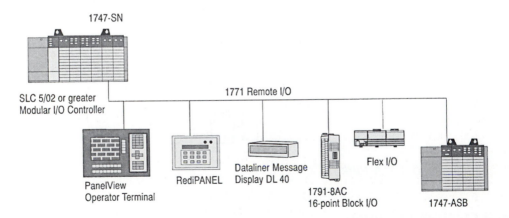

Figure 6-20. Example of the use of remote I/O modules. Courtesy of Allen-Bradley Company, Inc., a unit of Rockwell International.

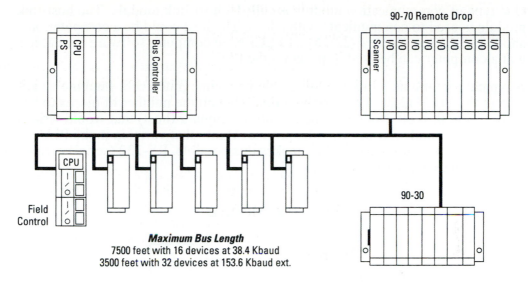

Figure 6-21. Example of a GE Fanuc Genius I/O bus. Courtesy GE Fanuc Inc.

Figure 6-21 shows an example of a GE Fanuc Genius I/O bus. The Genius I/O bus consists of a twisted-pair shielded cable that connects the Genius bus controller to up to 31 other devices. The devices might be Genius I/O blocks, hand-held monitors, bus controllers, or remote I/O scanners.

The use of modules that allow I/O to be remotely located will be expanding rapidly. I/O devices and the buses they connect to are making rapid advances in capabilities, lowered costs, and ease of use. These will be covered in more detail in Chapter 11.

COMMUNICATION MODULES

Communications are becoming a much larger part of the PLC's task. As systems are integrated, data must be shared throughout the system. PLCs must be able to communicate with computers, CNC machines, robots, and even other PLCs. Much of this communication involves file or data transfer.

For example, in a flexible system, a PLC might pass a program to a CNC machine. The CNC machine receives the program and then runs the program. The PLC was not originally designed to perform these duties. Various communication modules are now available to meet most of these needs.

The two main types are host-link and peer-to-peer modules.

Host-Link Modules

One type of communication module is called a host-link module. The host-link module is used to communicate with a host. The host could be a computer or another PLC (see Figure 6-22). Most PLCs have computer software available so that a computer can be used to program the PLC.

A typical use of a host-link module might be an integrated cell. Imagine a variable in the PLC ladder. The variable might contain the number of pieces to be produced in the cell. Now imagine a central computer (host) that would write a number (the number of pieces to be produced) into that variable in the PLC (download). The host-link module makes this possible. RS-232 (serial) communication devices are normally used.

Host Link Module

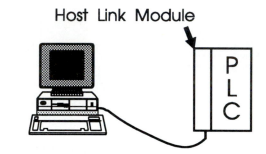

Figure 6-22. Use of a host-link module.

The host link can also be used to send data from the PLC to the computer (upload). The computer can then track process times, completed parts, and so on.

Peer-to-Peer Modules

Peer-to-peer modules are used by PLCs to communicate with other PLCs of the same brand (see Figure 6-23). This is usually done by assigning each PLC a different number. For example, some manufacturers allow 256 PLCs to be on the same network. Each PLC has a unique number (name) between 0 and 255. They communicate over a proprietary network.

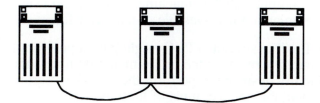

Figure 6-23. Peer-to-peer communications.

ASCII Modules

There are modules available that will transmit and receive ASCII files. These ASCII files are usually programs or manufacturing data. These modules are normally programmed with basic commands. The user writes a program in the basic language. The basic program works with the ladder logic as the program runs.

The modules are designed to make it easier for the PLC user to communicate with various devices. These modules typically have 2 to 24 kilobytes of memory. These modules can also be used to create an operator interface. Basic modules can be used to output text to a printer or terminal in order to update an operator. Communication can be done in several ways.

POSITION CONTROL MODULES

There are modules available for open- and closed-loop position control. Closed-loop systems have feedback devices to ensure that the command is completed.

Open-Loop Position Control

These are modules available to drive stepper motors (see Figure 6-24). Stepper motors can be used to control position in low-power, low-speed applications. They can be used to control positioning of axes very precisely. These modules are designed to make stepper motor integration easy. There are typically acceleration and deceleration functions, as well as teach functions, available. There are also inputs on the module that can be used to home the motor and for overtravel protection.

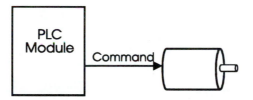

Figure 6-24. Block diagram of an open-loop servo. The PLC module issues a command to the motor and assumes it follows the command. There is no feedback. This is a very common configuration with stepper motors.

Figure 6-25 shows an example of a PLC stepper application. Note that the PLC outputs pulses and a direction signal (high or low) to a stepper drive.

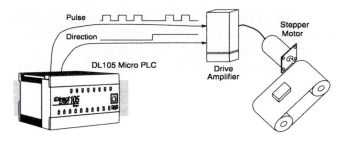

Figure 6-25. A stepper control system. Courtesy PLC Direct by Koyo.

Closed-Loop Position Control

There are many complex control applications. Robots and CNC machines are typical examples of closed-loop position control applications. The applications typically involve AC or DC motors to drive tables or axis of motion. The position loop is typically closed (monitored) by encoders. The velocity loop is normally closed by a tachometer. The modules are available to monitor and control velocity and position.

Study Figure 6-26. There are two closed loops in this figure.

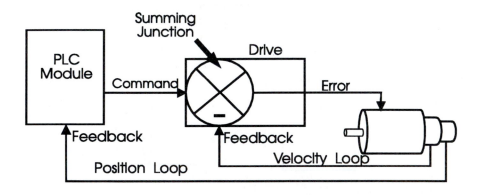

Figure 6-26. Closed-loop servo system.

There is a velocity loop and a position loop. The control device (in this case a PLC) closes the position loop. The feedback device for the position loop is normally an encoder or resolver. The motor drive closes the velocity loop. The feedback device for velocity is a tachometer.

The control device issues a command voltage. It is a velocity command. The drive receives the analog command (typically -10 to +10 V). The summing junction compares the command and the feedback from a tachometer on the motor and generates an error command to the motor. The velocity is continually monitored and adjusted by the drive. The position is monitored and controlled by the PLC. The minus sign in the summing junction shows that the negative-feedback principle is used.

Figure 6-27 shows an example of the use of a GE Fanuc axis positioning module (APM). This module is available in one- and two-axes versions. The two-axis module can be used in a normal mode so that two axes can be controlled independently or in follower mode. Follower mode is used to relate two axes of motion. Follower mode is often called master-slave mode. It can be used to establish an electronic "geared" relationship between two axes so that they are precisely coordinated.

Figure 6-28 shows a GE Fanuc servo motor and two digital AC drive modules. In a digital drive, all parameters are set digitally; there is no retuning required. The velocity and position loops are closed in the controller. These drives can act as "electronic gearboxes." The drives take input from a 8K absolute encoder that is standard on each motor. The absolute encoder eliminates the need for rehoming.

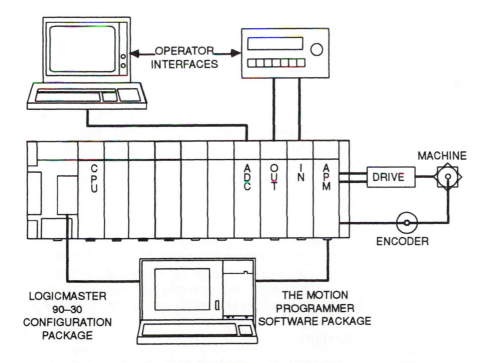

Figure 6-27. A GE Fanuc servo system. Courtesy GE Fanuc Inc.

Figure 6-28. Servo motor and two drives. Courtesy GE Fanuc Inc.

VISION MODULES

Vision is becoming more and more widely used in industry. Inspection costs industry billions of dollars each year. Most of the inspection occurs after the product is made. That means that the product is already scrap. Automated vision allows for in-process 10 percent inspection. It also allows automatic adjustment of the process while it is running. This means that the process can be corrected before it makes scrap. This can drastically improve the quality of parts and productivity of processes.

Vision began to enter industry in the early 1980s. There were a multitude of new vision companies. Industry immediately saw the benefit that was possible through vision. Unfortunately, industry tried to apply vision to the wrong applications. Instead of trying to apply vision to the many tasks that were appropriate and easy to implement, industry decided that they needed vision for automated welding, and automated guidance, and other applications. The applications they chose were too complex for the technology at the time.

The horror stories began. Programming easily cost more than the vision equipment. The systems were not very good when misapplied. Vision achieved a bad name. Industry backed off on planned expenditures for vision. Many vision companies failed. In the process, however, the surviving vision companies retrenched. They chose niches of applications that were appropriate for their product. Industry learned from its early vision failures, too. Industry has now begun to choose applications more appropriately.

Vision systems can inspect at rates of thousands of parts per minute, at extremely high rates of accuracy. They are particularly suited to tasks at which people are not efficient. People are not very good at inspecting when the tasks are very quick, tedious, or involve rapid identification and/or measurement. Vision systems do not get bored. For example, pharmaceutical manufacturers are required to inspect the date and lot code on every package they manufacture. There are no people, no matter how dedicated, that can inspect the high volumes that are required. A vision system is very well suited to this application.

The cost of vision is decreasing. This is due mainly to the fact that vision manufacturers have improved their software. They have made them much easier to program and apply. Thus huge reductions have occurred in application costs.

PLC manufacturers have seen the increased interest and now offer vision modules for their PLCs. Allen-Bradley, for example, offers a full range of vision equipment: stand-alone and PLC vision modules.

Allen-Bradley calls their product a Vision Input Module (VIM). The VIM, which can be plugged into several of the Allen-Bradley PLC families, can inspect pieces at up to about 1800 pieces per minute. The VIM can be used for part presence sensing, measurement, alignment, and other inspection tasks.

The VIM module does not utilize a programming language. The programmer just "teaches" the system. The programmer places a good part in front of the camera. The camera provides a picture of the image to the video monitor. There are symbols (called icons) across the bottom of the screen. Each icon represents a different type of inspection task. The programmer just chooses the icons. Once the programmer has chosen icons to match the steps in the inspection application, the vision system is ready.

Omron also offers vision modules and stand-alone systems for visual inspection.

A vision system typically consists of the following components: one or more cameras, lighting, a video monitor, and the vision processor (see Figure 6-29). The lighting can often be the most difficult part of the application. The story is told of one vision company that designed and installed an application. The application worked perfectly. It continued to work perfectly for several months. Then, all of a sudden, it seemed to act up every day. The company contacted the vision company to come and fix the system.

The vision technician struggled for several days and could not find any problem with the system, although every day it seemed to get erratic in the middle of the afternoon. Finally, the technician happened to glance at the ceiling. He noticed that there were some skylights that the sun was shining through. The company covered the windows and the system worked perfectly again. The time of year was just right for the sun to shine on the application in the middle of the afternoon. This story illustrates that lighting is a crucial part of every application. Many systems today will compensate for normal variations in ambient lighting.

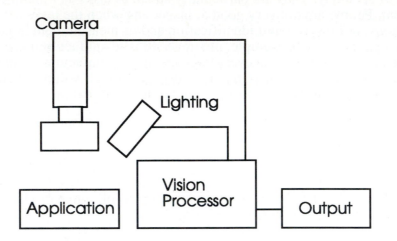

Figure 6-29. Block diagram of a typical vision system.

BAR-CODE MODULES

Automatic identification is becoming more widely used in industry. There are several types of automatic identification: vision, bar-coding, and radio frequency identification (RFID). Bar-coding is a rapidly growing technology. We see it in stores, markets, industry, and so on. It has become commonplace.

Bar-code modules are available for PLCs. A typical application might involve a bar-code module reading the bar code on boxes as they move along a conveyor line. The PLC is then used to divert the boxes to the appropriate product lines. They are also used extensively to monitor inventories in factories and businesses. A factory is able to have real-time inventory counts. Imagine the information the grocery store has from bar-coding every product and price into the database. They can measure very accurately the effect of advertising vs. sales of specific products, for example. They also use the information to learn which products and quantities must be ordered. The bar-coding also saves the cost of marking each and every product. There are several standard bar-coding systems. The grocery business has standardized on the universal product code (UPC) standard.

PROPORTIONAL, INTEGRAL AND DERIVA-TIVE MODULES (PID)

Proportional, integral, and derivative control is used to control processes. PLC manufacturers have approached PID control in two ways: Some offer special-purpose processing modules and some utilize standard analog/digital I/O with

PID software in the CPU (see Figure 6-30). PID can be used to control physical variables such as temperature, pressure, concentration, and moisture content. It is widely used in industrial control to achieve accurate control under a wide variety of process conditions. Although PID often seems complex, it need not be. PID is essentially an equation that the controller uses to evaluate the controlled variable. The controlled variable (temperature, for example) is measured and feedback is generated. The control device receives this feedback. The control device compares the feedback to the setpoint and generates an error signal. The error is examined in three ways. It is looked at with proportional, integral, and derivative methodology. Each of the three factors can be thought of as a gain. Each can affect the amount of response to a given error. We can control how much of an effect each has. The controller then uses these gains (proportional gain, integral gain, and derivative gain) to calculate an command (output signal) to correct for any measured error.

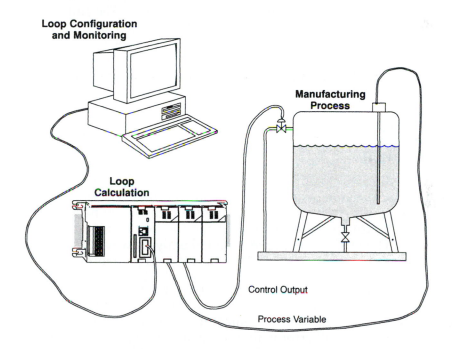

Figure 6-30. Block diagram of a process that is controlled by a PID controller. The system is a tank that controls temperature. A PLC is used to control this system. The CPU takes input from the analog input module and performs the PID calculation. The CPU generates an error signal and send it to the output module (digital or analog). The output would control the fuel valve. Courtesy PLC Direct by Koyo.

Proportional Control

The proportional gain portion of PID looks at the magnitude of the error. The proportional response to an error has the largest effect on the system. Proportional control reacts proportionally to errors. A large error receives a large response. (In the case of a large temperature error, the fuel valve would be opened a great deal.) A small error receives a small response. Figure 6-31 shows the PID equation.

Imagine a furnace that can be heated to 1500 degrees. There is a portion of this range of temperature where the response of the system will be proportional. For example, let's assume that between 1000 and 1500 degrees the system adjusts the valve opening in proportion to the error. Below 1000 degrees the valve is open 100 percent. Above 1500 degrees the valve is 0 percent open. The *proportional band* in this case is 500 degrees. Proportional band is normally given as a percentage. The percentage is calculated by dividing the proportional band in degrees by the full controller range and multiplying by 100. The full controller range is simply the range of temperatures that the furnace can control. In this case let's assume that the full controller range is 1440 (1500 - 60). Are you wondering why 60 was subtracted? The furnace cannot control below room temperature, so that is excluded in our calculation. In this case the proportional band in degrees is equal to (500/1440)*100 or 34.7 percent. The technician can adjust the width of the proportional band to make the system more or less responsive to an error. The narrower the proportional band, the greater the response to a given error.

Integral Control

Proportional control does not solve all of our problems. The first problem is that proportional control cannot correct for very small errors. These small errors are called offset or steady-state errors. An example of this might be driving your car with the cruise control on a level road. The car does not maintain the exact speed that you chose. The proportional term is quite happy, however. The small variations, or steady-state error, are handled by the integral gain term.

The integral gain portion attempts to correct for the small error (offset) that proportional cannot. Integral looks at the error over time. It increases the importance of even a small error over time. Integral is error multiplied by the time the error has persisted. A small error at time zero has zero importance. A small error at time 10 has an importance of 10 X error. In this way integral increases the response of the system to a given error over time until it is corrected. Integral can also be adjusted. The integral adjustment is called *reset rate*. Reset rate is a time factor. The shorter the reset rate the quicker the correction of an error. In hardware-based systems the adjustment will be done by a potentiometer. The potentiometer essentially adjusts the time constant of an *RC* circuit. Too short a reset rate can cause erratic performance.

Derivative Control

The second problem with proportional control is that it cannot adjust its output based on the rate of change in the error (remember that the proportional term looks at magnitude of error, not rate of change in the error). For example, if you are driving your car on the highway at 55 miles per hour and the car in front of you stops, you may respond with 50 percent brake pressure. This is a proportional response. The car ahead is stopping, so you respond with 50 percent brake. This will not work if the car stops too quickly. We, of course, would give much more brake pressure when the rate of stopping of the car in front is greater. In other words, we thought we applied enough brake pressure but the distance between our cars keeps getting smaller. We naturally respond to rate of error change. Proportional control systems do not. Proportional systems only respond to the magnitude of the error, not to the error's rate of change.

The derivative gain portion of control attempts to look at the rate of change in the error. Derivative will cause a greater system response to a rapid rate of change that to a small rate of change. Think of it this way: If a system's error continues to increase, the control device must not be responding with enough correction. Derivative senses this rate of change in the error and causes a greater response. Derivative is also adjusted as a time factor. Derivative is also called *rate time*.

The derivative term helps "damp" a system. If it is desired to have a system that will be very responsive and correct quickly for errors it is necessary to have a relatively high proportional gain. A high proportional gain will cause the system to overshoot. The derivative gain can help damp the overshoot. When a system overshoots the commanded value (position, or velocity, or temperature, etc.) the error rapidly increases.

$$Co = K(E + 1/TI \int_0^t E \, dt + KD(E-E(n-1))/dt) + bias$$

Co	The output of the equation
K	The overall controller gain
1/Ti	Reset gain (integral factor)
KD	Rate gain (derivative factor)
dt	Time between samples
bias	Output bias
E	System error. This is equal to the setpoint minus the measured value.
E(n-1)	This is the error of the last sample.

Figure 6-31. PID equation.

The derivative term responds to this rapid rate of change in the error and damps the proportional response. The derivative term allows us to have a much higher proportional gain and still have a stable system. The derivative term also adds "stiffness" to a system. If we were to turn the shaft on a small motor with proportional but no derivative term it would move relatively easily. When we add derivative gain the shaft becomes much more difficult to move. The system fights more to maintain position or zero velocity.

If it is not possible to watch the actual errors on a screen, an oscilloscope can be used. In general a command signal is issued to the drive. The signal is often a square wave. By comparing the square wave command on the oscilloscope screen to the system response, adjustments can be made to the PID parameters. The system response can be traced by looking at the tachometer feedback or the current.

Figure 6-32 shows an example of how a PLC implements the PID algorithm. Note the loop table in the diagram. The loop table will hold the user parameters for each of the gains.

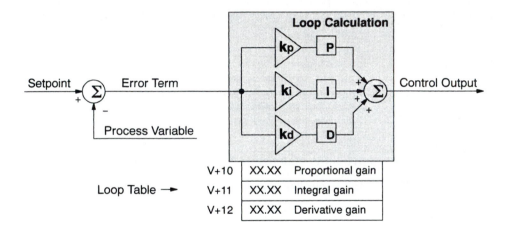

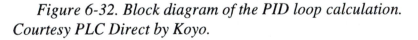

Figure 6-32. Block diagram of the PID loop calculation.
Courtesy PLC Direct by Koyo.

Tuning a PID System

Tuning a PID system involves adjusting potentiometers or software parameters in the controller. Many systems now employ software-based PID control in which the technician simply changes a software parameter value for each factor. The goal of tuning the system is to adjust the gains so that the loop will have optimal performance under dynamic conditions. Tuning is a trial and error process.

The technician should follow the procedures in the technical manual for the system being adjusted. Care must be exercised whenever one makes changes to a system. The following example is based on a system in which the technician can watch the actual error terms while the system is operating. In general the tuning procedure will be as follows:

1. Turn all gains down to zero.

2. Adjust the proportional gain until the system begins to oscillate.

3. Reduce the proportional gain until the oscillation stops and then reduce it by about 20 percent more.

4. Increase the derivative term to improve the system stability.

5. Next raise the integral term until the system reaches the point of instability and back it off slightly.

Feed Forward

Feed forward is an enhancement to PID control. It can be used to improve system response when there is a predictable error when setpoints or commands are changed. For example, assume that we have adjusted the PID parameters as well as is possible and there is still a predictable following error when a new position command is sent.

Feed forward can be used to compensate for these errors. Figure 6-33 shows a block diagram of feed forward. The feed forward term is "fed forward" around the PID equation and summed with its output. Without feed forward when a new command is issued the loop does not really know what the new operating point will be. The loop has to essentially increment/decrement its way until the error disappears. When the error disappears the loop has found the new operating point. If the error is somewhat predictable (known from previous testing) when a new command is issued we can change the output directly (feed forward). Feed forward can also be referred to as a bias. It is a term that can be added in many controllers to help improve system response. If used correctly it can also help reduce integral gain and improve system stability.

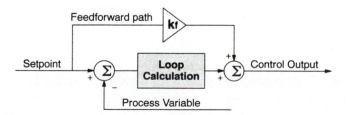

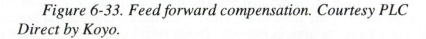

Figure 6-33. Feed forward compensation. Courtesy PLC Direct by Koyo.

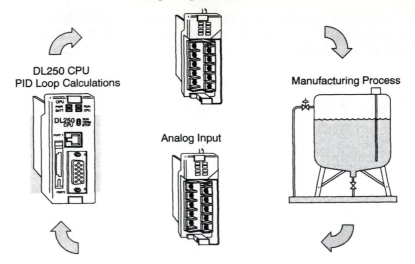

Figure 6-34. PLC system to control the tank process. Courtesy PLC Direct by Koyo.

Most larger PLCs offer PID capability through the use of software and analog I/O modules. Some offer the capability through special PID modules. The PLC Direct 250 CPU offers extensive PID capability. Figure 6-34 shows an example of how a PLC could be used to control a tank process using PID control.

Note that there is an analog input module so that the process can be monitored. The output can be either analog or digital. In this application a PLC Direct 250 processor was used. The 250 DirectSOFT programming software allows the user to use dialog boxes to create a forms-like editor to set up the loops. DirectSOFT's PID Trend View can then be used to view and tune each PID loop.

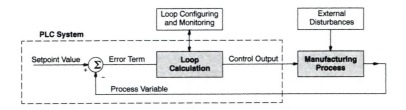

Figure 6-35. PID controlled tank process. Courtesy PLC Direct by Koyo.

Figure 6-35 shows a diagram of the system. The manufacturing process in this case would be the level of the fluid in the tank. Note the external disturbances that act on the process. They are always present. Temperature changes can cause the

Chapter 6: Input/Output Modules and Wiring

viscosity of the fluid to change, the pressure of the supplied fluid into the tank will vary, and so on. There are many disturbances and their individual effects are constantly changing. The PID control system must overcome these.

FUZZY LOGIC MODULES

Fuzzy logic promises to become a much larger part of industrial control. (Some would say that most of us have always used "fuzzy logic.") Fuzzy logic is a very new technology in electronic control. It is being widely used in consumer electronics.

Much of industrial decision making is based on binary logic: for example, the temperature is either right or wrong, the device is either in position or out of position. The motor is on or off. The heating coil is on or off. PID control is used to control variables more closely. PID still relies on fixed formulas. The operator or programmer can adjust the gains of the PID, but then the controller still works with a fixed formula.

Fuzzy logic attempts to make decisions more like those of human beings. People make decisions based on multiple inputs. They do not make hard-and-fast decisions based on equations. Humans are able to vary decisions based on the relative importance of inputs and other influences. They also understand concepts such as warm, cool, lukewarm, and so on.

People tend to make decisions in a "fuzzy" way. They can take in many inputs and decide on the relative importance of each based on present conditions. Imagine carrying a hot paper cup of coffee that was filled too full. A person can change how it is carried to adjust for speed of travel, going around a corner, going up/down stairs, making a quick stop, and so on. We adapt readily to new situations. We do this seemingly without thinking. We are, however, adjusting our acceleration/deceleration, speed, inclination of the cup, and so on. Fuzzy logic attempts to give hardware this capability.

One example of fuzzy logic control is the video camera. Some video recorders are now equipped with electronic circuitry to eliminate the unwanted movement of the camera that makes the video "jumpy." This is a relatively tough problem to solve. A person cannot hold a camera still. If the camera could compensate for the unwanted movement the jumpiness could be eliminated. How does the camera differentiate between desired movement and undesired movement? Fuzzy logic is used.

The process of fuzzy logic can be divided into two functions: inference and conclusion. Even a very complex decision is really just a set of simpler decisions. If each of the simpler questions is answered, the overall decision can be made based on the sum of the conclusions. In fuzzy logic the user sets up some parallel decisions.

For the fuzzy logic controller these simple decisions are called *rules*. Each rule is analyzed and a decision is made. The conclusion from each rule is then summed to achieve a logical sum. This logical sum is then "defuzzed" mathematically. This result is then used to calculate a value that the control device can use for output. Omron offers fuzzy logic modules for PLCs. They can be programmed to control a wide variety of industrial applications. Fuzzy logic is covered in greater detail in Chapter 7.

RADIO-FREQUENCY MODULES

The use of radio-frequency (RF) modules is rapidly expanding. They are used for identification, data collection, production control, product tracking, and more. They eliminate some of the problems associated with bar codes. Bar codes must be kept clean. The air must be relatively clean, too. Radio-frequency modules are unaffected by dirt, oil, or any other common contaminant. They are also unaffected by plant noise. Manufacturing environments are filled with electronic noise. RF is unaffected by it. RF has achieved fairly wide use in automatic-guided vehicles (AGVs). It is used to send instructions to the vehicles as they "roam" factories. Present AGVs typically follow wires in the floor. Newer systems do not require a wire for guidance.

RF is also widely used in retail stores. The small plastic tags on expensive items such as leather jackets and tennis rackets are RF tags. They are single-bit tags. They only need to be sensed by the RF module to set off an alarm if you take the item out of the store with the tag.

Another advantage of RF is that it features bidirectional communication. It can be used to read and write. Imagine an RF tag on a car as it moves through production. The tag could easily contain all of the specifications for the car, including the options that the customer had ordered. As it moves through production, each station reads the RF tag so that the correct options are installed (see Figure 6-36).

The same tag can be written to. So as the car moves along, quality tests are run and the results are written to the tag. When the car reaches the end of the line a complete production history is on the tag. The tags are capable of large amounts of storage. The typical tags available for PLCs feature 100 to 2000 bytes of memory.

There are two types of tags, active and passive. *Active* tags contain a battery; *passive* tags do not. As a passive tag passes the transmitter/receiver, the field that the transmitter radiates is used to power the tag. The only disadvantage to the passive tag is that its sensing distance is much less than the active tag's. The batteries in active tags can last for several years. Active tags can work at much greater distances.

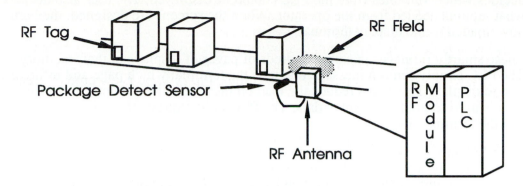

Figure 6-36. Typical RF application. As the pallets travel down the conveyor, they enter the RF field. A product sensor detects the product. The sensor is connected to the antenna. The antenna is triggered by the product sensor to read/write to/from the tag. The use of a product sensor helps avoid "false reads."

Radio frequency is catching on fairly rapidly. It is an "invisible" technology. With a bar code we can see the code, whereas we cannot see RF systems communicate. This has in some ways hampered the acceptance of RF technology. People do not understand things that they cannot see. They are also somewhat fearful or distrustful of invisible technology.

OPERATOR INPUT/OUTPUT DEVICES

As systems become more integrated and automated, they become more complex. Operator information becomes crucial. There are many devices available for this information interchange.

Operator Terminals

Many PLC makers now offer their own operator terminals. They can be very simple to very complex. The simpler ones are able to display a short message. The more complex models are able to display graphics and text in color while taking operator input from touch screens (see Figure 6-37), bar codes, keyboards, and so on. These display devices can cost from a couple hundred dollars up to several thousand. If we remember that the PLC has most of the valuable information about the processes it controls in its memory, we can see that the operator terminal can be a window into the memory of the PLC.

The greatest advances have been in the ease of use. Many PLC manufacturers have software available that runs on an IBM personal computer. The software essentially writes the application for the user. The user draws the screens and decides which variables from the PLC should be displayed. The user also decides what input is needed from the operator. When the screens are designed, they are downloaded to the display terminal.

These smart terminals can store hundreds of pages of displays in their memory. The PLC simply sends a message that tells the terminal which page and information to display. This helps reduce the load on the PLC. The memory of the display is used to hold the display data. The PLC only requests the correct display, and the terminal displays it.

The PLC only needs to update the variables that may appear on the screen. The typical display would include some graphics showing a portion of the process, variables showing times or counts, and any other information that might aid an operator in the operation or maintenance of a system.

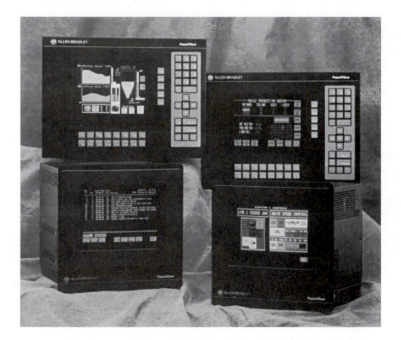

Figure 6-37. Operator display terminal. Courtesy of Allen-Bradley Company, Inc., a unit of Rockwell International.

Speech Modules

One of the more interesting and unique operator information systems are speech modules. These modules are used to output messages to operators (see Figure 6-38).

Speech modules typically are used to digitize a human voice pronouncing the desired word, phrase, or sentence. The digitized sound is stored in the module's memory. Each word, phrase, or sentence is given a number. Ladder logic is used to output the appropriate message at the appropriate time. The sound from such modules is remarkably good.

There are two modes that can be utilized for output. Sentence mode would play a recorded sentence. Phrase mode would play selected words in the proper order to form a message. For example, assume that there are five machines in a cell. We might need a message to maintain the machine, refill the machine, and so on. Instead of storing a complete message for each case for each machine, phrases are combined to form a complete sentence (see Figure 6-39).

Special-purpose modules are available for just about any application. These modules can be used to greatly simplify the task of developing a system. As PLCs gain more capability, more and more special-purpose modules will become available. The emphasis will also shift to making the modules easier to use. The modules that will become more widely used are those that provide information to and from systems (networking and integration). The data these modules can provide can be used to improve the efficiencies of manufacturing processes.

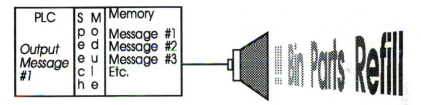

Figure 6-38. How a message is converted and stored in the memory of the speech module. The human voice is an analog signal. The analog acoustic signal is converted to an electrical analog signal by a microphone. The speech module uses an analog to digital converter to change the signal to a digitized file. The digitized message is given an address and stored in memory. The explanations about speech modules are based on Omron Electronic's speech module.

Message Address	Message
100	Maintenance required on machine number
101	Refill parts bin on machine number
102	Shut down machine number
103	One
104	Two
105	Three
106	Four
107	Five
108	Thank You

Figure 6-39. How the phrase mode of a speech module would be used. Individual messages would be combined to form a complete order. For example, if the message order was 102, 106, the complete message would be "Shut down machine number four." After the operator shut down number four and acknowledged it to the PLC, the PLC might output message number 108, "Thank you." (Machines can be more polite than people.)

Questions

1. What voltages are typically available for I/O modules?

2. If an input module was sensing an input from a load-powered sensor when it should not, what might be the possible problem, and what could you do about it?

3. What types of output devices are available in output modules? List at least three.

4. Explain the purpose of A/D modules and how they function.

5. Explain the purpose of D/A modules and how they function.

6. What type of module could be used to communicate with a computer?

7. If noise were a problem in an application, what types of changes might help alleviate the problem?

8. What are remote I/O modules?

9. Explain the term *resolution*.

10. With information becoming more important, which modules do you think will become more widely used? Explain your rationale.

11. What are the differences between RF and bar-code systems? Why might one be chosen over another?

12. Name the typical components in an RF system, and describe the fundamentals of its operation.

13. What is the difference between passive and active tags? Why might one be chosen over the other?

Chapter 7

Arithmetic Instructions

Arithmetic instructions are vital in the programming of systems. They can simplify the task of the programmer. In this chapter we cover compare, add, subtract, multiply, and divide instructions. Several brands of PLC instructions will be covered.

OBJECTIVES

Upon completion of this chapter, the student will be able to:

>*Describe typical uses for arithmetic instructions.*

>*Explain the use of compare instructions.*

>*Explain the use of arithmetic instructions such as add, subtract, multiply, and divide.*

>*Write ladder logic programs involving arithmetic instructions.*

>*Explain the similarities and differences among the various brands of PLC instructions.*

INTRODUCTION

There are many times that contacts, coils, timers, and counters fall short of what the programmer needs. There are many applications that require some mathematical computation. For example, imagine a furnace application that requires the furnace to be between 250 and 255 degrees (see Figure 7-1). If the temperature variable is between 250 and 255 degrees, we might turn off the heater coil. If the temperature is below 250 we would turn on the heater coil. If the temperature is between 250 and 255 degrees, we turn on a green indicator lamp. If the temperature falls below 240 degrees, we might sound an alarm. (*Note*: Industrial temperature control is normally more complex than this. Complex process control is covered in Chapter 8.)

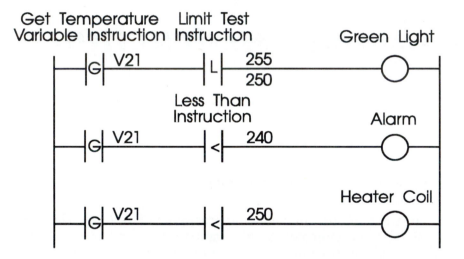

Figure 7-1. How comparison instructions could be used to program a simple application. In this application the heater coil will be on if the temperature is below 250 degrees. An alarm would sound if the temperature drops below 240 degrees. A green indicator lamp is on if the temperature is between 250 and 255 degrees.

This simple application required the use of many relational operators (arithmetic comparisons). The application involved tests of equality, less than, and greater than. The use of arithmetic statements makes this a very easy application to write. Many of the small PLCs do not have arithmetic instructions available. All of the larger PLCs offer a wide variety of arithmetic instructions.

There are also many cases when numbers need to be manipulated. They need to be added, subtracted, multiplied, or divided. There are PLC instructions to handle all of these computations. For example, we may have a system to control the temperature of a furnace. In addition to controlling the temperature we would like to convert the Celsius temperature to degrees Fahrenheit for display to the operator. If the temperature was to rise too high, an alarm would be triggered. Arithmetic instructions could do this very easily. An example of this is shown later in this chapter.

ALLEN-BRADLEY INSTRUCTIONS

Allen-Bradley PLC-2 Arithmetic Instructions

Allen-Bradley PLC-2 controllers offer many arithmetic instructions. The standard instructions include addition, subtraction, multiplication, and division. There are many more instructions available.

Addition -(+)-

The addition instruction adds two values that are "fetched" by two get instructions in the rung. A get instruction is used to "get" a word (a word in this case is a number that is 16 bits long) from a memory location. A get instruction is used in a rung to "get" data from memory. A get instruction "gets" 16 bits of data from one location in memory. Although the instruction looks like a contact, it does not determine a rung's condition. Study Figure 7-2. If contact 11501 is true, the whole rung is true. The get instructions merely tell the processor to fetch two numbers from memory. Get instructions can be programmed at the beginning of the rung or other conditions can precede it. The number that is fetched by a get can then be used in some way by another instruction. In this case it is used to get two values to be added together. The decimal number fetched by the get will be displayed below the get instruction. The result of the add is stored in the word address of the add instruction. In this case it will be stored in address 010. The add instruction is programmed in the output position of the rung.

Figure 7-2. Allen-Bradley add instruction. If the first contact (11501) is true, the add instruction will be performed. In this example the numbers (450 and 500) are "fetched" from addresses 030 and 031. They are then added. The result is stored in address 010.

Subtraction -(-)-

The subtraction instruction subtracts the two values that are "fetched" by two get instructions (see Figure 7-3). The second word value is subtracted from the first word value. The result is stored in the word address given by the subtraction instruction. Only positive values should be used.

Figure 7-3. Allen-Bradley subtraction instruction. If the first contact (11501) is true, the instruction will be performed. In this example the numbers (450 and 471) are "fetched" from addresses 030 and 031. They are then subtracted. The result is stored in address 010.

Multiplication -(X)-(X)-

The multiplication instruction is used to multiply two values that are "fetched" by get instructions in the rung. The result is stored in the two addresses specified by the multiplication instruction.

The multiplication instruction is programmed in the output position of the rung. The addresses to which the result will be stored should be consecutive. Note that the result is stored in the format XXX,XXX (see Figure 7-4).

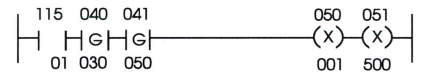

Figure 7-4. Allen-Bradley multiply instruction. If the first contact (11501) is true, the multiplication instruction will be performed. In this example the numbers (30 and 50) are "fetched" (by two get instructions) from addresses 040 and 041. They are then multiplied. The result is stored in addresses 050 and 051.

Division

The division instruction is used to divide two values. The two values are "fetched" by two get instructions in the rung. The result of the division is stored in two word addresses that are specified by the division instruction. The addresses used for storing the result should be consecutive.

Study Figure 7-5 to see how the result is stored. The whole portion of the result is placed in the first memory address (50). The decimal portion of the result is placed in the second address (51).

Chapter 7: Arithmetic Instructions

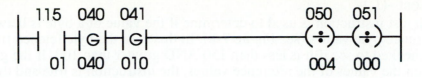

Figure 7-5. Allen-Bradley division instruction. If the first contact (11501) is true, the division instruction will be performed. The numbers (40 and 10) are "fetched" from addresses 040 and 041. Forty is then divided by 10. The result is stored in addresses 050 and 051. The division is shown to three-decimal-place precision (XXX.XXX).

Allen-Bradley PLC 2 Compare Instructions

The compare instruction compares the data in a memory location with data from another location. The result of the compare determines the rung condition. The get instruction is used to get the first data value to be compared.

Equality Comparison -[=]-

Two values can be compared to check if they are equal. The result determines the rung condition. If true, the rung is true and the output will be energized; if false, the rung is false and the output is deenergized (see Figure 7-6).

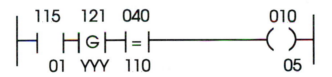

Figure 7-6. Use of a comparison instruction to check for equality.

Less Than Instruction -|<|-

The less than instruction is used to compare two values (see Figure 7-7). If the get value is less than the reference value stored in the less than instruction, the instruction is true and the output is energized.

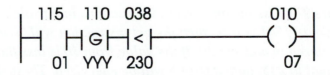

Figure 7-7. Less than comparison. The get instruction is used to get the value that is held in register 110. If the number held in 110 is less than 230, output 01007 will be energized.

Limit Test -(L)-

The limit test instruction is used to determine if the value of a byte is between two reference byte values (see Figure 7-8). In this case the byte fetched from 110 will be checked to see if it is less than 150 AND greater than 100. If the get value is between the values of the reference values, the instruction is true and the output energizes. Note that this instruction uses only one byte. Study Figure 7-8. Instead of the "G" we used to get a word from memory, a "B" is used. The "B" fetches one byte from memory. The normal get instruction "gets" a word (two bytes) from memory. The get byte instruction must be used. The get byte instruction -|B|- gets one byte from one word of memory. The data are shown in octal. In fact, the limit test and the get byte instruction work only in octal.

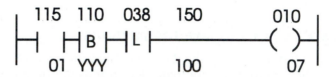

Figure 7-8. Use of a limit instruction. If input condition 11501 is true, a byte from address 110 will be "fetched." The value from 110 will then be compared to the limit values. If it is less than 150 and more than 100, output 01007 will be energized.

The programmer can use arithmetic instructions such as compares in series or in parallel in ladder logic. The programmer is then able to test for multiple conditions. It is almost like constructing your own instruction to accomplish a new task.

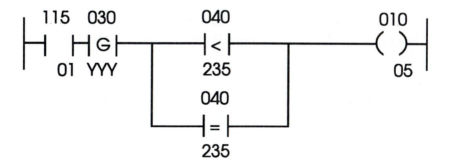

Figure 7-9. Use of comparison instructions in parallel. In this case if input 11501 is true, then the get instruction will "fetch" the number stored in address 30. If the value from 30 is less than 235, OR is equal to 235, output 01005 will be energized. These instructions can be programmed in series or parallel or in combination as long as the screen limits are not exceeded. This is a good way to test for multiple conditions.

Study Figure 7-9. The programmer needs to determine if a number was less than OR equal to 235. There is no less than or equal to instruction, so the programmer created one. In this case a less-than instruction and an equal to instruction were used in parallel. If either of them is true, the rung is true and the output will be energized.

The rungs shown in Figure 7-10 operate as follows:

Rung 1: The get instruction at address 200 is used to multiply the Celsius temperature 100 by 9. The result (900) is stored in address 203.

Rung 2: The get instruction at address 203 is used to divide 900 by 5. The result (180) is stored in address 205. Remember that had there been a decimal remainder, it would have been stored in address 206.

Rung 3: The get instruction at 207 is used to add 32 to the value 180, which is located at get address 205. The sum (212) is stored at address 210. The ladder so far has converted 100 degrees Celsius to 212 degrees Fahrenheit.

Rung 4: This rung gets the temperature from address 210. If the value is less than 190, the timer (T33) begins timing for 3 seconds.

Rung 5: After 3 seconds has elapsed, the output at address 01115 will energize. The output is a heating coil. The heater will bring the temperature back into the desired range.

Rung 6: The counter in this rung (counter 034) is used to count the number of times that the system temperature falls below 190 degrees. Input 03317 (timer 33, status bit 17) is used to increment the counter. Every time the timer energizes, bit 17 of the timer energizes.

Rung 7: When the temperature is equal to 212 degrees, a latching output (01116) energizes. The output could be an alarm, light, or any other device.

Rung 8: Rung 8 enables an operator to acknowledge the alarm condition. If an operator closes input switch 11014, output 01116 is unlatched and is deenergized.

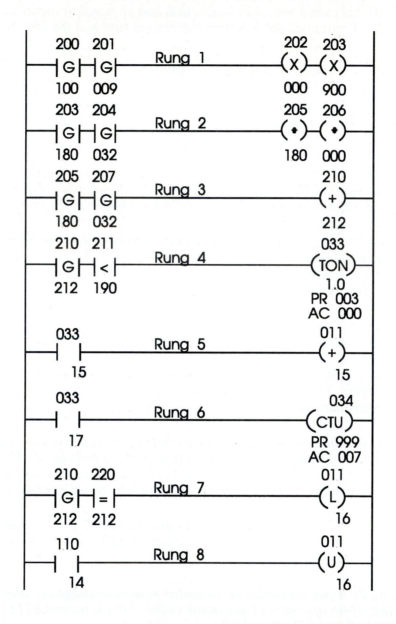

Figure 7-10. Ladder diagram used to control temperature. The logic first converts 100 degrees Celsius to 212 degrees Fahrenheit. If the temperature falls below 190 degrees Fahrenheit a heating coil is turned on. If the temperature reaches 212 an alarm is latched on until an operator acknowledges the alarm condition.

Chapter 7: Arithmetic Instructions

Instruction	Function	Description of Operation and Use
ADD	Add	Adds source A to source B and stores the result in the destination.
SUB	Subtract	Subtracts source B from source A and stores the result in the destination.
MUL	Multiply	Multiplies source A by source B and stores the result in the destination.
DIV	Divide	Divides source A by source B and stores the result in the destination and the math register.
DDV	Double Divide	Divides the contents of the math register by the source and stores the result in the destination and the math register.
CLR	Clear	Sets all bits of a word to zero.
SQR	Square Root	Calculates the square root of the source and places the integer result in the destination.
SCP	Scale with Parameters	Produces a scaled output value that has a linear relationship between the input and scaled values.
SCL	Scale Data	Multiplies the source by a specified rate, adds to an offset value and stores the result in the destination.
ABS	Absolute	Calculates the absolute value of the source and places the result in the destination.
CPT	Compute	Evaluates an expression and stores the result in the destination.
SWP	Swap	Swaps the low and high bytes of a specified number of words in a bit, integer, ASCII, or string file.
LN	Natural Log	Takes the natural log of the value in the source and stores the result in the destination.
LOG	Log to the Base 10	Takes the log base 10 of the value in the source and stores the result in the destination.
XPY	X to the power of Y	Raises a value to a power and stores the result in the destination.

Figure 7-11. AB arithmetic instructions.

AB PLC-5 and SLC 500 and Micrologix 1000 Arithmetic Instructions

There are a multitude of arithmetic instructions available in the AB-5 and SLC-500 controllers. A few of them will be covered in this section. Figure 7-11 shows the arithmetic functions that are available. A few of the more common instructions are covered next.

Compute

The compute instruction (CPT) can be used to convert between number systems, manipulate numbers, or perform trigonometric functions.

Remember that floating-point numbers are 32-bit numbers and integer numbers are 16 bits long. The instruction converts numbers from the expression to the type required by the destination address. If the conversion requires a 32-bit number to be changed to a 16-bit number and the result is too large for a 16-bit number, the CPU will set an overflow bit in S2:01 and will also set a minor fault bit (bit 14). The resulting converted number could cause problems. You should monitor these status bits in your program.

Order	Operation	Description
1	**	Exponential (x to the power of y)
2	-	Negate
	NOT	Bitwise Complement
3	*	Multiply
	/	Divide
4	+	Add
	-	Subtract
5	AND	Bitwise AND
6	XOR	Bitwise Exclusive OR
7	OR	Bitwise OR

Figure 7-12. Precedence (order) in which math operations are performed. When precedence is equal the operations are performed left to right. Parentheses can be used to override the order.

Operations are performed in a prescribed order. Operations of equal order are performed left to right. Figure 7-12 shows the order in which operations are performed. The programmer can override precedence order by using parentheses.

Figure 7-13 shows the use of a CPT instruction. The mathematical operations are performed when contact I:012/10 is true. The result of the operation is stored in the destination address (N76:20).

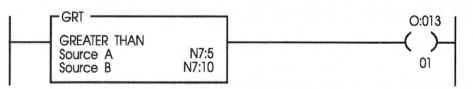

Figure 7-13. Use of a compute (CPT) instruction.

Add

The add instruction (ADD) is used to add two values together (source A + source B). The result is put in the destination address. Figure 7-14 shows an example of an add instruction. If contact I:012/10 is true, the add instruction will add the number from source A (N7:3) and the value from source B (N7:4). The result will be stored in the destination address (N7:20). The effects on the status bits are shown in Figure 7-15.

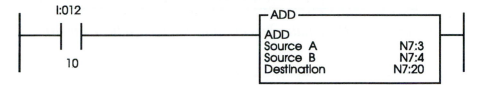

Figure 7-14. Use of an add instruction.

Bit	CPU Action
Carry (C)	1 if a carry occurs, otherwise a 0
Overflow (V)	1 if an overflow occurs, otherwise a 0
Zero (Z)	1 if result = zero, otherwise a zero
Sign (S)	1 if result is negative, otherwise a 0

Figure 7-15. Effects of an add instruction on the status bits.

Subtraction

The subtraction instruction (SUB) is used to subtract two values. The subtract instruction subtracts source B from source A. The result is stored in the destination address. Figure 7-16 shows the use of a subtract instruction (SUB). If contact I:012/10 is true, the SUB instruction is executed. Source B is subtracted from source A; the result is stored in destination address N7:20. The effect on status bits is shown in Figure 7-17.

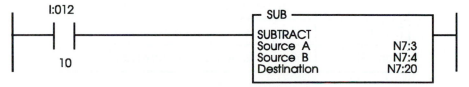

Figure 7-16. This figure shows the use of a subtract (SUB) instruction.

Bit	CPU Action
Carry (C)	1 if a borrow occurs, otherwise a 0
Overflow (V)	1 if an underflow occurs, otherwise a 0
Zero (Z)	1 if result = zero, otherwise a zero
Sign (S)	1 if result is negative, otherwise a 0

Figure 7-17. How status bits are affected by a subtract instruction.

Multiply

The multiply instruction (MUL) is used to multiply two values. The first value, source A, is multiplied by the second value, source B. The result is stored in the destination address. Source A and source B can be either values or addresses of values. Figure 7-18 shows the use of a multiply instruction. If contact I:012/10 is true, source A (N7:3) is multiplied by source B (N7:4) and the result is stored in destination address N7:20. Status bits are set according to Figure 7-19.

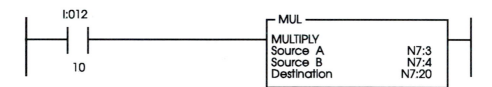

Figure 7-18. Use of a multiply (MUL) instruction.

Bit	CPU Action
Carry (C)	Always resets
Overflow (V)	1 if an overflow occurs, otherwise a 0
Zero (Z)	1 if result = zero, otherwise a zero
Sign (S)	1 if result is negative, otherwise a 0

Figure 7-19. How a multiply instruction affects status bits.

Divide

The divide instruction (DIV) is used to divide two values. Source A is divided by source B and the result is placed in the destination address. The sources can be

　　　　　　　　　　　　　　　　　　Chapter 7: Arithmetic Instructions

values or addresses of values. Figure 7-20 shows the use of a divide instruction. If contact I:012/10 is true, the divide instruction will divide the value from source A (N7:3) by the value from source B (N7:4). The result is stored in destination address N7:20. Figure 7-21 shows how a divide instruction affects the status bits.

Figure 7-20. Use of a divide (DIV) instruction.

Bit	CPU Action
Carry (C)	Always resets
Overflow (V)	1 if an overflow occurs or if division by 0, otherwise a 0
Zero (Z)	1 if result = zero, otherwise a zero, undefined if overflow is set
Sign (S)	1 if result is negative, otherwise a 0, undefined if overflow is set

Figure 7-21. How the status bits are affected by the divide instruction.

Negate

The negate instruction (NEG) is used to change the sign of a value. If it is used on a positive number, it makes it a negative number. If it is used on a negative number, it will change it to a positive number. Remember that this instruction will execute every time the rung is true. Use transitional contacts if needed. The use of a negate instruction is shown in Figure 7-22.

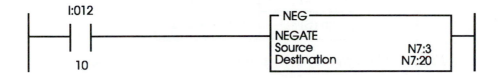

Figure 7-22. Use of a negate (NEG) instruction.

Bit	CPU Action
Carry (C)	Sets if the operation generates a carry
Overflow (V)	1 if an overflow occurs, otherwise a 0
Zero (Z)	1 if result = zero, otherwise a zero
Sign (S)	1 if result is negative, otherwise a 0

Figure 7-23. Effect of a negate instruction on status bits.

If contact I:012/10 is true, the value in source A (N7:3) will be given the opposite sign and stored in destination address N7:20. Figure 7-23 shows how status bits are affected by a negate instruction.

Clear Instruction

The clear instruction (CLR) is used to set all bits of a word to zero. The destination must be an address. An example is shown in Figure 7-24. If contact I:012/10 is true, the negate instruction will clear the value in address N7:3. All bits in word N7:3 will be set to zero. The effect on status bits is shown in Figure 7-25.

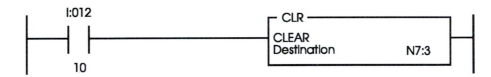

Figure 7-24. Use of a clear (CLR) instruction.

Bit	CPU Action
Carry (C)	Always resets
Overflow (V)	Always resets
Zero (Z)	Always sets
Sign (S)	Always resets

Figure 7-25. Effect of a clear instruction on status bits.

Square Root

The square root instruction is used to find the square root of a value. The result is stored in a destination address. The source can be a value or the address of a value. Figure 7-26 shows the use of a square root instruction. If contact I:012/10 is true, the SQR instruction will find the square root of the value of the number found at the source address N7:3. The result will be stored at destination address N7:20. The effect on status bits is shown in Figure 7-27.

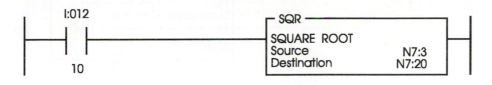

Figure 7-26. Use of a square root (SQR) instruction.

Bit	CPU Action
Carry (C)	Always resets
Overflow (V)	Sets if an overflow is generated during floating-point to integer conversion, otherwise resets
Zero (Z)	Sets if result is zero, otherwise resets
Sign (S)	Always resets

Figure 7-27. Effect of an SQR instruction on the status bits.

Standard Deviation

The standard deviation instruction (STD) is used to find the standard deviation of a set of values. It stores the result in a destination address. It is used extensively in process control applications. The instruction is executed only on a false to true transition.

Figure 7-28 shows an example of the use of an STD instruction. There are several values that the programmer must enter. File is an address that contains the first value of the set to be included in the calculation. The destination is the address where the result of the calculation will be stored. Control is the address of the control structure in the control area (R) of the CPU memory. The CPU uses this information to perform the instruction. Length is the number of values that will be used in the calculation (number of file elements, 0 to 1000). Position points to the element that the instruction is currently using.

In Figure 7-28, if contact I:012/10 is true, the STD instruction will execute. When the STD instruction is enabled, the .EN bit for R6:0 is set. This turns on output O:010/05. The instruction uses the values from elements N7:1, N7:2, N7:3, and N7:4 to calculate the standard deviation. The result is stored in destination address N7:0. When the instruction is done the done bit (.DN) is set, turning output O:010/07 on. The effect of the STD instruction on the status bits is shown in Figure 7-29.

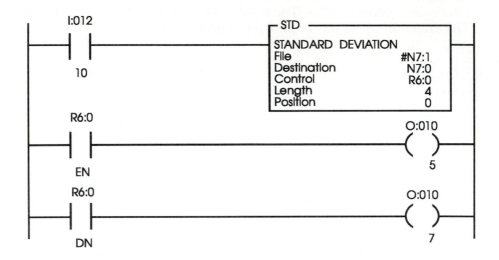

Figure 7-28. Use of a standard deviation instruction.

Bit	CPU Action
Carry (C)	Always resets
Overflow (V)	Sets if an overflow occurs, otherwise resets
Zero (Z)	Sets if result is zero, otherwise resets
Sign (S)	Always resets

Figure 7-29. Effect of the STD instruction on the status bits.

Average Instruction

The average instruction (AVE) is a file instruction. It is used to find the average of a set of values. The AVE instruction calculates the average using floating-point regardless of the type specified for the file or destination. If an overflow occurs, the CPU aborts the calculation. In that case the destination remains unchanged.

The position points to the element that caused the overflow. When the .ER bit is cleared, the position is reset to zero and the instruction is recalculated. Every time there is a low-to-high transition the value of the current element is added to the next element. The next low-to-high transition causes the current element value to be added to the next element, and so on. Every time another element is added, the position field and the status word are incremented.

Figure 7-30 shows how an average instruction (AVE) is used in a ladder diagram. The programmer must provide certain information while programming. The file is the address of the first element to be added and used in the calculation. The destination is the address where the result will be stored. This address can be floating-point or integer. The control is the address of the control structure in the control area (R) of CPU memory. The CPU uses this information to run the instruction. Length is the number of elements to be included in the calculation (0 to 1000). Position points to the element that the instruction is currently using.

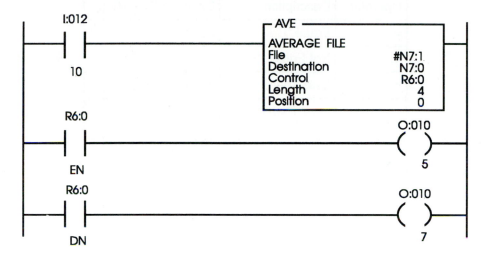

Figure 7-30. Use of an average (AVE) instruction.

Relational Operators

Compare

The compare instruction (CMP) is used to compare two values or expressions to check on their relationship.

The compare instruction (CMP) is an input instruction that uses mathematical operators to perform comparisons of values or expressions (see Figure 7-31). When the comparison is true, the rung is true. The compare instruction execution time is longer than other instructions, such as equal to (EQU). The compare instruction should be used when the programmer needs to perform some mathematical manipulation on the values before comparing them.

The expression portion of the instruction defines what operations the programmer wants to perform. The programmer can use constants or addresses. An expression can be up to 80 characters long. Program constants can be integer or floating-point. If octal numbers are entered, &O must precede the number so that the CPU knows that it is an octal number. If a hexadecimal number is used, it must be preceded by &H. If binary numbers are used, they must be preceded by &B.

The use of a compare instruction (CMP) is shown in Figure 7-32. The compare (CMP) instruction tells the CPU to compare the sum of the values in N7:0 and N:7:1 to the difference of the values in N7:2 and N7:3. If the sum of the first two values is greater than the result of the subtraction of the other two values, the rung will be true and output O:013/01 will be turned on.

Operator	Description	Example Operation
=	Equal	if a=b, then...
<>	Not Equal	if a<>b, then...
<	Less Than	if a<b, then...
<=	Less Than or Equal to	if a<=b, then...
>	Greater Than	if a>b, then...
>=	Greater Than or Equal to	if a>=b, then...

Figure 7-31. Operators that can be used with a compare instruction.

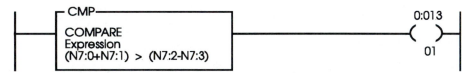

Figure 7-32. Use of a compare instruction.

Equal To

The equal to instruction (EQU) is used to test if two values are equal. The values tested can be actual values or addresses that contain values. An example is shown in Figure 7-33. Source A is compared to source B to test to see if they are equal. If the value in N:7:5 is equal to the value in N7:10, the rung will be true and output O:013/01 will be turned on.

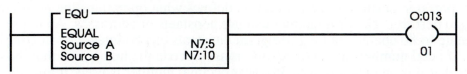

Figure 7-33. An equal to instruction (EQU).

Greater Than or Equal to

The greater than or equal to instruction (GEQ) is used to test two sources to determine whether or not source A is greater than or equal to the second source. The use of a GEQ instruction is shown in Figure 7-34. If the value of source A (N7:5) is greater than or equal to source B (N7:10), output O:013/01 will be turned on.

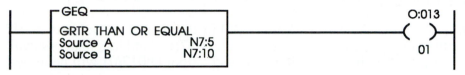

Figure 7-34. Use of a greater than or equal to (GEQ) instruction.

Greater Than

The greater than instruction (GRT) is used to see if a value from one source is greater than the value from a second source. An example of the instruction is shown in Figure 7-35. If the value of source A (N7:5) is greater than the value of source B (N7:10), output O:013/01 will be set (turned on).

Figure 7-35. Use of a greater than (GRT) instruction

Less Than

The less than instruction (LES) is used to see if a value from one source is less than the value from a second source. An example of the instruction is shown in Figure 7-36. If the value of source A (N7:5) is less than the value of source B (N7:10), output O:013/01 will be set (turned on).

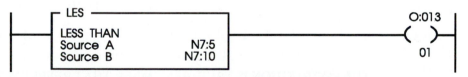

Figure 7-36. Use of a less than (LES) instruction.

Limit

The limit instruction (LIM) is used to test a value to see if it falls in a specified range of values. The instruction is true when the tested value is within the limits. This could be used, for example, to see if the temperature of an oven was within the desired temperature range. In this case the instruction would be testing to see if an analog value (a number in memory representing the actual analog temperature) was within certain desired limits.

The programmer must provide three pieces of data to the LIM instruction when programming. The programmer must provide a low limit. The low limit can be a constant or an address that contains the desired value. The address will contain an integer or floating-point value (16 bits). The programmer must also provide a test value. This is a constant or the address of a value that is to be tested. If the test value is within the range specified, the rung will be true. The third value the programmer must provide is the high limit. The high limit can be a constant or the address of a value.

Figure 7-37 shows the use of a limit (LIM) instruction. If the value in N7:15 is greater than or equal to the lower limit value (N7:10) and less than or equal to the high limit (N7:20), the rung will be true and output O:013/01 will be turned on.

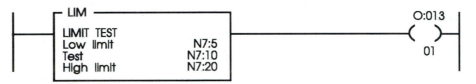

Figure 7-37. Use of a limit test (LIM) instruction.

Mask Compare Equal To

The mask compare equal to (MEQ) instruction is used to compare two values to check for equality (see Figure 7-38). The difference between this instruction and a regular equal to (EQU) instruction is that the MEQ instruction permits the masking of bits so that they are not considered in the comparison. This can permit the programmer to consider parts of words when making a ladder decision. Bits that do not matter can be masked. The mask value is what determines whether or not a particular bit is compared. If the bit in the mask is a 1, the bit is used in the comparison. If the bit of the mask is a 0, it is not used in the comparison.

Source	0	1	0	1	0	1	0	1	0	1	0	1	1	1	1	1
Mask	1	1	1	1	1	1	1	1	1	1	1	1	0	0	0	0
Reference	0	1	0	1	0	1	0	1	0	1	0	1	x	x	x	x

Result   THE INSTRUCTION IS TRUE. REMEMBER THAT REFERENCE BITS xxxx ARE NOT COMPARED BECAUSE THE FIRST FOUR BITS IN THE MASK ARE A ZERO AND ARE NOT COMPARED.

Figure 7-38. MEQ comparison. The instruction would be true in this case.

The actual use of a MEQ instruction is shown in Figure 7-39. The source (N7:5) is compared to the compare value N7:10. Any bits that are a 1 in the mask value (N7:6) will be compared; any bits that are a zero will not be compared. If the compared bits are all the same, the instruction is true. All values used in a MEQ instruction must be 16 bits. The values used are unchanged by the instruction. If the user wishes to change the value of the mask, an address should be used. The value of the address can then be modified by other instructions. A hex value (constant) can also be used for a mask value. When entering a hex value that begins with a letter such as D170, you must enter the value with a leading zero (0D170).

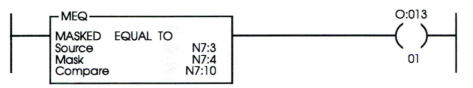

Figure 7-39. Use of a mask compare equal to (MEQ) instruction.

Not Equal To

The not equal to instruction (NEQ) is used to test two values for inequality. The values tested can be constants or addresses that contain values. An example is shown in Figure 7-40. If source A (N7:5) is not equal to source B (N7:10), the instruction is true and output O:013/01 is turned on.

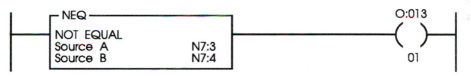

Figure 7-40. Use of a not equal to (NEQ) instruction.

Logical Operators

There are several logical instructions available. They can be very useful to the innovative programmer. They can be used, for example, to check the status of certain inputs while ignoring others.

And

There are several logical operator instructions available. The AND instruction is used to perform an AND operation using the bits from two source addresses. The bits are ANDed and a result occurs. See Figure 7-41 for a chart that shows the results of the four possible combinations. An AND instruction needs two sources (numbers) to work with. These two sources are ANDed and the result is stored in a third address (see Figure 7-42). Figure 7-43 shows an example of an AND instruction. Addresses D9:3 and address D10:4 are ANDed. The result is placed in destination address D12:3. Examine the bits in the source addresses so that you can understand how the AND produced the result in the destination.

Source A	Source B	Result
0	0	0
1	0	0
0	1	0
1	1	1

Figure 7-41. Results of an AND operation on the four possible bit combinations.

Source A D9:3	0	0	0	0	0	0	0	0	1	0	1	0	1	0	1	0
Source B D10:4	0	0	0	0	0	0	0	0	1	1	1	0	1	0	1	1
Destination D12:3	0	0	0	0	0	0	0	0	1	0	1	0	1	0	1	0

Figure 7-42. Result of an AND on two source addresses. The ANDed result is stored in address D12:3.

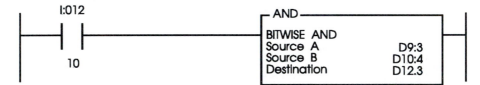

Figure 7-43. Use of an AND instruction. If input I:012/10 is true, the AND instruction executes. The number in address D9:3 is ANDed with the value in address D10:4. The result of the AND is stored in address D12:3.

NOT

NOT instructions are used to invert the status of bits. A one is made a zero and a zero is made a 1. See Figures 7-44, 7-45, and 7-46.

Source	Result
0	1
1	0

Figure 7-44. Results of a NOT instruction on bit states.

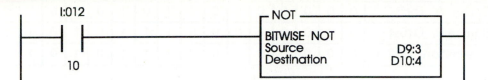

Figure 7-45. Use of a NOT instruction. If input I:012/10 is true, the NOT instruction executes. The number in address D9:3 is NOTed with address D10:4. The result is stored in destination address D12:3.

Source A D9:3	0	0	0	0	0	0	0	0	1	0	1	0	1	0	1	0
Source B D10:4	0	0	0	0	0	0	0	0	0	1	0	1	0	1	0	1

Figure 7-46. What would happen to the number 0000000001010101010 if a NOT instruction were executed? The result is shown in destination address D10:4.

OR

Bitwise OR instructions are used to compare the bits of two numbers. See Figures 7-47, 7-48, and 7-49 for examples of how the instruction functions.

Source A	Source B	Result
0	0	0
1	0	1
0	1	1
1	1	1

Figure 7-47. Result of an OR instruction on bit states.

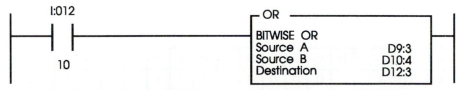

Figure 7-48. How an OR instruction can be used in a ladder diagram. If input I:012/10 is true, the OR instruction executes. Source A (D9:3) is ORed with source B (D10:4). The result is stored in the destination address.

Source A D9:3	0	0	0	0	0	0	0	0	1	0	1	0	1	0	1	0
Source B D10:4	0	0	0	0	0	0	0	0	1	1	1	0	1	0	1	1
Destination D12:3	0	0	0	0	0	0	0	0	1	1	1	0	1	0	1	1

Figure 7-49. Result of an OR instruction on the numbers in address D9:3 and address D10:4. The result of the OR is shown in the destination address D12:3.

Exclusive Or

Bitwise exclusive or instructions (XOR) are used to compare the bits of two numbers. The result of the XOR is placed in the destination address. Figures 7-50, 7-51, and 7-52 show how the instruction is used.

Source A	Source B	Result
0	0	0
1	0	1
0	1	1
1	1	0

Figure 7-50. How the XOR instruction evaluates bit states.

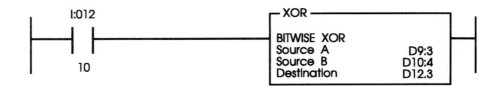

Figure 7-51. Use of an XOR instruction. If contact I012/10 is true, source A is exclusive ORed with source B. The result is stored in address D12:3. Source A and source B are not modified.

Source A D9:3	0	0	0	0	0	0	0	0	1	0	1	0	1	0	1	0
Source B D10:4	0	0	0	0	0	0	0	0	1	1	1	0	1	0	1	1
Destination D12:3	0	0	0	0	0	0	0	0	0	1	0	0	0	0	0	1

Figure 7-52. How an XOR instruction would evaluate two sources. The result is stored in destination address D12:3. The source addresses are not modified.

Number System Conversion

From BCD

The from BCD instruction (FRD) is used to convert a BCD number to its binary equivalent (see Figure 7-53). The source number (a BCD value) is converted and its binary equivalent is stored in the destination address.

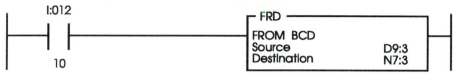

Figure 7-53. Use of a FRD instruction. If contact I:012/10 is true, the instruction executes. Source address (D9:3) contains a BCD value. The FRD instruction converts it to its binary equivalent. The result is stored in the destination address (N7:3).

To BCD

The to BCD instruction (TOD) is used to convert a binary number to a BCD equivalent (see Figure 7-54). The source address contains a binary number. The TOD converts it to a BCD value and stores it to the destination address.

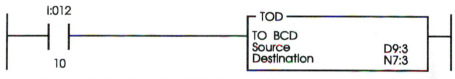

Figure 7-54. Use of a TOD instruction.

Instruction	Function	Description of Operation and Use
ASN	Arc Sine	Takes the arc sine of a number and stores the result as radians in the destination.
ACS	Arc Cosine	Takes the arc cosine of a number and stores the result as radians in the destination.
ATN	Arc Tangent	Takes the arc tangent of a number and stores the result as radians in the destination.
COS	Cosine	Takes the cosine of a number and stores the result in the destination.
SIN	Sine	Takes the sine of a number and stores the result in the destination.
TAN	Tangent	Takes the tangent of a number and stores the result in the destination.

Figure 7-55. AB trigonometric functions.

SIN

The SIN instruction is used to take the sine of a number (source in radians) and store the result in the destination (see Figure 7-55). The sin function is shown in Figure 7-56. The source must be greater than or equal to -205887.4 and less than or equal to 205887.4. The result is placed in the destination. This instruction can be used with SLC 5/03 OS302 and SLC 5/04 OS401 processors. Figure 7-57 shows how the instruction affects the bit states.

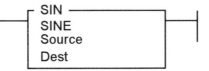

Figure 7-56. Use of a SIN instruction.

Bit	CPU Action
Carry (C)	Always resets
Overflow (V)	Sets if an overflow occurs or an unsupported data is detected; otherwise resets.
Zero (Z)	1 if result = zero, otherwise a zero
Sign (S)	1 if result is negative, otherwise a 0

Figure 7-57. Bit conditions for the SIN instruction.

There are a wealth of other instructions available for the AB-5 family of processors. These examples are intended to make the technician comfortable with some of the basic instructions. The PLC programming manual provides information on many additional instructions.

GE FANUC FUNCTIONS

GE Fanuc Arithmetic Functions

GE Fanuc uses a function block style for arithmetic instructions. There are many arithmetic functions available (see Figure 7-66). Only a few are covered here. The allowed data types are shown with each description. INT stands for a signed integer, DINT stands for a double precision signed integer, BIT for bit, BYTE for byte, WORD stands for 16 bits of consecutive memory, BCD-4 is a 4-digit binary-coded decimal, and REAL stands for a floating-point number.

Add Function (INT, DINT, REAL)

The GE Fanuc add instruction is shown in Figure 7-58. If input 1 (%I0001)

Chapter 7: Arithmetic Instructions

becomes energized it enables the add function. The value in I1 (the number in register %R0004) will then be added to the number in I2 (the number in register %R0007). The result will be put in register 3 (%R0003). If the add does not result in an overflow the add function will pass power and output 5 (%Q0005) will be turned on.

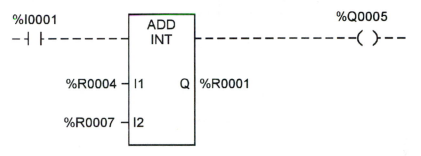

Figure 7-58. Example of the use of an add (ADD) function.

Subtract Function (INT, DINT, REAL)

The GE Fanuc subtract function is shown in Figure 7-59. If input 1 (%I0001) becomes energized it enables the subtract function. The value in I2 (constant value 6 in this case) will then be subtracted from the number in I1 (the number in register %R0005). The result will be put in register 2 (%R0002). If the subtract does not result in an overflow, the subtract function will pass power and output 5 (%Q0005) will be turned on.

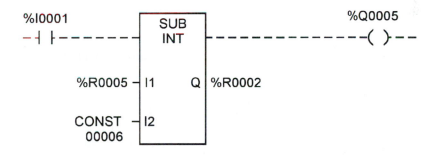

Figure 7-59. Example of the use of a subtract (SUB) function.

Multiply Function (INT, DINT, REAL)

The GE Fanuc multiply function is shown in Figure 7-60. If input 1 (%I0001) becomes energized it enables the multiply function. The value in I1 (the number in register %R0002) will then be multiplied by the number in I2 (constant value of 3). The result will be put in register 6 (%R0006). If the multiply does not result in an overflow the function will pass power and output 5 (%Q0005) will be turned on.

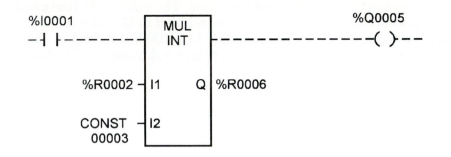

Figure 7-60. Example of the use of a multiply (MUL) function.

Divide Function (INT, DINT, REAL)

Divides one number by another, which yields a quotient. An example of the use of a DIV function is shown in Figure 7-61. The quotient is stored in Q:, in this case %R0006. The divide function passes power if the operation does not result in an overflow and if there is no attempt to divide by zero.

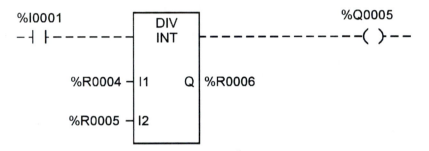

Figure 7-61. Example of the use of a divide (DIV) function.

Square Root Function (INT, DINT, REAL)

The square root function is used to find the square root of an input value. An example of the use of a square root function is shown in Figure 7-62. When the function receives power flow (%I0003), the function takes the square root of the value found in input IN (%R0002) and puts the result in Q (%R0004).

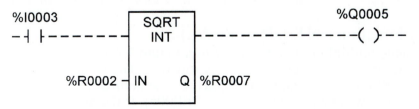

Figure 7-62. Example of the use of a square root (SQRT) function.

Modulo Function (INT, DINT)

The modulo function is used to find the remainder from a division. An example of the use of a modulo function is shown in Figure 7-63. When the function receives power flow (%I0002), the value found in input IN (%R0002) is converted to degrees and the result is placed in Q (%R0008).

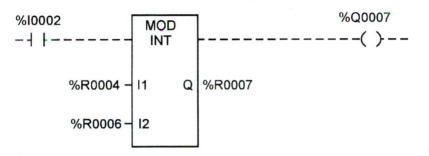

Figure 7-63. Example of the use of a modulo (MOD) function.

Sine Function

The sine function is used to find the trigonometric sine of the input value at IN. An example of the use of a SIN function is shown in Figures 7-64 and 7-67. If input 1 (%I0001) is true the sine function will find the trigonometric sine of the value held in input IN. In this case a register was used for the input value (%R0004).

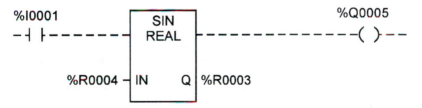

Figure 7-64. Example of the use of a sine (SIN) function.

Degree Function (REAL)

The degree function is used to convert radians to degrees. An example of the use of a DEG function is shown in Figure 7-65. When the function receives power flow (%I0004), the radian value found in input IN (%R0002) is converted to degrees and the result is placed in Q (%R0008).

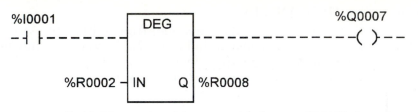

Figure 7-65. Example of the use of a degree (DEG) function.

Instruction	Function	Description of Operation and Use
add	Addition	Adds two numbers. Add functions pass power if the operation does not result in an overflow.
sub	Subtraction	Subtracts one number from another. Sub functions pass power if the operation does not result in an overflow.
mul	Multiplication	Multiplies two numbers. The mul function passes power if the operation does not result in an overflow.
div	Division	Divides one number by another, yielding a quotient. The div function passes power if the operation does not result in an overflow and if there is no attempt to divide by zero.
mod	Modulo Division	Divides one number by another, yielding a remainder. The div function passes power if the operation does not result in an overflow and if there is no attempt to divide by zero.
sqrt	Square Root	Finds the square root of an integer or real value. When the function receives power flow, the value of the output Q is set to the square root of the input IN.
log 10	Base 10 Logarithm	When the function receives power flow, it finds the base 10 logarithm of the real value in input IN and places the result in output Q.
ln	Natural Logarithm	When the function receives power flow, it finds the natural logarithm base (e) of the real value in input IN and places the result in output Q.
exp	Power of e	When the function receives power flow, it finds the natural logarithm base (e) raised to the power specified by IN and places the result in Q.
expt	Power of X	When the function receives power flow, X is raised to the power specified by IN and places the result in Q.

Figure 7-66. Arithmetic instructions for the GE Fanuc PLC.

Instruction	Function	Description of Operation and Use
sin	Sine	Finds the sine of the input. When the function receives power flow, it computes the sine of the value of IN, whose units are radians, and stores the result in output Q.
cos	Cosine	Finds the cosine of the input. When the function receives power flow, it computes the cosine of the value of IN, whose units are radians, and stores the result in output Q.
tan	Tangent	Finds the tangent of the input. When the function receives power flow, it computes the tangent of the value of IN, whose units are radians, and stores the result in output Q.
asin	Inverse Sine	Finds the inverse sine of the input. When the function receives power flow, it computes the inverse sine of the value of IN and stores the result in output Q.
acos	Inverse Cosine	Finds the inverse cosine of the input. When the function receives power flow, it computes the inverse cosine of the value of IN and stores the result in output Q.
atan	Inverse tangent	Finds the inverse tangent of the input. When the function receives power flow, it computes the inverse tangent of the value of IN and stores the result in output Q.
deg	Convert to Degrees	When the function receives power flow, a RAD_TO_DEG conversion is performed on the real radian value of IN and the result is placed in output degree real value Q.
rad	Convert to Radians	When the function receives power flow, a DEG_TO_RAD conversion is performed on the real degree value in input IN and the result is placed in output real radian value Q.

Figure 7-67. Trigonometric instructions for the GE Fanuc PLC.

Exponential Function (REAL)

The GE Fanuc exponential function is shown in Figure 7-68. If input 1 (%I0001) becomes energized it enables the exponential function. The value in I1 (the number in %AI001) will then be raised to the power of the number in I2 (constant of 2.5 in this case). The result will be put in register 1 (%R0001). If the exponential function does not result in an overflow, the function will pass power and output 5 (%Q0005) will be turned on unless an invalid operation occurs and/or IN is not a number or is negative.

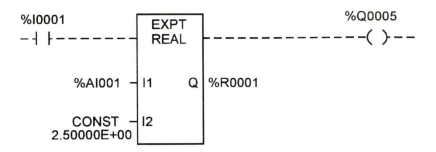

Figure 7-68. Example of the use of a exponential (EXPT) function.

GE Fanuc Relational Functions

Equal To (EQ) Function (INT, DINT, REAL)

GE Fanuc has many relational functions available (see Figure 7-71). The GE Fanuc EQUAL TO function is shown in Figure 7-69. If input 1 (%I0001) becomes energized it enables the equal to function. The value in I1 (the number in register %R0004) will then be added to the number in I2 (the number in register %R0007). The result will be put in register 3 (%R0003). If the add does not result in an overflow the add function will pass power and output 5 (%Q0005) will be turned on.

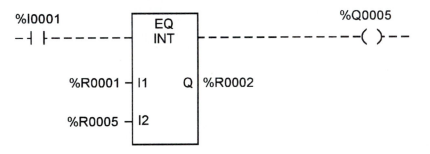

Figure 7-69. Example of the use of an equal to (EQ) function.

Range Function (INT, DINT, WORD)

The GE Fanuc range function is shown in Figure 7-70. If input 1 (%I0001) becomes energized it enables the function. The range function compares the value in input parameter IN against the range specified by limits L1 and L2, inclusive. When IN is within the range specified by L1 and L2, output Q will be turned on.

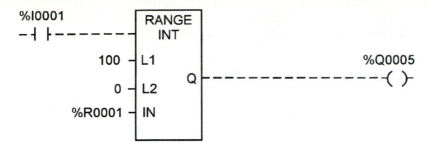

Figure 7-70. Example of the use of a range function.

Instruction	Function	Description of Operation and Use
eq	Equal	This function tests for equality between two numbers. The EQ function passes power if the two inputs are equal.
ne	Not Equal	This function tests for inequality between two numbers. The NE function passes power if the two inputs are not equal.
gt	Greater Than	This function tests to see if one number is greater than another number. The GT function passes power if the first number is greater than the second.
ge	Greater Than or Equal To	This function tests to see if one number is greater than or equal to another number. The GE function passes power if the first number is greater than or equal to the second.
lt	Less Than	This function tests to see if one number is less than another number. The LT function passes power if the first number is less than the second.
le	Less Than or Equal To	This function tests to see if one number is less than or equal to another number. The LE function passes power if the first number is less than or equal to the second.
range	Range	This function tests the input value against a range of two numbers. This instruction is only available for release 4.50 or higher CPUs (4.02 of the 341) and in Logic master 90-30/20 Version 4.5 and above.

Figure 7-71. GE Fanuc comparison instructions.

GE Fanuc Bit Functions

And Function (WORD)

GE Fanuc has many bit functions available (see Figure 7-76). The GE Fanuc AND function is shown in Figure 7-72. If input 1 (%I0001) becomes energized it enables the function. If a bit string in I1 and the corresponding bit in bit string I2 are both 1, a 1 is placed in the corresponding location in output string Q (%R0001 in this case). The function will pass power and output 5 (%Q0005) will be turned on whenever power is received by the function.

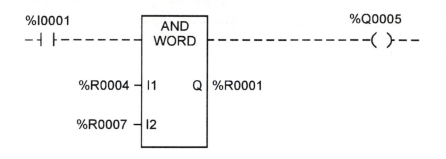

Figure 7-72. Example of the use of an AND function.

Not Function (WORD)

The GE Fanuc NOT function is shown in Figure 7-73. If input 1 (%I0001) becomes energized it enables the function. The NOT function is used to set the state of each bit in the output bit string Q to the opposite of the state of the corresponding bit in bit string I. All bits are altered on each scan that power is received. Output string Q will be the logical complement of I1. The function will pass power whenever it receives power.

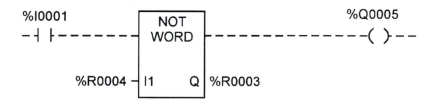

Figure 7-73. Example of the use of a not (NOT) function.

Shift Left Function (WORD)

The GE Fanuc shift left function is shown in Figure 7-74. If input 1 (%I0001) becomes energized it enables the function. The function is used to shift all the bits in a word or group of words to the left by a specified number of places. The specified number of bits is shifted out of the output string to the left. As the bits are shifted out of the high end of the string, the same number of bits are shifted into the low end. The function will pass power whenever it receives power. A string length of 1 to 256 words can be used.

The number of bits to be shifted is placed in N. The bit to be shifted in located at B1. The bit to be shifted out is located at B2. The word to be shifted is located at

IN. LEN is the length of the array. Output Q contains a shifted copy of the input string. If you want to actually shift the input string you would use the same address for IN and Q.

Study Figure 7-74. If input %I0001 becomes true, the output string in %R0001 will be made a copy of IN (%R0005) left shifted by the number of bits specified in N (8 in this example). The status of the input bits will be taken from input B1. The bits shifted out will appear at B2.

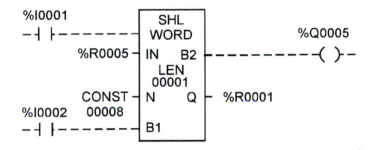

Figure 7-74. Example of the use of a shift left (SHL) function.

Rotate Left Function (WORD)

The GE Fanuc rotate left function is shown in Figure 7-75. If input 1 (%I0001) becomes energized it enables the function. The function is used to rotate all the bits in a string to the left by a specified number of places. The specified number of bits is shifted out of the output string to the left and back into the string on the right. The function will pass power unless the number of bits specified to be rotated is greater than the total length of the string or less than zero. A string length of 1 to 256 words can be used.

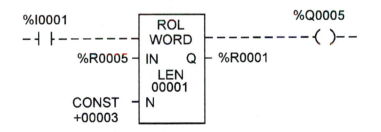

Figure 7-75. Example of the use of a rotate left (ROL) function.

Instruction	Function	Description of Operation and Use
and	Logical AND	This instruction is a logical AND of two bit strings.
or	Logical OR	This instruction is a logical OR of two bit strings.
xor	Logical Exclusive OR	This instruction is a logical Exclusive OR of two bit strings.
not	Logical Invert	This instruction is a logical inversion of a bit string.
shl	Shift Left	This instruction shifts a bit string left.
shr	Shift Right	This instruction shifts a bit string right.
rol	Rotate Left	This instruction rotates a bit string left.
ror	Rotate Right	This instruction rotates a bit string right.
bittst	Bit Test	This instruction tests a bit within a string.
bitset	Bit Set	This instruction sets a bit within a string to true.
bitclr	Bit Clear	This instruction tests a bit within a string to false.
bitpos	Bit Position	This instruction locates a bit set to true within a bit string.
maskcmp	Masked Compare	This instruction performs a masked compare of two arrays (available only for release 4.5 or higher CPUs and in Logicmaster 90-30/20 Version 4.5 and above).

Figure 7-76. Ge Fanuc instructions.

GE Fanuc Data Move Functions

There are a wide variety of data move functions available. They are shown in Figures 7-77 and 7-78.

Move Function (BIT, INT, WORD, REAL)

The GE Fanuc move function is shown in Figure 7-77. If input 1 (%I0001) becomes energized it enables the function. The function will then move 48 bits from memory location %M0001 to memory location %M0034. LEN specifies the number of words to be moved in this case because move-word was used. Three words * 16 bits is equal to 48 bits. The data is copied in bit format so the new location does not have to be the same data type. The function passes power whenever it receives power.

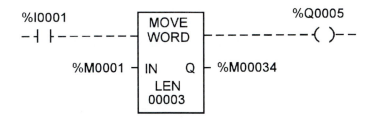

Figure 7-77. Example of the use of a move function.

Instruction	Function	Description of Operation and Use
MOVE	Move	Copy data as individual bits and move to another location. The maximum length is 256 words unless the MOVE_BIT function is used. The move bit limit is 256 bits. Data can be moved to a different data type without prior conversion.
BLKMOVE	Block Move	Copies a block of seven constants to a specified memory location. The constants are input as a part of the function.
BLKCLR	Block Clear	Replaces the content of a block of data with all zeroes. This can be used on bit and word memory. Maximum length is 256 words.
SHFR	Shift Register	Shifts one or more data words into a table. The maximum length is 256 words.
BITSEQ	Bit Sequencer	Performs a bit sequence shift through an array of bits. The maximum length allowed is 256 words.
COMMREQ	Communications Request	Allows the program to communicate with an intelligent module, such as a Genius Communications Module or a Programmable Coprocessor Module.

Figure 7-78. GE Fanuc data move instructions.

GOULD MODICON INSTRUCTIONS

Arithmetic Instructions

Gould Modicon uses a function block style for arithmetic instructions. There are many arithmetic instructions available. Only a few are covered here.

Addition Instruction

The addition instruction is used to add two numbers and store the result to an address (see Figure 7-79). The addition instruction requires three entries from the programmer. The top entry is added to the middle entry and the result is stored in the register shown by the bottom entry. The top two entries can be an actual number or a 4XXX or a 3XXX register. If actual numbers are used, they can be between 0 and 999. If registers are used for the top two entries, the instruction gets the numbers from those two addresses and adds them. The result is stored in the register shown by the bottom entry. The programmer must also supply an input condition. The instruction is executed every time the rung is scanned. The use of a transitional input can assure that the instruction is executed only for transitions, instead of continuously if the input condition is true.

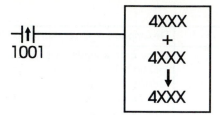

*Figure 7-79. Gould Modicon addition instruction. The X's
represent digits.*

See Figure 7-80 for an example of the use of an addition instruction. Imagine that
a PLC is controlling a bottling line. The manufacturer would like to count the
number of bottles produced by the line. An addition instruction is used to keep
track of the total number produced. Every time a case is full, a sensor is triggered
as the case leaves the packing station. The output from the sensor (input 1001) is
used as an input coil to trigger the addition instruction. The addition instruction
then adds 24 to the total count. The total count is being held in register 4002.
Note that one actual value was used and two addresses (both the same address).
The same address was used for both so that a running total of bottles can be kept.
(If input 1 becomes true, 24 is added to the current total from address 4002 and
then the new total is put back in 4002.)

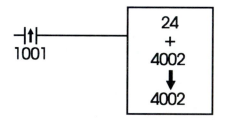

Figure 7-80. Gould Modicon addition instruction.

Subtraction Instruction

The subtraction instruction is very similar to the addition instruction. The subtraction instruction requires three values. The middle value is subtracted from the top
value and the result is stored in the register shown in the bottom entry (see Figure
7-81). The subtraction instruction is much more versatile than the addition instruction. The subtraction instruction has three outputs available. The subtraction
instruction can actually be used to compare the size of numbers. The top output is
energized when the value of the top element is greater than the middle element.
The middle value is energized when the top and middle elements are equal. The
bottom output is energized when the top value is less than the middle value.

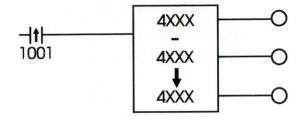

Figure 7-81. Gould subtraction instruction.

Multiplication Instruction

The multiplication instruction also uses a block format. Three entries are required. The top and middle entries are multiplied by each other and the result is placed in the register shown by the bottom entry. The top and middle entries can be an actual number or a register (see Figure 7-82). The third entry the programmer makes is the register where the answer is to be stored. The Gould Modicon uses two addresses to store the result in case the number gets large. For example, if the programmer designated register 4002 for the bottom entry, the PLC would store the result in registers 4002 and 4003. The programmer should keep this in mind when using registers. Do not use the same register for two different purposes unless there is a valid reason.

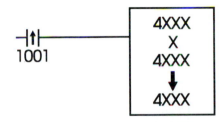

Figure 7-82. Gould multiplication instruction.

Division Instruction

The division instruction also uses a block format (see Figure 7-83). There are three entries. The top and middle elements can be actual numbers or registers. The third entry is where the answer will be stored. It must be a register. The division instruction divides the top number by the middle number. Note that a transitional input was used for the input to this instruction. This instruction uses two registers for the top entry. This means that if register 4002 was programmed for the top value, the number to be divided would be retrieved from registers 4002 and 4003. The output of the instruction is energized any time the result is too large for the storage register. The output could be used for error checking by the programmer.

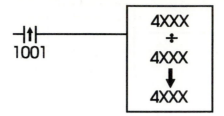

Figure 7-83. Gould divide instruction.

OMRON ARITHMETIC INSTRUCTIONS

Omron offers a wide variety of arithmetic instructions for binary and BCD numbers. For all of these instructions the programmer provides numbers, or sources of numbers, and the address where the result will be stored. There are instructions available for binary and BCD numbers. A few of these are described in the next few pages.

Omron C200H Memory Areas

Omron divides memory into areas (see Figure 7-84). There are several types of areas, each with a specific use.

Omron Data Areas	
HR	Holding Relay Area
TR	Temporary Relay Area
AR	Auxiliary Relay Area
LR	Link Relay Area
TC	Timer/Counter Area
DM	Data Memory Area
IR	I/O and Internal Relay Area
SR	Special Relay Area
UM	Program Memory

Figure 7-84. This figure shows the memory areas for the Omron C200H series.

Holding Relay Area

The HR (holding relay) area is used to store and manipulate numbers. It retains the values even when modes are changed or during a power failure. The address range for the HR area is 0000 to 9915. This memory area can be used by the programmer to store numbers that need to be retained.

Temporary Relay Area
The TR (temporary relay) area is used for storing data at program branching points.

Auxiliary Relay Area
The AR (auxiliary relay) area is used for internal data storage and manipulating data. A portion of this data area is reserved for system functions. The AR bits that can be written to by the user range from 0700 to 2215.

Link Relay Area
The LR (link relay) area is used for communications to other processors. If it is not needed for communications, it can be used for internal data storage and data manipulation. The address range for the LR area is 0000 to 6315.

Timer/Counter Area
The TC (timer/counter) area is used to store timer and counter data. The timer counter area ranges from 000 to 511. Note that a number can be used only once. The same number cannot be used for a timer and a counter. For example, if the programmer uses the numbers 0 to 5 for timers, they cannot be used for counters.

Data Memory Area
The DM (data memory) area is used for internal storage and manipulation of data. It must be accessed in 16-bit channel units. If a multiplication sign is used before the DM (*DM), it means that indirect addressing is being used. Data memory ranges from addresses 0000 to 1999. The user can only write to addresses 0000 to 0999.

I/O and Internal Relay Area
The IR area is used to store the status of inputs and outputs. Any of the bits that are not assigned to actual I/O can be used as work bits by the programmer. Channels 000 to 029 are allocated for I/O. The remaining addresses up to 24615 are for the work area. The IR area is addressed in bit or channel units. The addresses are accessed in channels or channel/bit combinations. Channel/bit addresses are 5 bits long. The two least significant digits are the bit within the channel. For example, address 01007 would be channel 10, bit (or terminal) 7.

Special Relay Area
The SR (special relay) area can be used to monitor the PLC's operation. It can also be used to generate clock pulses and to signal errors. For example, bit 25400 is a 1-minute clock pulse. Bit 25500 is a 0.1-second clock pulse. These special bits can be used by the programmer in ladder logic. There are many bits provided to signal errors that could occur. The programmer can use these in ladder logic to indicate problems.

Program Memory

The UM area is where the user's program instructions are stored. Memory is available in various sizes (RAM and ROM). These data areas are used simply by using the prefix (such as HR) followed by the appropriate address within that data area.

Omron Arithmetic Instructions

Binary Addition

To program a binary add instruction the user chooses the ADB instruction (function 50). The programmer enters the function number to specify which instruction will be used. The programmer then must supply the Au (augend) value, the Ad (Addend) value, and the R (result channel). If @ is used in front of the ADB (@ADB), the instruction is transitional. It will be active only when there is a transition on the input to it. The @ can be used with most of the instructions.

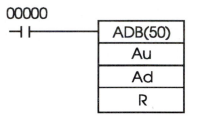

Figure 7-85. Binary add instruction.

Figure 7-86 shows the data types that can be used in an add instruction. The # means that actual data are being entered. A hex number would be entered.

Au and Ad	R
IR,SR,HR,AR,LR,TC,DM,*DM,#	IR,HR,AR,LR,DM,*DM

Figure 7-86. Data types that can be used with a binary add instruction.

The use of the ADB instruction in a ladder diagram is shown in Figure 7-87. This figure shows that a ADB (function 41) is used. If input 00000 comes true, then the CLC (clear carry) instruction clears the carry bit. Input 00000 also causes the ADB instruction to execute. In this case the binary data from channel 1 will be added to the hex number #F8C5. The result will be stored in holding relay 21. The clear carry should be used to ensure that the addition is correct. The clear carry instruction should be used with any addition, subtraction, or shift instruction to ensure the correct result. Note that it can be programmed using the same input conditions as the actual add or subtract instruction. Figure 7-88 shows how the add instruction works. The Au and Ad are added and the result is placed into the R address. In this case the binary data for Au comes from actual input channel 1.

These data are added to the hex number F8C5. The result is FA4C. The result is stored in HR21 (holding relay 21).

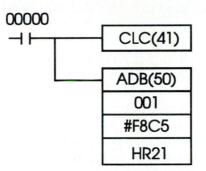

Figure 7-87. Use of the ADB (binary addition) instruction. Note the use of the clear carry instruction.

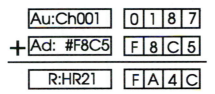

Figure 7-88. Results of a binary add instruction. The result is stored in address HR21.

Binary Subtraction

A binary subtraction instruction is programmed just like an add instruction. The instruction's mnemonic is SBB (see Figure 7-89).

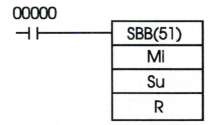

Figure 7-89. A binary subtract instruction.

If it is entered as @SBB it will only be active when there is a transition on its input condition. The programmer enters a Mi (minuend), a Su (subtrahend), and an R address where the result will be stored. The same data types may be used for numbers to be subtracted as with the addition instruction. See Figure 7-86 for the possible types. The function code for this instruction is a 51. A BCD subtraction instruction is also available.

The Su (subtrahend) will be subtracted from the Mi (minuend). The result is stored in the area specified by the R value. The same types of areas can be used for each of these as was used for the addition instruction. Remember that a clear carry [CLC(41)] should be used to ensure that the correct answer is obtained.

Multiply Instructions

The binary multiply instruction (MLB) can be used to multiply two numbers. The function number for this instruction is 52 (see Figure 7-90). The programmer supplies a Md (multiplicand) and a Mr (multiplier) and the result is stored in the area specified by R. The R specifies the first of two addresses that will be used to store the result of the multiplication. This instruction multiplies two 16-bit numbers together and stores a 32-bit answer at channel R and R+1. The same data types can be used for the numbers to be multiplied as for the addition instruction. See Figure 7-86 for a list of the possible data types. A BCD multiply instruction is also available.

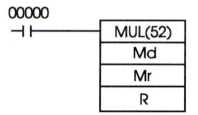

Figure 7-90. Binary multiply instruction.

Binary Division

The binary division instruction [DVB (53)] is programmed like the multiplication instruction. The programmer supplies a Dd (dividend), a Dr (divisor), and an R address. The R is the beginning address of the result. The quotient will be stored in channel R, and the remainder of the division problem will be stored in channel R+1. The same data types can be used as for the addition instruction. (See Figure 7-86 for the possible data types.) See Figure 7-91 for an example of the instruction.

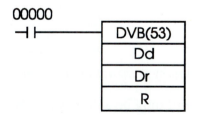

Figure 7-91. Binary divide instruction.

SQUARE D ARITHMETIC INSTRUCTIONS

One of the simplest math instructions available is the "let" instruction. The let instruction can be used to assign a value to a variable: For example, let S46 = 100. It can also be used to perform arithmetic instructions such as addition, subtraction, multiplication, division, and square root (see Figures 7-92 to 7-96). The larger Square D models also perform sine, cosine, log, and absolute value operations.

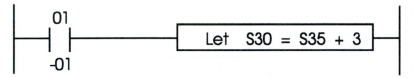

Figure 7-92. Let instruction used to add 3 to the number stored in register S35. The result is then stored in register S30. Note that the number subtracted could also have been a value stored by a register. The number held in register S35 does not change.

Figure 7-93. Let instruction used to subtract two numbers. The number in register S40 is subtracted from the value held in S35. The result is then stored in register S30. Note: The values on the right of the equal sign could have been actual numbers instead of registers that hold a number. The number held in S35 does not change.

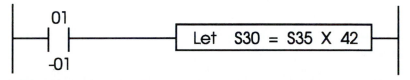

Figure 7-94. Use of a let instruction to multiply two numbers. In this case the number held by register S35 is multiplied by 42. The result is stored in register S30. Note that the number held in register S35 does not change.

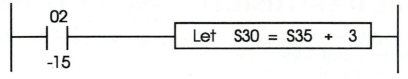

Figure 7-95. Use of a let instruction to divide two numbers. In this case the number held by register S35 is divided by 3. The result is stored in register S30. Note that the number held in register S35 does not change.

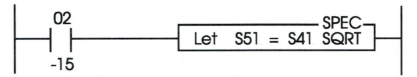

Figure 7-96. Let instruction used to find the square root of the number held in register S41. The result is then stored in register S51. The number held in register S41 does not change.

Let instructions can also be used to convert from one number system to another. The let instruction could be used to convert a BCD number to a binary number (see Figures 7-97, 7-98, and 7-99). This could be used to convert inputs from BCD thumbwheel switches to a binary equivalent. Thumbwheel switches typically output BCD numbers. The BCD must then be converted to a binary for the PLC to understand it.

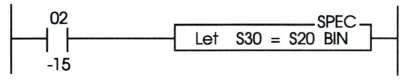

Figure 7-97. Use of a let instruction to convert the number held in register S20 to its binary equivalent. The result is stored in register S30. Note that the number held in register S20 does not change.

The let instruction can also be used to convert a binary number to a BCD number (see Figure 7-98). This is very useful when we need to convert a binary number to a decimal number for output to an operator. This could be used to output a BCD number that could drive a numeric display.

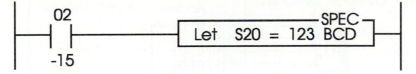

Figure 7-98. Use of a let instruction to convert a number to a BCD number. If input 0215 is true, 123 will be converted to BCD and stored in register S20. Note that instead of using an actual value the address of a register could have been used.

Figure 7-99. (Shown on page 272) How BCD thumbwheel switches could be used to input decimal numbers into a normal PLC input module. The thumbwheels output BCD. Each thumb-wheel switch outputs 4 bits. Each bit corresponds to a bit in storage register S1. Note that S1 is the register that holds the status of real-world inputs 1-1 through 1-16. The BCD number is stored in register S1. A let instruction is used to convert the BCD to its binary equivalent and is then stored in register S2. The CRT would then display the proper decimal equivalent.

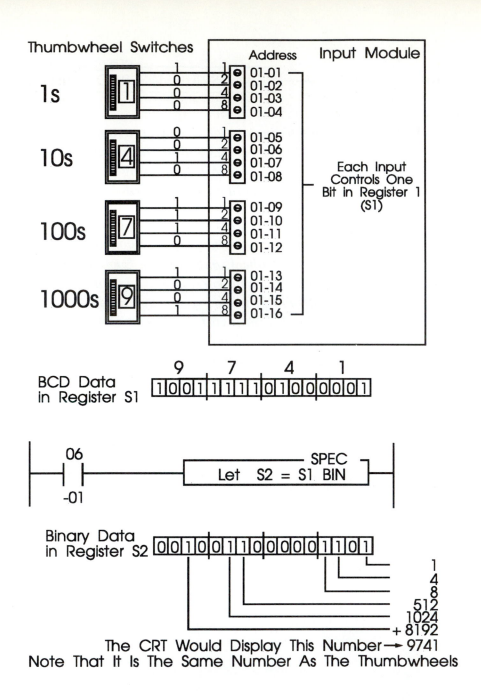

Figure 7-99. Thumbwheel example.

Square D Compare Instructions

The IF statement is used with Square D controllers to perform comparisons (see Figures 7-100 to 7-103). The IF instruction can be used to compare the number in a storage register to another storage register, a constant value, or the result of a math instruction. The IF can be used to test for equal, not equal, greater than or equal to, less than, or greater than.

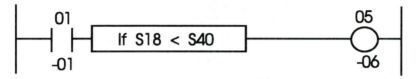

Figure 7-100. IF statement being used to test for a less-than condition. If the number held in storage register S18 is less than the number held in storage register S40, coil 5-06 will be energized.

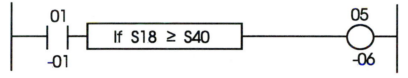

Figure 7-101. IF statement being used to test for a greater than or equal to condition. If the number held in storage register S18 is greater than or equal to the number held in storage register S40, coil 5-06 will be energized.

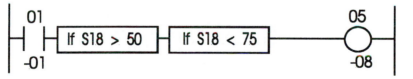

Figure 7-102. The use of multiple IF statements in series (AND condition). If storage register S18 is greater than 50 AND less than 75, coil 5-08 will be energized.

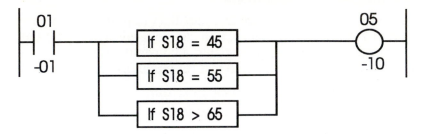

Figure 7-103. Use of IF statements in parallel. If storage register S18 is equal to 45 OR equal to 55 OR greater than 65, coil 5-10 will be energized.

Let instructions can also be used for Boolean logic. Let instructions can perform ANDs, ORs, and XORs (exclusive ORs).

TEXAS INSTRUMENTS AND PLC DIRECT ARITHMETIC INSTRUCTIONS

There are many arithmetic instructions available for the programmer's use. There are binary and BCD instructions. Many of these instructions make use of the accumulator. The accumulator is a memory location that is used to store numbers temporarily. Think of it as a scratch pad that can be used to hold a number. Numbers can be "loaded" into it, or the number in the accumulator can be stored to a different memory location. For example, the states of inputs can be loaded into the accumulator. They could then be manipulated and sent to actual outputs. There are many uses for the accumulator (see Figures 7-104 and 7-105).

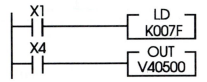

K007F = 0000000001111111

Figure 7-104. Use of a load instruction. If input X1 becomes true, the load instruction will load the accumulator with the hex number 007F. The K preceding the number stands for constant. If X4 becomes true, the number in the accumulator (007F) will be output to variable V40500. Variable V40500 is the word that controls the states of the first 16 outputs. The number below the ladder logic shows that the first 7 outputs would be energized.

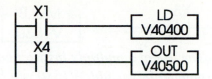

Figure 7-105. Use of a load instruction. If input X1 becomes true, the first 16 input states are loaded. Variable V40400 is a word that contains the states of the first 16 inputs. If input X4 becomes true, the number (word) held in the accumulator will be output to variable V40500. Variable V40500 is the word that controls the states of the first 16 outputs.

Omron arithmetic instructions utilize the accumulator. The use of an add instruction is shown in Figure 7-106. When input X5 becomes true the hex number 1324 will be added to whatever number is in the accumulator. The number that was in the accumulator is replaced by the result. If we wanted to add the 5 and 7 we would first use a load instruction to load 5 into the accumulator. We would then use an add instruction with a value of 7. The result of the instruction would be placed into the accumulator. We can use variables or constants with these instructions. In fact, the actual states of inputs, outputs, timers, and counters are all stored in variables. This means that we can utilize this data in instructions also.

Study Figures 7-107, 7-108, and 7-109 for examples of other arithmetic instructions.

Figure 7-106. Use of a add instruction. If input X5 becomes true, the hex number 1324 will be added to the value in the accumulator. The result is then stored back in the accumulator. The value that was in the accumulator is lost when the new value is stored there.

Figure 7-107. Use of a subtract instruction. If input X4 becomes true, the number in variable V1500 will be subtracted from the number in the accumulator. The result is stored in the accumulator.

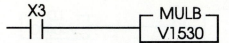

Figure 7-108. The use of a multiply instruction. If input X3 becomes true, the number in variable V1530 will be multiplied by the number in the accumulator. The result is stored in the accumulator.

Figure 7-109. Use of a division instruction. If input X3 becomes true, the number in the accumulator will be divided by the number in variable V2320. The result is stored in the accumulator.

Questions

1. Explain some of the reasons why arithmetic instructions are used in ladder logic.

2. What are let instructions used for?

3. What are comparison instructions used for?

4. Why might a programmer use an instruction that would change a number to a different number system?

5. Write a rung of ladder logic that would compare two values to see if the first is greater than the second. Turn an output on if the statement is true.

6. Write a rung of logic that checks to see if one value is equal to a second value. Turn on an output if true.

7. Write a rung of logic that checks to see if a value is less than 20 or greater than 40. Turn on the output if the statement is true.

8. Write a rung of logic to check if a value is less than or equal to 99. Turn on an output if the statement is true.

9. Write a rung of logic to check if a value is less than 75 or greater than 100 or equal to 85. Turn on an output if the statement is true.

10. Write a ladder logic program that accomplishes the following. A production line produces items that are packaged 12 to a pack. Your boss asks you to modify the ladder diagram so that the number of items is counted and the number of packs is counted. There is a sensor that senses each item as it is produced. Use the sensor as an input to the instructions you will use to complete the task. (*Hint*: One way would be to use a counter and at least one arithmetic statement.)

11. Write a ladder diagram program to accomplish the following. A tank level must be maintained between two levels (see Figure 7-110). An ultrasonic sensor is used to measure the height of the fluid in the tank. The output from the ultrasonic sensor is 0 to 10 volts (see Figure 7-111). This directly relates to a tank level of 0 to 5 feet. It is desired that the level be maintained between 4.0 and 4.2 feet. Output 1 is the input valve. Output 2 is the output valve. The sensor output is an analog input to an analog input module.

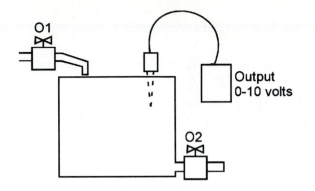

Figure 7-110. Tank level application.

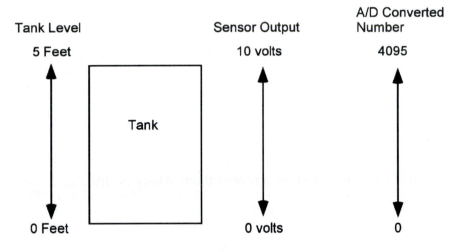

Figure 7-111. A comparison of tank level and A/D output.

Chapter 8

Advanced Programming

There are many instructions available that can make the programming systems easier. In this chapter we examine a few of them. Many of these instructions were designed to "look" like the mechanical control devices they replaced. The purpose was to make them easy to understand.

OBJECTIVES

Upon completion of this chapter, the student will be able to:

> *Explain a few of the advanced instructions that can make programming a complex application easier.*
>
> *Explain such terms as* **drum controller, interlocking, sequencers, stage programming,** *and* **step programming.**
>
> *Write simple sequencing programs using an appropriate advanced instruction.*
>
> *Choose an appropriate advanced instruction for an application.*

SEQUENTIAL CONTROL

Before PLCs there were many innovative ways to control machines. One of the earliest control methods for machines was punched cards. These date back to the earliest automated weaving machines. Punched cards controlled the weave. Until a relatively few years ago the main method of input to computers was punched cards.

Most manufacturing processes are very *sequential*, meaning that they process a series of steps, from one to the next. Imagine a bottling line. Bottles enter the line, are cleaned, filled, capped, inspected, and packed. This is a very sequential process. Many of our home appliances work sequentially. The home washer, dryer, dishwasher, breadmaker, and so on, are all examples of sequential control. Plastic injection molding, metal molding, packaging, and filling, are a few examples of industrial processes that are sequential.

Many of these machines were (and some still are) controlled by a device called a drum controller. A drum controller functions just like an old player piano. The player piano was controlled by a paper roll with holes punched in it. The holes represent the notes to be played. Their position across the roll indicates which note should be played. Their position around the roll indicates when they are to be played. A drum controller is the industrial equivalent.

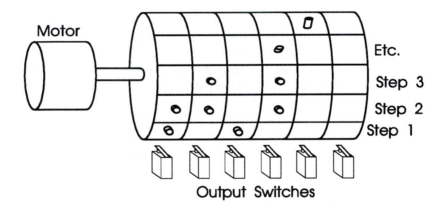

Figure 8-1. Drum controller. Note the pegs that activate switches as the drum is turned at slow speed by the motor. See Figure 8-2, which shows the output conditions for the steps.

The drum controller is a cylinder with holes around the perimeter with pegs placed in the holes (see Figure 8-1). There are switches that the pegs hit as the drum turns. The peg turns, closing the switch that it contacts and turning on the output to which it is connected. The speed of the drum is controlled by a motor. The motor speed can be controlled. Each step must, however, take the same amount of time. If an output must be on longer than one step, consecutive pegs must be installed.

Chapter 8: Advanced Programming

The drum controller has several advantages. It is easy to understand, which makes it easy for a plant electrician to work with. It is easy to maintain. It is easy to program. The user makes a simple chart that shows which outputs are on in which steps (see Figure 8-2). The user then installs the pegs to match the chart.

Step	Input Pump	Heater	Add Cleaner	Sprayer	Output Pump	Blower
1	on		on			
2	on	on		on		
3		on		on		
4					on	
5						on

Figure 8-2. Output conditions for the drum controller shown in Figure 8-1.

Many of our home appliances are also controlled with drum technology. Instead of a cylinder they utilize a disk with traces (see Figure 8-3). Think of a washing machine. The user chooses the wash cycle by turning the setting dial to the proper position. The disk is then moved very slowly as a synchronous motor turns. Brushes make contact with traces at the proper times and turn on output devices such as pumps and motors.

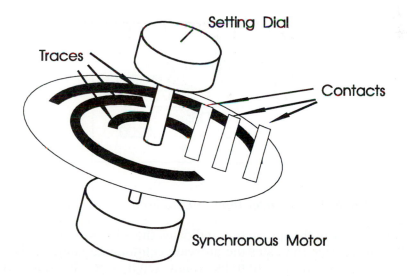

Figure 8-3. Disk-type drum controller. It is commonly used in home appliances. Note that when you turn the dial you are actually turning the disk to its starting position.

Although there are many advantages to this type of control, there are also some major limitations. The time for individual steps cannot be controlled individually. The sequence is set and it does not matter if something goes wrong. The drum will continue to turn and turn devices on and off. In other words, it would be nice if the step would not occur until certain conditions are met. PLC instructions have been designed to accomplish the good traits of drum controllers and also to overcome the weaknesses.

SEQUENCER INSTRUCTIONS

Sequencer instructions can be used for processes that are cyclical in nature. Sequencer instructions can be used to monitor inputs to control the sequencing of the outputs. Sequencer instructions can make programming many applications a much easier task. The sequencer is very like the drum controller.

Allen-Bradley Sequencer Instructions

Allen-Bradley PLCs use function blocks to program sequencers. Three function block instructions are available: sequencer input, sequencer output, and sequencer load. Each sequencer block has up to 999 steps available. Each step can control up to 64 outputs. As you can see, some very complex applications can be programmed with these. Examine Figure 8-4. This figure shows a typical sequencer input instruction block. The data that the programmer inputs to the instruction are shown in Figure 8-5.

```
SEQUENCER INPUT
COUNTER ADDRESS: 0200
CURRENT STEP:        001
SEQ LENGTH:          006
WORDS PER STEP:        2
FILE:           0400- 0413
MASK:           0070- 0071

INPUT WORDS:
1:    0110     2:    0200
3:             4:
```

Figure 8-4. Sequencer input instruction.

The counter address in the instruction indicates the address where the counter is stored. The current step indicates the present accumulated value of the counter. This instruction uses a counter to track which step it is in. The sequence length determines how many steps the sequencer has. (This is also the preset value of the counter.)

The words per step determines how many inputs are used. (Remember that one word is 16 bits, or inputs in this case.) The file input determines the address of the file. The mask input determines the address of the file that will be used as a mask. The last input shows which words will be used for the inputs. As this instruction is incremented through its steps, the inputs will be examined through the mask.

Counter address:	Address of the instruction
Current step:	Accumulated value of the counter (present step)
Seq. length:	Present value of the counter (number of steps)
Words per step:	This is the width of the sequencer table
File:	Starting address of the sequencer table
Mask:	Address of the mask file
Input words:	These are the input words that are examined by the instruction

Figure 8-5. Values required for a sequencer input instruction.

In actual use the input sequencer will be used as the input to an output sequencer instruction. For example, assume that the sequencer input instruction is in step one. If the input to the input instruction becomes true, the sequencer input instruction will increment to the next step. The inputs will be evaluated through the mask. (Remember that if the mask bit is a zero, the actual input condition for that bit is ignored.) If states of the inputs are correct, the input instruction will be true. The true output is used as an input to the output sequencer instruction. The output sequencer then increments to the next step. It outputs the new word of output conditions to the actual outputs. When the input sequencer instruction is enabled again, it will examine the states of the inputs that correspond to the new input step and continue the process.

The inputs that the programmer provides the output instruction are just like those of the sequencer input instruction. An example of a sequencer output instruction is shown in Figure 8-6. Figure 8-7 gives an explanation for each value that the programmer must input.

```
SEQUENCER  OUTPUT
COUNTER  ADDRESS:  0200
CURRENT  STEP:            001
SEQ  LENGTH:              006
WORDS  PER  STEP:          2
FILE:              0400- 0413
MASK:              0070- 0071

OUTPUT  WORDS:
1:     0110     2:     0200
3:                4:
```

Figure 8-6. Format for a sequencer output block for an Allen-Bradley PLC.

Counter address:	Address of the instruction
Current step:	Accumulated value of the counter (present step)
Seq. length:	Present value of the counter (number of steps)
Words per step:	This is the width of the sequencer table
File:	Starting address of the sequencer table
Mask:	Address of the mask file

Figure 8-7. Entries for an Allen-Bradley PLC-2.

The mask values are used to enable and disable certain inputs and outputs. A mask is a means of selectively screening out data. Study Figure 8-8. The mask value will be used for each step. Look at step 1; it contains all ones. This means that step 1 says that all outputs should be on for both words. The mask value must be considered, however. The mask value and the step bits are "ANDed." The result becomes the actual output status for each bit. Only if the mask bit AND the step bit is a 1 will the actual output be turned on. One reason for this might be that each bit in the step might be a processing station in a system. The mask value might be the status of presence sensors for each station. We would only want to process the stations that had parts present. The mask value would allow us to do that.

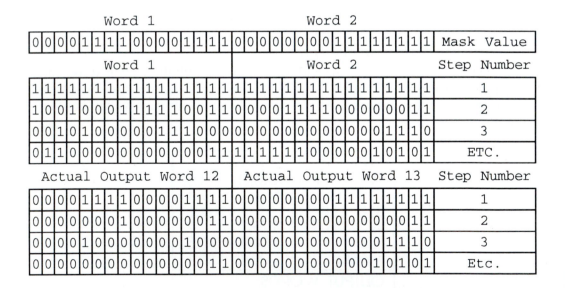

Figure 8-8. Use of a mask to determine whether or not an output is active. Note that a 1 in the mask bit means that the output is active. If the output word bit is a 1 and the mask bit is a 1, the output bit is set.

Chapter 8: Advanced Programming

The real power of these instructions can be seen when the sequencer input and sequencer output instructions are used together. Figure 8-9 shows how a sequencer input and a sequencer output instruction might be used in a ladder diagram. The programmer can put values into the input, output, and mask tables or a sequencer load instruction can be used to place data in the sequencer file. The sequencer load instruction is programmed as an output instruction. A false-to-true transition enables this instruction. It would then place its data into a sequencer file. This could be used so that one ladder diagram program could be used to produce different products. The proper data could be loaded into the sequencer files when needed.

Sequencers are very powerful tools when programming sequential-type processes. There are some limitations, however. Two events cannot run simultaneously and if one sequencer instruction is out of order, the process stops.

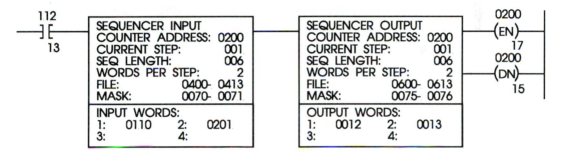

Figure 8-9. Sequencer input and sequencer output instruction used in a ladder diagram. Note that input I1213 is used to increment the sequencer input instruction. When the input sequencer instruction is true it is used to increment the output sequencer instruction. The output sequencer instruction then increments to the next step and sends that step's output conditions to the actual outputs.

SHIFT RESISTER PROGRAMMING

A shift register is a storage location in memory. These storage locations can typically hold sixteen bits of data, that is, 1's or 0's. Each 1 or 0 could be used to represent good or bad parts, presence or absence of parts, or the status of outputs (see Figure 8-10). Many manufacturing processes are very linear in nature. Imagine a bottling line. The bottles are cleaned, filled, capped, and so on. This is a very linear process. There are sensors along the way to sense for the presence of a bottle, there are sensors to check fill, and so on. All of these conditions could easily be represented by 1's and 0's.

Shift registers essentially shift bits through a register to control I/O. Think of the bottling line. There are many processing stations, each represented by a bit in the shift register. We want to run the processing station only if there are parts present. As the bottles enter the line a one is entered into the first bit. Processing takes place. The stations then release their product and each moves to the next station. The shift register also increments. Each bit is shifted one position. Processing takes place again. Each time a product enters the system a 1 is placed in the first bit. The 1 follows the part all the way through production to make sure that each station processes it as it moves through the line. Shift register programming is very applicable to linear processes.

Station #	1	2	3	4	5	6	7	8
Part Present	1	0	0	0	1	0	0	1

Figure 8-10. What a shift register might look like when monitoring whether or not parts are present at processing stations. A 1 in the station location would be used to turn an output on and cause the station to process material. In this case the PLC would turn on outputs at station 1, 5, and 8. After processing has occurred, all bits would be shifted to the right. A new 1 or a 0 would be loaded into the first bit depending on whether a part was or was not present. The PLC would then turn on any stations that had a 1 in their bit. In this way processing only occurs when parts are present.

STAGE PROGRAMMING

Stage programming is a new concept in PLC programming. The concept is to make programming complex systems easier. This concept involves breaking the program into logical steps or stages. The stages can then be programmed individually without concern for how they will affect the rest of the program. Stage programming is used by PLC Direct PLCs.

This method of programming can reduce programming time by up to about 70 percent. It can also drastically reduce the troubleshooting time by up to 85 percent. Much of the time and effort in writing a ladder logic program is spent programming interlocks to be sure that one part of the ladder does not adversely affect another. Stages help eliminate this problem.

A process or a manufacturing procedure is simply a sequence of tasks or stages. The PLC ladder diagram that we write must make sure that the process executes in the correct order. We have to design the ladder very carefully so that rungs execute only when they should. To do this we program interlocks. The process of designing the interlocks can take the majority of programming and debugging time.

A ladder logic program that deals with process control may require as much as 35 percent of the ladder dedicated to interlocking. A ladder written to control a sequential process may require that up to 60 percent be devoted to interlocking.

Let's consider a simple example. The process is shown in Figure 8-11. The process involves a conveyor and press operation. Figure 8-12 shows one ladder logic solution for the process. Remember that there are as many possible ladder logic programs as there are programmers. No two programmers would write the ladder in the same way. This makes it a little difficult when we are asked to troubleshoot or modify someone else's ladder. Even this simple process requires a fairly complex program and would require considerable time to understand. Also remember that a substantial portion of this ladder is devoted to interlocking.

What if we could program in English-like, logical blocks: for example, a block for starting the process, a block for checking for part presence, a block for locking the part, and so on. Then all we would have to do is break down any process into logical steps and the steps (or stages) would be our program. Figure 8-13 shows a solution for the press process. It is much easier to understand than a ladder program.

> *When we turn on the PLC it will start in stage 0 (see Figure 8-14). When the start button is pushed, the PLC changes to stage 1.*
>
> *Stage 1 checks for part presence. When contact X2 closes, the PLC moves into stage 2.*
>
> *Stage 2 locks the part by turning on coil Y1. When contact X3 closes, the PLC moves into stage 3.*
>
> *Stage 3 turns on coil Y2, which activates the press. When the lower limit is reached, it closes contact X4 and the PLC enters stage 4.*
>
> *Stage 4 raises the press by turning on coil Y3. When the press reaches the upper limit, it activates the upper limit contact X5. The PLC then moves into stage 5.*
>
> *Stage 5 unlocks the part by turning coil Y4 on. When the confirm unlock contact X6 closes, the PLC moves into stage 6.*
>
> *Stage 6 moves the conveyor by turning on coil Y5. When X7 closes the PLC will move into stage 7.*

Let's add a stop stage to our program (see Figure 8-14). Note that now when contact X1 is closed, the PLC enters stage 1 and stage 10. Both stages are active. In fact, stage 10 will be active during all of the stages. This allows us to stop the program at any point during operation. If X20 is then closed, stages 0 through 7 are reset and the PLC is told to jump to stage 0. Stage 10 is deactivated when the PLC is in step 0.

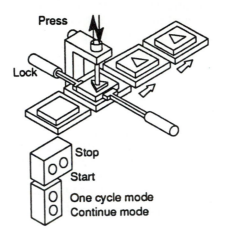

Figure 8-11. Simple press process. Courtesy Siemens Industrial Automation, Inc.

Rules of Stage Programming

Only instructions in active stages are executed. This eliminates the need for all of the complex and sometimes devious interlocking.

Stages are activated by one of the following:

1. Power flow makes contact with a stage label.

2. A "jump to stage" is executed.

3. The stage status bit is turned on by a "set" instruction.

4. The initial stage is executed when the PLC enters run mode.

Stages are deactivated by:

a. Power flow transitions from the stage.

b. Jumping from the stage.

c. A "reset" instruction.

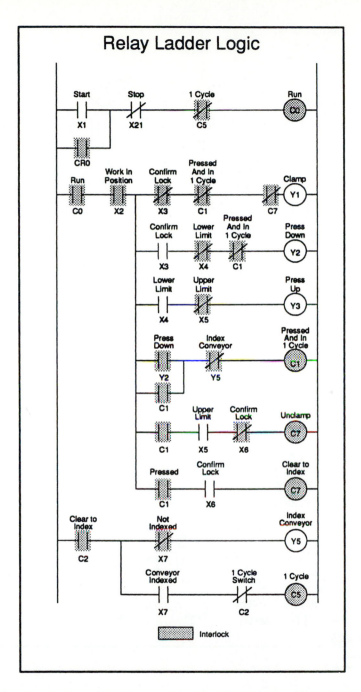

Figure 8-12. Ladder logic required to run the simple system shown in Figure 8-11. Courtesy of Siemens Industrial Automation, Inc.

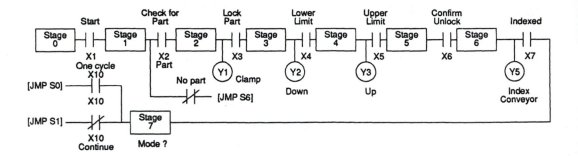

Figure 8-13. Same process as in Figure 8-11 but programmed using stage programming. Courtesy of Siemens Industrial Automation, Inc.

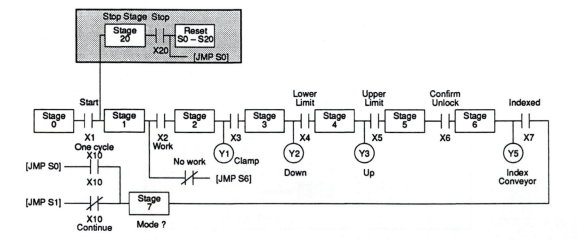

Figure 8-14. How a stop stage could be added to the stage program. Courtesy of Siemens Industrial Automation, Inc.

STEP PROGRAMMING

Omron PLCs have a similar type of programming available. Omron calls it "step" programming. It is very similar to stage programming. The application to be written is broken into logical steps. This is very much the way that manufacturing processes actually function. The production of a product can be broken down into logical processing steps: step A, then step B, then step C "or" step D, then step E, and so on. The beauty of step programming is that only the desired steps are active at that time. That means that there is almost no interlocking required. The individual blocks (steps) can be written and will not affect other steps.

Figure 8-15. Two types of step instruction.

There are two types of instruction blocks: step and step next (SNXT) (see Figure 8-15). The step next is used to move from one block of instructions to the next. The step instruction is used to show the beginning of a block of ladder logic. The block of instructions is then marked by a step next (SNXT) instruction.

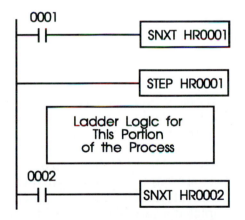

Figure 8-16. How a step instruction is used in a ladder diagram. Courtesy of Omron Electronics.

Input conditions are used with step next instructions to switch between blocks (steps) (see Figure 8-16). If input 0001 becomes true, the PLC will move to step HR0001. It will only evaluate the logic between step HR0001 and the following step next instruction. It will stay in this block of instructions until input 0002 becomes true. When that happens the PLC will move to step HR0002 and evaluate the logic for that step. Note that only the logic between the step and the step next is evaluated. This eliminates all of the interlocking that normally occupies about two-thirds of the ladder and two-thirds of the programmer's time.

Figure 8-17 shows the diagram of a two-process manufacturing line. Product enters and is weighed on the input conveyor. It is then routed to one of two possible processes, depending on the weight of the product. After the individual processing all product is printed at the last processing station. Note that sensors are used to sense when product enters and leaves a process.

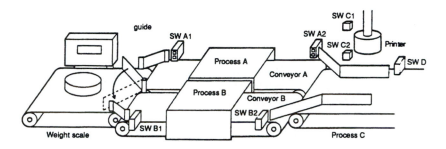

Figure 8-17. Conveyor system. The processing of the product is dependent on the weight of the product. Courtesy of Omron Electronics.

Figure 8-18 shows a block diagram of the process. Note that sensor input conditions have been used to route the product through the various possible production steps. Also note that parallel processing is possible. Steps do not have to be serial.

Figure 8-19 shows an example of what the ladder logic would look like for this process. Switches A1 and B1 have been used as input conditions to step next instructions. If switch A1 is true, the PLC will jump to step HR0000. This is process A. The logic of step 0000 is then active until switch A2 senses the product leaving process A. Switch A2 is then true, which makes the conditions true for step next to jump to HR0002. The PLC then evaluates step HR0002 (process C). When the product leaves process C it makes switch D true. Switch D makes the step next 24614 true. This sets a bit in memory, which indicates that process C is complete and has completed a part. This bit could be used by the programmer to allow further processing.

If the product weight had indicated that the product was a process B type, the product would have been routed through process B and then C. While this is a relatively simple example, it would require quite a lengthy ladder diagram if it were written with normal ladder logic. Which would you rather troubleshoot, a regular ladder or a ladder written in logical steps? The answer is that in addition to being logically organized into processing steps, the step ladder logic is probably one-fifth to two-thirds shorter.

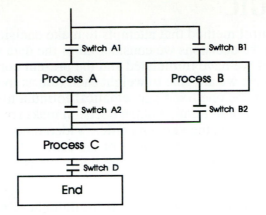

Figure 8-18. This figure shows a block diagram of the conveyor processing system, including the input conditions. Courtesy of Omron Electronics.

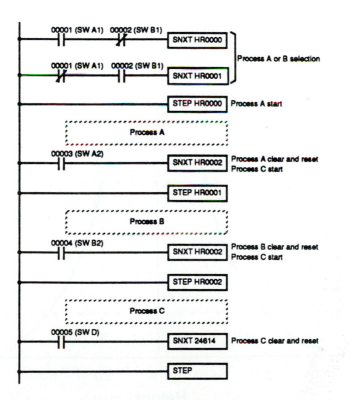

Figure 8-19. Ladder diagram for the conveyor processing system. Note that the actual ladder logic for each process has not been shown. Courtesy of Omron Electronics.

FUZZY LOGIC

Fuzzy logic is a control method that attempts to make decisions as a human being would. When we make decisions we consider all of the data we have available to us based on the rules we have formulated and the present conditions. We do not use hard and fast rules; we are able to weight each rule as to its importance. This means that we do not use one fixed mathematical formula to make our decision. Fuzzy logic is an attempt to mimic human decision making. The example mentioned earlier in the book is the video recorder. Video recorders with fuzzy logic are able to differentiate wanted movement of the camera from unwanted movement and thus stabilize the picture. There are many industrial applications that are very appropriate for fuzzy logic.

The basis of fuzzy logic is the "fuzzy" set. If we thought of people's height or weight, we could easily see an image of normal height and weight. To us it seems straightforward. If you think about it, though, what is normal? A bell curve can be used to show the relationship of people's heights (see Figure 8-20). Let's assume for the sake of discussion that 5 feet 9 inches is a normal height. We are not trying to say that someone who is 5 feet 1 inch or 6 feet 5 inches is not of normal height, however. Let's assume that we can readily agree that below 5 feet 0 inches is short and above 6 feet 6 inches is tall. As you can see, the definition of normal height is actually quite complex.

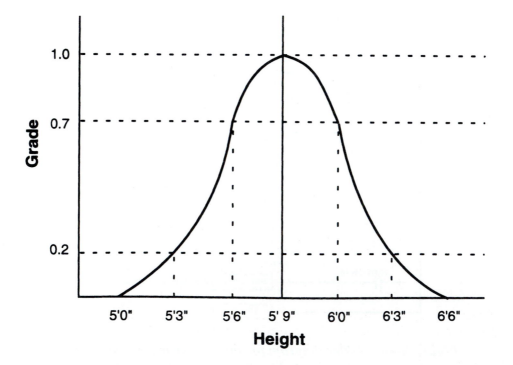

Figure 8-20. Chart of people's heights. Courtesy of Omron Electronics.

Consider Figure 8-20 again. A grade between 1.0 and 0.0 has been assigned to show how strongly we feel that the height is "normal." For example, 5 feet 9 inches has been assigned the value 1.0. We do not feel quite as strongly that 5 feet 3 inches is normal height. We only assign a value of 0.2 to it. The graph (bell curve) can also be called a *membership function*.

Fuzzy logic decision making can be divided into two steps: the inference step and the "defuzzifier" step (see Figure 8-21).

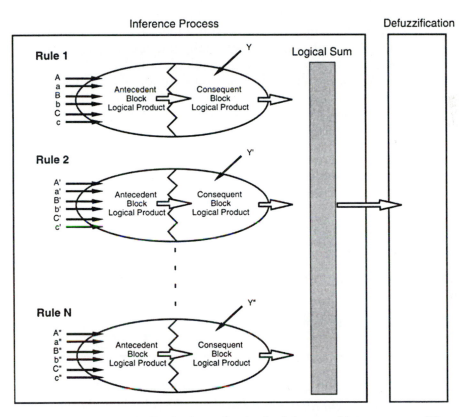

Figure 8-21. Basic fuzzy logic decision-making process. The rules are shown on the left. Decisions are made based on the individual rules. These decisions are then summed to provide a logical sum. The sum is then "defuzzed" and used to control the system. Courtesy of Omron Electronics.

The inference process is made up of several rule-based decisions that are summed to a single logical sum. Each rule is composed of input conditions (antecedent block) and a conclusion (consequent block). The logical products of each of these are summed.

Each rule is independent. They are analyzed independently. They are separate until they are combined to produce the logical sum. This combination of simple rules allows very complex decisions to be made. The rules are analyzed in parallel and a logical sum is generated. This produces a much better result than a simple formula to control all situations.

Imagine a cart with a stick (see Figure 8-22). The cart must be moved to balance the stick. Earlier when we thought about height we used terms such as *short*, *normal*, and *tall*. We could call these words *codes* or *labels*. We use these labels to express degree, such as moderate, almost, or a little. This example will utilize seven labels to express degree. More or fewer labels can be used.

Figure 8-23 will at first seem complex. It is not. The figure shows the antecedent membership functions and their labels. Note that the figure actually shows a series of triangle shapes. A triangular membership function is used instead of the bell curve we used in the height example. Originally the bell-shaped membership function was used, but due to the complexity of calculation the triangular function is now used most often. The results of both are very comparable. These triangular membership functions (antecedent membership) are composed so that the labels overlap. This permits reliable readings even when the level is not distinct or when the input from sensors is continually changing.

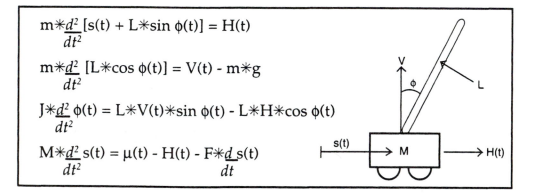

Figure 8-22. System controlled in this example. The cart must be moved just the right amount and at the right velocity to balance the stick. Diagram and example courtesy of Omron Electronics.

Developing the Rules in Code

The inclination of the stick from vertical is θ and the speed with which the inclination is changing (the angular velocity) is $d\theta$. Both θ and $d\theta$ are inputs from sensors. The sensors might be an encoder and a tachometer. The tachometer would give input on the angular velocity and the encoder would give input on the inclination.

These two variables can be used to write the "production" rules. The change in speed of the platform on which the stick is mounted is ΔV.

The antecedent block is usually composed of more than one variable linked by ANDs. The rules are linked by ORs. In this example we use seven rules. Refer to Figure 8-23. There are seven possible states for each input (NL: negative large, NM: negative medium, etc.).

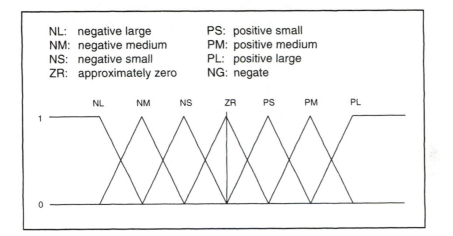

Figure 8-23. *The triangular membership functions used to represent this system. Note that there are seven triangular functions shown.*

Rule	Antecedent Block	Consequent Block
Rule 1	If the stick is inclined moderately to the left and is almost still	then move the hand to the left quickly
Rule 2	If the stick is inclined a little to the left and is falling slowly	then move the hand to the left slowly
Rule 3	If the stick is inclined a little to the left and is rising slowly	then keep the hand as it is
Rule 4	If the stick is inclined moderately to the right and is almost still	then move the hand to the right quickly
Rule 5	If the stick is inclined a little to the right and is falling slowly	then move the hand moderately to the right slowly
Rule 6	If the stick is inclined a little to the right and is rising slowly	then keep the hand as it is
Rule 7	If the stick is almost vertical and is almost still	then keep the hand as it is

Figure 8-24. *The 7 rules that will be used for the decision-making process. Note the consequent blocks. The consequent block shows what should be done if the rule is true.*

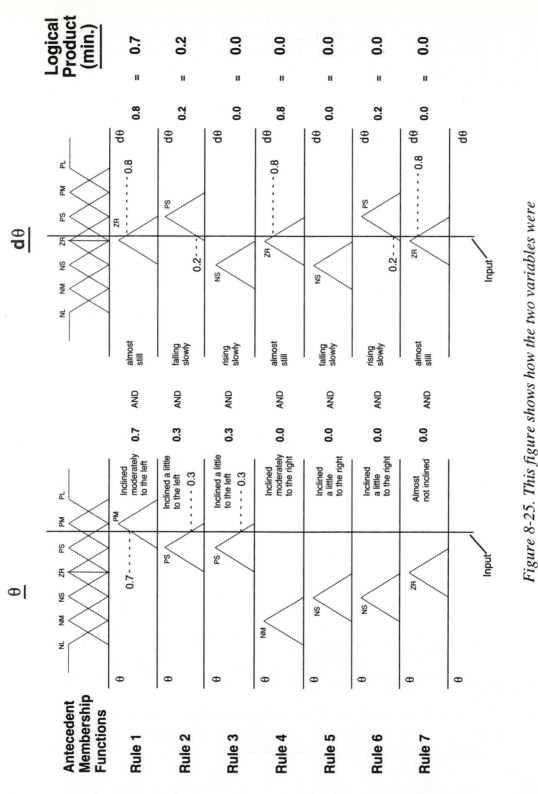

Figure 8-25. This figure shows how the two variables were evaluated by each rule at one point in time. Courtesy of Omron Electronics.

In our example there are two inputs: inclination and angular velocity. This means that if there are seven rules, we could have 49 potential combinations. (The number of possible combinations was calculated by raising 7 to the second power.) Each combination could be a rule. For this example seven rules are used. Only important rules need to be defined.

Examine Figure 8-24. Only five of the seven states describe the angle of inclination. Only three describe the angular velocity. This reduces the possible combinations to 15 (3X5) antecedent blocks. Of the 15 remaining combinations only seven are useful in describing how the system must operate. This is very similar to the way we think. We discard the rules that are not applicable to a particular decision and use only the relevant ones.

The following explanation of fuzzy logic evaluation will at first seem very confusing. You must study the graphs carefully as you read to gain an understanding of the process. Then it will seem simple and straightforward. Study Figure 8-25. This figure shows the two variables (inclination and angular velocity) at one point in time. The input value at that precise point in time is shown by the vertical line.

Each rule evaluates the input based on its membership function and assigns a value. For example, rule 1 evaluates the input based on the PM label (positively medium inclination to the right.) The rule assigns a value of 0.7 based on where the input intersects the triangular membership function (see Figure 8-25). Rule 1 is also evaluated for the angular velocity and a value of 0.8 is assigned.

The values of rule 1 inclination and rule 1 angular velocity are then evaluated to find the logical product (minimum). This means that the minimum value is used. In this case rule 1 inclination was 0.7 and rule 1 angular velocity was 0.8. The logical product is the smallest value, or 0.7 for rule 1. Note that each of the seven rules is evaluated based on the two inputs (inclination and angular velocity). Logical minimums are found for each rule. Note that each rule is evaluated based on where the input intersects (or does not intersect) its membership function.

The next step is to find the logical sum. The logical sum is the combination of the results of the rule evaluations. Study Figure 8-26. This figure shows how the logical sum is derived. Again it may look complex at first. Study rule 1. The logical product of rule 1 was 0.7. The area equating to 0.7 is filled in the triangular membership function for the rule 1 consequent block. The result of this first evaluation would say that the cart should be moved moderately to the left quickly. Rule 2 shows that the cart should be moved to the left a little quickly (value of 0.2). The products of rules 3 to 7 were zero, so they do not affect the outcome.

The logical sum of the consequent blocks is shown on the bottom of the figure. The membership functions for rules 1 and 2 are simply combined to give the result shown at the bottom of the figure. A decision must now be made on how to move the cart (how far and how fast). This is done by calculating the center of gravity for the logical sum of the rules. The result becomes the output value.

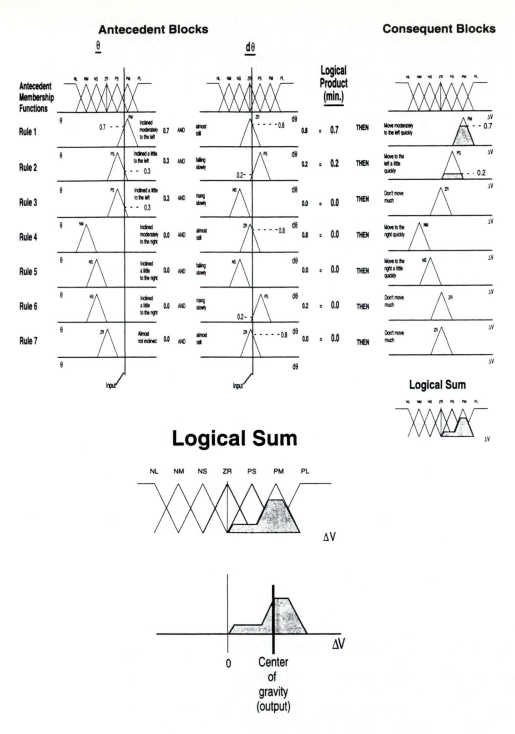

Figure 8-26. How the logical sum is derived.

As you have seen, seven rules and two inputs were used for this simple example. The evaluation that was done resulted in an output value that was based on an evaluation of the rules and the status of the inputs at that particular time. The advantage of fuzzy logic is that it is more flexible than mathematical models of systems. PID is a mathematical model. With the wide variety of industrial systems and operating conditions, it is very difficult to develop an accurate mathematical model for industrial systems.

Fuzzy logic helps solve the problem. Fuzzy logic breaks any system down into more humanlike rules. These simple rules are evaluated and compensate for the actual conditions at that time. Fuzzy logic is also easier to understand because the operator's own knowledge and thought process is reflected in the control process. PID, on the other hand, is often difficult for people to understand.

Fuzzy theory was born about 25 years ago. In 1965, Professor L. A. Zadeh of the University of California, Berkeley presented a paper that outlined fuzzy theory. At first the theory was met with indifference and even hostility.

In about 1970 fuzzy logic began to be used in Japan, Europe, and China. It began to show positive results. There will be room for fuzzy logic and other control systems, such as PID. Each will have its own niche. The future of fuzzy logic does appear bright, however. Although fuzzy technology is still in its infancy, it has already contributed significantly to commercial and industrial control applications. Fuzzy technology can be applied through software, dedicated controllers, or through fuzzy microprocessors in products. This flexibility and simplicity of fuzzy logic may make fuzzy logic a basic control and information-processing technology in the twenty-first century.

> *Here are a few examples of applications that could benefit from fuzzy logic control:*
>
> *Nonlinear systems such as process control, tension control, and position control.*
>
> *Systems with gross input deviations or insufficient input resolution.*
>
> *Difficult-to-control systems that require human intuition and judgment.*
>
> *Systems that require adaptive signal processing to overcome changing environmental or process conditions.*
>
> *Processes that must balance multiple inputs or that have conflicting constraints.*

STATE LOGIC

State logic is one of many new approaches to programming manufacturing systems. There are many new languages being developed to program control systems. They are an attempt to make programming an easier task. They do not use ladder logic. The commands typically used in these new languages are very straightforward English statements. State logic is presented here as an example of one of these emerging languages. State logic represents a different approach to control logic. State logic does not use ladder logic. State logic breaks processes into states and tasks. The actual logic is written in plain English. The resulting program is very easy to understand.

Study Figures 8-27 and 8-28. The system is shown in Figure 8-27 and the logic is shown in Figure 8-28. Note that the logic consists of small sections called tasks. Each task is one logical portion of the entire process. Each task is then divided into states that control a portion of the task. Each state has a statement associated with it to make decisions based on real inputs and/or values and control outputs. Note how English-like the statements are.

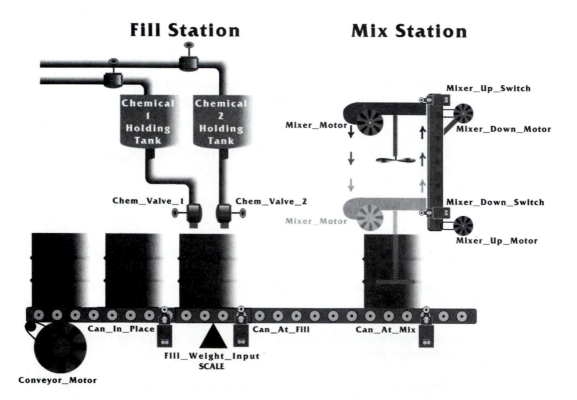

Figure 8-27. Example of a fill and mix system. Figure and description courtesy of Adatek.

Example System

Cans are filled with 2 chemicals and mixed as they move along a conveyor belt. When a can is placed on the conveyor line, it trips the can_in_place limit switch, and if neither the fill nor the mix tasks are currently active, the conveyor will start. The conveyor will run until a can arrives at either the fill or mix stations, at which time the conveyor will stop and wait for another can and any filling or mixing tasks to be completed.

When a can arrives at the fill station, it trips the can_at_fill limit switch. Chemical valve 1 will open until the fill weight is above 20 lbs. Then chemical valve 2 will open until the fill weight is above 30 lbs. Then the fill station will wait for another can to arrive to begin another cycle.

When a can arrives at the mix station, it trips the can_at_mix limit switch. The mixer down motor will run until the mixer down limit switch is tripped. The mixer motor will then start. After the mixer motor has run for 30 seconds the mixer up motor will run until the mixer up switch is tripped. A counter is incremented to keep track of the number of completed cans. (The counter is called Can_Inventory.) Then the message "Batch Cycle Complete" is written to the operator panel every time a can is mixed. The actual logic (program) is shown in Figure 8-28.

State logic is a very high-level programming language for programming systems. It is based on finite state machine theory. State logic is essentially a framework for modeling and real-world processes. It is a language designed to control systems.

Project: Batching System

Task: Fill_Station
> State: PowerUp
>   When Can_At_Fill is on, go to the Batch_Chem_1 State.
> State: Batch_Chem_1
>   Open Chem_Valve_1
>   When Fill_Weight is above 20 pounds, go to Batch_Chem_2.
> State: Batch_Chem_2
>   Open Chem_Valve_2 until Fill_Weight is more than 30 lbs,
>   then go to the Batch_Complete State.
> State: Batch_Complete
>   When Can_At_Fill is off, go to the PowerUp State.

Task: Mix_Station
> State: PowerUp
>   If Can_At_Mix is on, go to the Lower_Mixer State
> State: Lower_Mixer
>   Run the Mixer_Down_Motor until the Mixer_Down_Switch is tripped,
>   then go to the Mix_Chemicals State.

State: Mix_Chemicals
 Start the Mixer_Motor.
 When 30 seconds have passed, go to the Raise_Mixer State
State: Raise_Mixer
 Run the Mixer_Up_Motor until the Mixer_Up Switch is tripped,
 then go to the Batch_Complete.
State: Batch_Complete
 When Can-At-Mix is off, go to the Update_Inventory State.
State: Update_Inventory
 Add 1 to Can_Inventory.
 Write "Batch Cycle Complete" to the Operator_Panel and go to PowerUp.
Task: Conveyor
State: PowerUp
 When Can_In_Place is on and
 (Fill_Station Task is in the PowerUp or Batch_Complete) and
 (Mix_Station Task is in the PowerUp or Batch_Complete),
 go to Start_Cycle
State: Start_Cycle
 Start the Conveyor_Motor.
 When Can_In_Place is off, Go to Index_Conveyor State.
State: Index_Conveyor
 Run Conveyor_Motor.
 When Can_At_Fill or Can_At_Mix is on, go to the PowerUp State.

Figure 8-28. This figure shows the actual logic for the system shown in Figure 8-27. The state logic control language can presently be used on personal computers and some PLCs. Logic and explanation courtesy of Adatek.

The State Logic Model

All real-world processes move through sequences of states as they operate. Every machine or process is a collection of real physical devices. The activity of any device can be described as a sequence of steps in relation to time. For example, a cylinder can exist in only one of three states: extending, retracting, or at rest. Any desired action for that cylinder can be expressed as a sequence of these three states. Even a continuous process goes through startup, manual, run, and shutdown phases. All physical activity can be described in this manner. It is not difficult to express an event or condition that could be used to cause the cylinder (or other device) to change states. For example: If the temperature is over 100 degrees F, turn the warning light on and go to the shutdown procedure. Time and sequence are natural dimensions of the state model just as they are natural dimensions of the design and operation of every control system, process, machine, and system. State logic control uses these attributes (time and sequence) as the components of program development. As a result, the control program is a clear snapshot of the system that is being controlled. State logic is a hierarchical programming system that consists of tasks, states, and statements.

Tasks

Tasks are the primary structural elements of a state logic program. A task is a description of a process activity expressed sequentially and in relation to time. If we were describing an automobile engine the tasks would include the starting system task, the fuel system task, the charging system task, the electrical system task, and so on. Almost all processes contain multiple tasks operating in parallel. Tasks operating in parallel are necessary because most machines and processes must do more than one thing at a time. State logic provides for the programming of many tasks that are mutually exclusive in activity yet interactive and joined in time.

States

States are the building blocks of tasks. The activity of a task is described as a series of steps called states. A state describes the status or value of an output or group of outputs. These are the outputs of the control system and thus are inputs to the process. Every state contains the rules that allow the task to transition to another state. A state is a subset of a task that describes the output status and the conditions under which the task or process will change to another state. The states when taken in aggregate provide a description of the sequence of activity of the process or machine under control. Further they provide an unambiguous specification of how that portion of the process will respond in all conditions.

Statements

Statements are the user's command set to create state descriptions. The desired output-related activity of each state can be described by using statements. Statements can initiate actions or can base an output status change on a conditional statement or a combination of conditions. Any input value or variable can be used in conditional statements. Variables can include state status from other tasks as well as typical integer, time, string, analog, and digital status variables.

State logic allows the programmer to write the control program in natural English statements. State logic and other types of system languages are sure to increase in acceptance and popularity because of their simplicity and ease of use.

State logic can be used to program computers and some PLCs. An example of a state logic module for a PLC is shown in Figure 8-29.

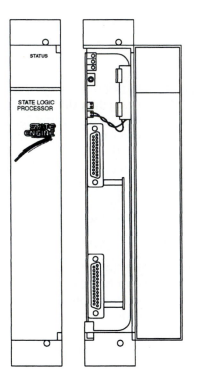

Figure 8-29. A GE Fanuc state logic processor module for a series 90-30 PLC. Courtesy GE Fanuc Inc.

Questions

1. What is a drum controller?

2. What are the benefits of a drum controller?

3. List at least three disadvantages of drum controllers.

4. How is a sequencing instruction different from a drum controller?

5. Construct a matrix of what a stoplight program would look like. Show the east/west outputs and north/south outputs as well as the steps.

6. Explain how the matrix you constructed in question 5 would be used to program a sequencer.

7. What is stage/step programming?

8. List at least three advantages of stage/step programming.

9. What is a shift register?

10. What types of applications are appropriate for the use of shift register programming?

11. What is fuzzy logic?

12. What are rules?

13. What types of applications are appropriate for fuzzy logic?

14. What is state logic programming?

IEC 1131-3 Programming

IEC 1131-3 is a standard for programming. It promises to revolutionize PLC and other controller programming.

OBJECTIVES

Upon completion of this chapter, the student will be able to:

Describe the purpose of the IEC 1131-3 standard.

List and explain each of the languages specified by IEC 1131-3.

Explain how the languages can be integrated in programs.

Explain terms such as: **function blocks, functions, statements,** *and so on.*

Explain how IEC 1131-3 will affect industrial automation.

OVERVIEW OF IEC 1131-3

The IEC 1131-3 standard has developed to meet a very natural need. Users have struggled for years with the problem of programming different brands of PLC. Every different brand has different ladder logic. Different models from the same manufacturer often have different languages. This has required users to have multiple programming packages. It has also required users to learn the software and logic differences. The other thing that has changed is that special-purpose controllers and languages emerged to fill the ladder logic gaps. Languages for more complex control and for special purposes like motion control emerged. Languages that attempted to simplify programming also emerged.

There were many weaknesses in ladder logic. It is hard to reuse ladder logic. Complex programs that have been developed are little use in developing new programs. Each application is essentially created from scratch. Ladder logic can be quite inconvenient and cumbersome to perform mathematical computations and comparisons. It is also difficult to segment ladder logic. Every line affects every other line. This also means that it is difficult to control the execution of the ladder diagram. Ladder diagrams are normally executed top to bottom. It would be advantageous to be able to control which sections execute and in which order.

IEC 1131-3 emerged to meet these needs and establish a standard for programming. IEC 1131-3 was really built using programming techniques that were already well established. The IEC standard specifies the following programming languages: ladder diagram, instruction list, function block diagram, structured text, and sequential function chart. The standard also allows the user to select any of the languages and mix their use in the program. The user can choose the best language for each part of the application. Three of the languages are graphics-based and two are text-based. The standard should ensure that the vast majority of software written for one PLC will run on PLCs from other manufacturers.

The first revision of IEC 1131-3 was published in 1993. Programs can be written using any of the IEC languages. A program is typically a collection of function blocks that are connected together. Programs can communicate with other programs and can control I/O.

The IEC 1131-3 standard defines a program as "a logical assembly of all programming elements and constructs necessary for the intended signal processing required for the control of a machine or process by a programmable controller system."

Function Blocks

Function blocks are one of the keys in IEC 1131-1 programming. Function blocks allow a program to be broken into smaller, more manageable blocks. Function blocks can even be used to create new, more complex function blocks. Function blocks take a set of input data, perform actions or an algorithm on the data and produce a new set of output data. Functions can hold data values between execution. You can think of a function block as a special-purpose integrated chip. We find special-purpose chips in many consumer products today. They take a set of inputs, process the data and produce outputs. Think about the processors in automobiles. Think about the processor used in antilock braking systems, for example. A function block is like that. Function blocks can be "plugged in" to programs and can perform a portion of the needed logic and control. Standard functions can be used "off the shelf" or users can develop and use their own functions.

Function blocks can be used to solve control problems such as PID control. Temperature and servo control are important in many industrial applications. The temperature and servo control is typically only a portion of the control problem, however. Function blocks can be used to develop a PID or fuzzy logic algorithms.

A function block defines the purpose of input and output data. This data can be shared with the rest of the control program. Only input and output data can be shared. A function block can also have an algorithm. These algorithms are run every time the block is executed. The algorithms process the current input data values and produce new output values.

Function blocks also have the ability to store data. Data can be used locally or may be used globally. Input or output data can be shared but internal variables are not accessible to the rest of the program. This is an important feature in that it allows function blocks to be modular and independent or to share their data.

The standard specifies some standard function blocks. It specifies counters, timers (see Figure 9-1), clocks, and so on. Function blocks can also be developed by the user. Users can build new blocks using existing function blocks and other software logic.

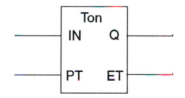

Figure 9-1. A TON function block timer.

Function blocks are crucial because they allow the user to develop very logical programs that are easy to understand and maintain.

Functions

IEC 1131-3 also specifies functions. Functions are different from function blocks. Functions are designed to perform common processing tasks such as trigonometric functions. They take a set of inputs and produce a result. Given the same input the output will always be the same.

All of the standard trigonometric calculations are available as functions. All of the common arithmetic calculations such as add, subtract, and so on are also available as functions. There are also a variety of bit functions available such as shifts and rotates, Boolean operators, and a set of selection type functions. Figure 9-2 shows an example of a limit function. A limit function limits the value of an input data value between the values at minimum (MN) and maximum (MX) and sends out the result.

There are also comparison functions such as greater than (GT), greater than or equal to (GE), equality (EQ), and so on. Character string function are also specified. There are also a wide variety of functions available for working with the time and date.

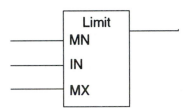

Figure 9-2. Example of a limit function.

STRUCTURED TEXT PROGRAMMING

Structured text (ST) is a high-level language that has some similarities to Pascal programming.

The IEC standard defines structured text as a language in which statements can be used to assign values to variables. The use of structured text is wider than this, however.

Expressions

Structured text uses expressions to evaluate values that are derived from other constants and/or variables. Expressions can also be used to calculate values based on values of other variables or constants. When using expressions the user must be sure to use a variable data type that will match the result of the expression.

Assignment Statements

Assignment statements are used to assign or change the value of a variable. An assignment statement in ST is very similar to an assignment statement in any programming language, particularly PASCAL.

An example of an assignment statement would be: TEMP := Z;. This would assign the value of whatever is in Z to TEMP. If the value of Z was equal to 78 the statement would put the value 78 into the variable called TEMP. We could have used a constant instead of Z.

Operators

There are many arithmetic operators available. Figure 9-3 shows a table that lists the arithmetic operators. They are shown in their order of precedence. The top of the table has the highest precedence.

Operator	Description
(....)	Parentheses used to group for precedence
Function(....)	Used for function parameter list and evaluation
**	Exponentiation
-	Negation
NOT	Boolean complement
*	Multiplication
/	Division
MOD	Modulus
+	Addition
-	Subtraction
<,>,<=,>=	Comparison
=	Equality
<>	Inequality
AND,&	Boolean AND
XOR	Boolean exclusive OR
OR	Boolean OR

Figure 9-3. Structured text arithmetic operators.

Statements

Structured text language has a variety of statements that can be used to call functions, perform iteration (loops), or perform conditional evaluation.

Function blocks can be run using statements. An example is shown below.

Temp (PV := 74, SP := 83);

In this example a statement is used to call the Temp function (see Figure 9-4). The Temp function requires two input values, PV and SP. Values are sent to the function by specifying them in the function. In this case 74 will be sent to the PV input of the function and 83 will be sent to the SP of the function. The function would then calculate the output value. The output value is always available. It can be used in any assignment statement. An example is shown below.

Tval := Temp.Out;

This would assign the output of the Temp function to variable Tval.

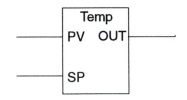

Figure 9-4. A function.

Conditional Statements

Conditional statements are used to control which statements are to be executed. IF, THEN, ELSE and CASE statements are allowed.

IF, THEN, ELSE Statements

These conditional statements evaluate a Boolean expression to determine whether to execute the logic they control. An example is shown below.

IF Tval = 87 THEN

    Var1 := 35;

    Var2 := 12;

ELSE

    Var1 := 54;

    Var2 := 17;

END_IF;

ELSIF can also be used.

IF Var1 < 83 THEN

    A:=5;

ELSIF VAR1= 83 THEN

    A:=6;

ELSIF VAR1 >83 THEN

    A:=7;

END_IF;

CASE Statements

The CASE statement is also conditional. It is a very useful statement when the user needs to execute a set of statements that are conditional on the value of an expression that returns an integer. In other words, a CASE statement evaluates an integer expression and executes the portion of the CASE code that matches the integer value.

CASE Var1 OF

1 : Var2 := 5;

2 : Var2 := 5; Pump1 := ON;

3,4 : Var2 := 5; Pump2 := ON;

5...7 : Var2 := 5; Alarm := ON;

END_CASE;

Loops

There are several types of loops available.

REPEAT ... UNTIL

The REPEAT UNTIL loop is used to execute one or more statements while a Boolean expression is true. The Boolean expression is tested after execution of the statements. If it is true, the statements are executed again. An example is shown below.

Count := 1;

REPEAT

   Count := Count + 1;

   Pack := 5;

   Pump1 := OFF;

UNTIL Count = 5

END_REPEAT

FOR ... DO Loop

The FOR Do loop allows execution of a set of statements to be repeated based on the value of a loop variable. An example is shown below.

FOR I := 1 TO I <= 50 BY 1 DO

   Temp (PV := Var1, SP := Var2);

END_FOR;

WHILE ... DO Loop

The WHILE DO loop permits repeated execution of one or more statements while a Boolean expression remains true. The expression is tested before executing the statements. If the expression is false, the statements are not executed. An example is shown below.

WHILE Var1 < 95 DO

    Temp (PV := Var1, SP := Var2);

END_WHILE

RETURN and EXIT Statements

RETURN and EXIT statements can be used to end loops prematurely. The return statement can be used within functions and function block bodies. RETURN is used to return from the code. An example is shown below.

FUNCTION_BLOCK TEMP_TEST

VAR_INPUT

    PV, SP : REAL;

END_VAR

VAR_OUTPUT

    OUT : REAL;

END_VAR

IF PV > 100 THEN

    ALARM := TRUE; RETURN;

END_IF;

IF SV > 100 THEN

    ALARM := TRUE; RETURN;

END_IF;

END_FUNCTION_BLOCK;

If any IF is true, ALARM is set to TRUE and the RETURN statement is executed to end the execution of this function block.

EXIT Statement

The EXIT statement can be used to end loops before they would ordinarily end. When an EXIT statement is reached, the execution of the loop is ended and execution continues from the end of the loop. An example is shown below. The example shows a loop that increments through a single-dimensional array named TEMP. As the loop increments the IF statement checks to see if the value of each element of the TEMP array is greater than 90. If it is the loop is exited. Program execution would then proceed to the next statement following this loop.

```
FOR I:= 1 TO 10 DO

    IF  TEMP[I] > 90 THEN

        EXIT;

    END_IF;

END_FOR;
```

FUNCTION BLOCK DIAGRAM PROGRAM-MING

Function block diagram (FBD) programming is one of the graphical languages that are specified by the standard. Function block diagrams can be used to show how programs, functions, and function blocks operate. An FBD looks very much like an electrical circuit diagram. In an electrical circuit diagram lines are used to show current flow to and between devices. There are typically inputs and outputs from each device. The diagram shows the overall system and the interrelationship of the components. This is very similar to function block diagrams. Many people will find function block diagrams more friendly to program than language-based languages like structured text.

Function block diagram programming is used when the application involves the flow of signals between control blocks. Remember that we examined functions and function blocks earlier in this chapter. Figure 9-5 shows an example of a function block. The function block always has inputs shown on the left and outputs on the right. The function block type name is shown inside the block, in this case TempControl. Remember that the name shown in the block is the type name. The name shown above the block in a function block diagram will be the name of the function block instance. The names of the inputs are shown on the left of the block and the output names are shown on the right.

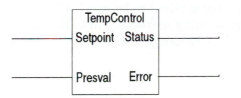

Figure 9-5. An example of a function block.

Figure 9-6 utilizes Boolean functions. Remember that we can utilize standard functions such as Boolean, comparison, string, selection, time, mathematical and numerical functions as well as user-created function blocks in FBDs.

Figure 9-7 shows an example of a function block diagram. This is a temperature control application. Two temperature control blocks are used. Note that the two blocks are of the same type. These blocks only needed to be created once. Note also that the name above each is different because each is a different use of the same function block type and the name identifies each particular instance. Note also that a block called TempMonitor has been used. It is providing the current temperature reading to the input called Presval in each of the TempControl blocks.

Variables or constants can be used as inputs to function blocks. Note also that the outputs from the function blocks are sent to variables. Outputs are used to supply values to variables and function block inputs (also variables).

The IEC 1131-3 standard also allows for values to be fed back to create feedback loops. Figure 9-8 shows an example of feedback. This allows values to be used as inputs to previous blocks.

This means that the output value of Status from MachControl1 is used as an input value to the input (ComVal) of LoopCont1. What happens if LoopCont1 is evaluated before MachControl1?

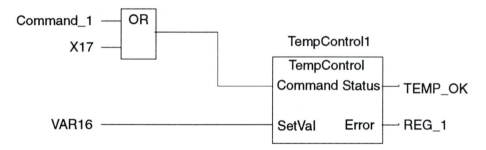

Figure 9-6. The use of Boolean operators in a simple FBD.

Figure 9-9 shows a list of operators that are reserved for use with standard function blocks. The standard states that implementations may provide a facility so that the order in which functions are evaluated in a network can be defined. The standard does not define the method to be used. This means that different companies that sell FBD programming software can define their own method of determining how this would be evaluated. One method would be a list that would specify evaluation order.

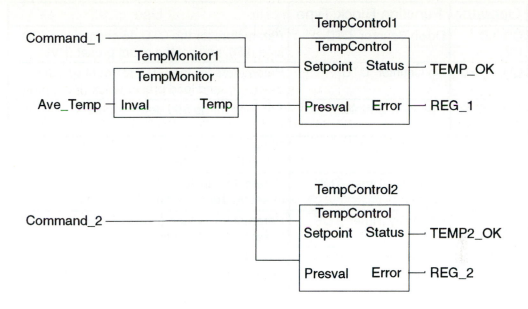

Figure 9-7. A small function block diagram.

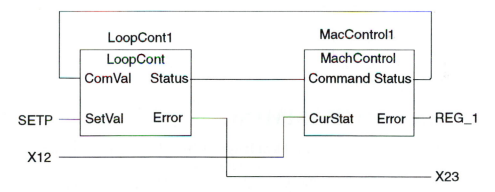

Figure 9-8. How a value can be used to create feedback.

Operator	Function Block Type	Use
CD, LD, PV	Down Counter CTD	Parameters for the CTD, to count down (CD), load (LD), and set the count preset (PV)
CU, R, PV	Up Counter CTU	Parameters for the CTU, to count up (CU), reset, (R), and load preset value (PV)
CU, CD, R, LD, PV	Up/Down Counter	Same as for up and down counters
IN, PT	Pulse Timer TP	Parameters are IN to start timing and PT to set up the pulse time
IN, PT	On-Delay Timer TON	Parameters include IN to start timing and PT to set up the delay time
IN, PT	Off-Delay Timer TOF	Parameters include IN to start timing and PT to set up the delay time
CLK	R_Trig, Rising Edge Detector	Clock input to the rising edge detector function block
CLK	F_Trig, Falling Edge Detector	Clock input to the falling edge detector function block
S1, R	SR Bi-stable	Set and reset the SR bi-stable
S, R1	RS Bi-stable	Set and reset the RS bi-stable

Figure 9-9. A list of operators that are reserved for use with standard function blocks.

LADDER DIAGRAMMING

Ladder diagramming is also a graphical language that IEC 1131-3 defines. Ladder diagramming was a logical choice as a programming language for one of the standard languages as it is the most widely used. Ladder diagramming can also be used in combination with all of the other programming methods that IEC 1131-3 specifies. Figure 9-10 shows the contacts that are specified in IEC 1131-3. Figure 9-11 shows the coils that are specified in IEC 1131-3. Ladder logic can also be used with the other IEC 1131-3 languages. Figure 9-12 shows an example of ladder logic used with function blocks.

Jumps

The standard for ladder logic programming specifies jumps. A jump can be used to jump from one section of a ladder diagram to another. This is based on the condition of a rung of logic. This is accomplished by having the rung of logic end in a label identifier. The label identifier is a name which is then used to identify the start of the logic the user wishes to jump to. Figure 9-13 shows an example of the use of a jump.

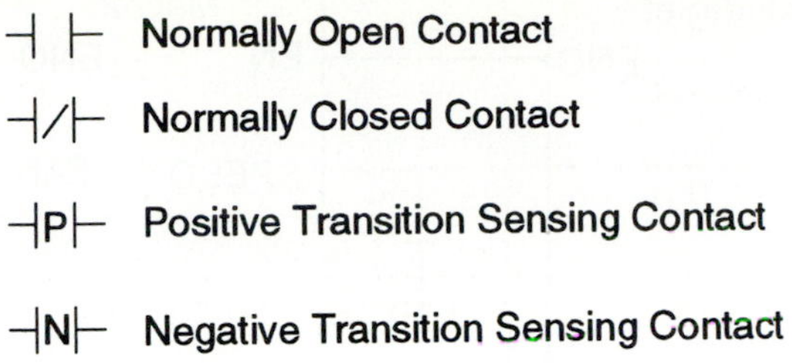

—| |— Normally Open Contact

—|/|— Normally Closed Contact

—|P|— Positive Transition Sensing Contact

—|N|— Negative Transition Sensing Contact

Figure 9-10. Contacts specified in IEC 1131-3.

—( )— Coil
—(/)— Negated Coil
—(S)— Set Coil
—(R)— Reset Coil
—(M)— Retentive Memory Coil
(SM)— Set Retentive Memory Coil
(RM)— Reset Retentive Memory Coil
—(P)— Positive Transition Sensing Coil
—(N)— Negative Transition Sensing Coil

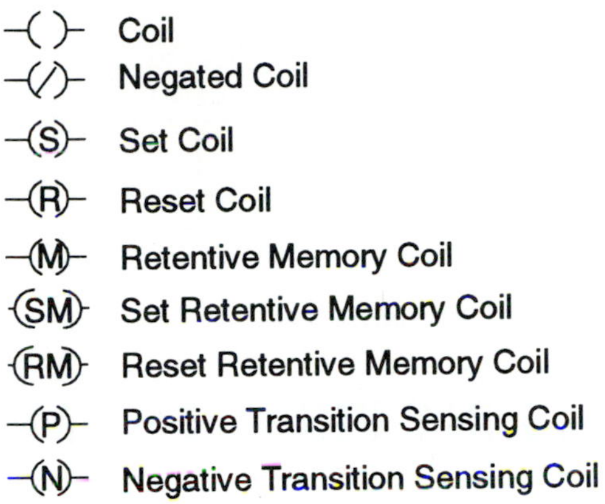

Figure 9-11. Coils specified in IEC 1131-3.

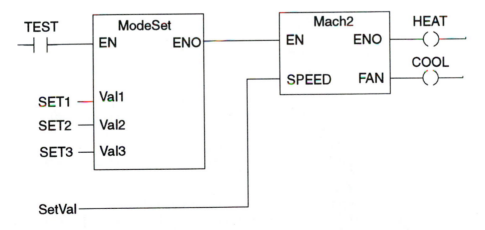

Figure 9-12. The use of ladder logic to combine function blocks.

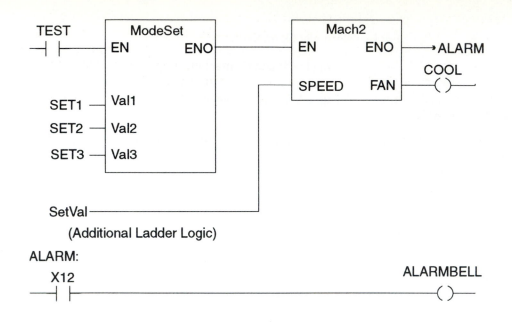

Figure 9-13. Jumps in ladder logic.

INSTRUCTION LIST PROGRAMMING

Instruction list (IL) programming is a very low-level programming language. Low level typically means that it is not very user-friendly. The higher level the language, the easier it is to use. Instruction list programming is very much like assembly language programming. Figure 9-14 shows an example of IL programming.

```
       Operator  Operand    Comments
          /        /           \
       LD       Temp       (* Load Temperature *)
       LT       90         (* Test if Temperature < 90 *)
       JMPCN    Temp_Hi    (* Jump to Temp_Hi if Temp is not < 90 *)
       LD       Count      (* Load Count *)

       DIV      6          (* Divide by 6 *)
       ST       CASES      (* Store Result of Division to Cases variable *)
Label  LD       0          (* Load 0 *)
   \   ST       Y12        (* Turns Output Y12 Off *)
    \
     \ ST       Y23        (* Turns Output Y23 Off *)
      \LD       1          (* Loads 1 *)
       ST       Y5         (* Turns on Output 5 *)
Temp_Hi: LD     1          (* Loads 1 *)
       ST       ALARM      (* Stores a 1 in ALARM Variable *)
```

Figure 9-14. Instruction list program example.

IL programs can be used by experienced programmers to develop very efficient, fast code. IL language is considered by many to be the base language for IEC-compliant PLCs. It is the language into which all of the other languages can be converted. IEC 1131-3 does not specify that any language is the base language, however. Figure 9-15 shows an example of arithmetic and Boolean instructions that can be used in IL programs. Figure 9-16 shows the comparison instructions that are available in IL programming.

Let's compare the IL program in Figure 9-14 to what it would look like in structured text.

IF Temp < 90 THEN

    CASES := Count / 6;

    Y12 := 0;

    Y23 := 0;

    Y5 := 1;

END_IF

ALARM : = 1;

Operator	Operand	Purpose
ADD	Any Type	Add
SUB	Any Type	Subtract
MUL	Any Type	Multiply
DIV	Any Type	Divide
LD	Any Type	Load
ST	Any Type	Store
S	Boolean	Set operand true
R	Boolean	Reset operand false
AND / &	Boolean	Boolean AND
OR	Boolean	Boolean OR
XOR	Boolean	Boolean exclusive OR

Figure 9-15. The arithmetic and Boolean instructions available in IL programming.

Operator	Operand	Purpose
GE	Any Type	Greater than or equal to
GT	Any Type	Greater than
EQ	Any Type	Equal
NE	Any Type	Not equal
LE	Any Type	Less than or equal to
LT	Any Type	Less than
CAL	Name	Call function block
JMP	Label	Jump to label
RET		Return from a function or function block
)		Execute last deferred operation

Figure 9-16. The comparison instructions that are available in IL programming.

Functions and function blocks can be called by using the CAL operator. An example is shown below.

CAL TEMP1(SETPT := 85, CYC := 5)

This would call and execute a function block called TEMP1. It would send the values of 85 to the SETPT input parameter and 5 to the CYC input parameter. These are the inputs that the TEMP1 function block requires. So as you can see IL can be combined with the other types of programming languages that IEC 1131-3 specifies.

There is another way to load parameters and make a function block call. In the example shown below the values are loaded and then stored to the input parameters for the function before the function is called. The LD function is used to load the value and then the ST is used to store the value into the parameter. The first ST stores 85 into the SETPT input parameter for the TEMP1 function block (TEMP1.SETPT). The same method is then used to load 5 into the CYC parameter. The function block is then called with the CAL operator.

LD 85.0

ST TEMP1.SETPT

LD 5

ST TEMP1.CYC

CAL TEMP1

Figure 9-17 shows an example of a function block diagram for a start sequence. The equivalent logic is shown below.

LD   Command_1

OR X17

ST StartRS.S

Load X12 Str\art.R1

LD StartRS.Q1

ST Start

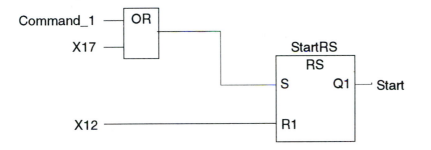

Figure 9-17. Example of the use of an OR.

SEQUENTIAL FUNCTION CHART PROGRAMMING

Sequential function chart (SFC) programming is a graphical programming method. Sequential function charts are a very useful method for describing sequential type processes. Figure 9-18 shows an example of a simple linear-type sequence. There are 5 steps in the sequence. There is a condition that must be fulfilled before we can move from one step to the next. For example, to move from step one to step 2 S1 must be true. S1 in this case might be a start switch in the process. It could be anything that would evaluate to either a true (1) or false (0). There are conditions between each step in this example. Transitions can be defined by name.

Sequential function charts can also be used for processes which have portions of the applications whose steps are dependent on conditions. For example, imagine a bottling application (see Figure 9-19). Two types of bottles come down a line at random.

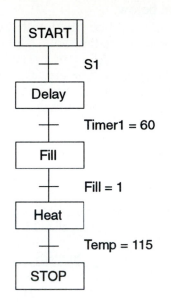

Figure 9-18. Example of a simple linear process.

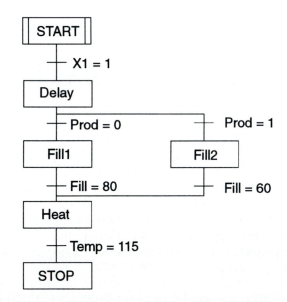

Figure 9-19. A simple process with alternate steps that are
dependent on a condition.

A sensor is mounted at the fill station to identify which bottle is present. The sensor has a tagname of Prod. If Prod is equal to 0, it should be filled with one product and if it is equal to 1 it should be filled with a different product. Most of the processes (steps) in Figure 9-19 are the same. The fill processes are the only ones that are different in this process. After the delay the bottles are either filled in the Fill1 process or the Fill2 process. The processes following the fill processes are the same in this application. Sequential function charts handle these types of applications very easily. It could have been much more complicated than this. There could have been many alternative processes. Each path could have had multiple steps, too.

Branching

Branches can also be used to alter the sequence processing during operation (see Figure 9-20). In this case, X17 is evaluated after the second step. If X17 is equal to 1 the sequence continues to the next step. If X17 is equal to 0 sequencing branches back to a position immediately following the start block.

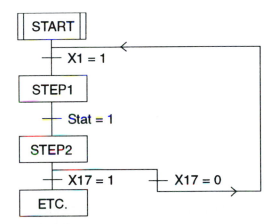

Figure 9-20. An example of branching in a sequence.

Concurrent Processing

There are many applications where the processing is not linear. There may be several processes occurring at the same time. An example is shown in Figure 9-21. In this case there are three processes that are all active at the same time. The first process has a Fill and a Heat step. The second process is an assembly process. The third contains two steps. Note that double lines are used to show where concurrent processes begin and end. Also note that there can be multiple steps in each process and there can be conditions in each to control movement between steps. This is a powerful tool. As processors gain more power and speed they have tremendous capability to control complex processes or even multiple processes.

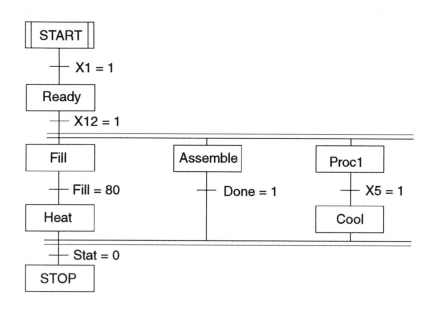

Figure 9-21. An example of concurrent processing.

There are some rules about how steps and transitions can be used. Transitions cannot be directly linked. There must be a step between any two transitions. If a transition leads to two or more steps, all of the steps are executed independently. They all execute simultaneously. Two steps cannot be linked directly, either. Steps must be separated by a transition.

Step Variables

There are two variables that are associated with every step. The programmer can utilize these variables. The first variable is set to a 1 while the step is active. This can be very useful for monitoring and for logic. It is called the step active flag. It is named .X. To use it the programmer uses the name of the step and adds the .X.

There was a step named *Ready* in Figure 9-22. If the programmer wanted to use the active flag he/she would simply use Ready.X. If the step is active, Ready.X would be equal to 1. If the step is inactive Ready.X would equal 0. The active flag can be used in another way also. Figure 9-23 shows how the active flag can be used directly. In this case Y21 is set to a one whenever this step is active.

The second type of step variable is a time variable. Every step has a variable that contains the time that the variable has been active. The elapsed time variable can be used by giving the step name and adding a .T. Figure 9-22 uses the elapsed time variable to control the transition between Step1 and Step2. The logic essentially says if the elapsed time of Step1 is greater than 15 minutes move to Step2 (Step1 > T#15m).

Ladder logic can be used as a transition condition between steps (see Figure 9-24). In this example when the rung is true the program will move from the START step to STEP1.

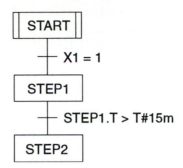

Figure 9-22. Use of the elapsed time flag for STEP1.

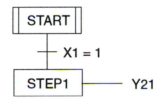

Figure 9-23. Use of the active flag to set Y21 when STEP1 is active.

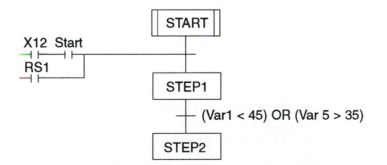

Figure 9-24. An example of ladder logic used as a transition condition.

Function blocks or function block diagrams can be used to control transitions between steps as long as their output is discrete (1or 0). Figure 9-25 shows an example of a function block being used to control the transition from START to STEP1.

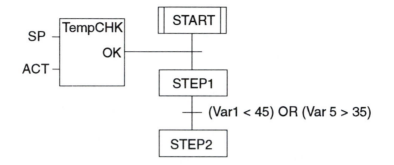

Figure 9-25. An example of a function block used as a transition condition between steps.

Ladder logic that ends in a transition connection can be used to control transitions between steps. Figure 9-26 shows a ladder diagram rung that ends in a transition connector named Alt1. Alt1 is also used as a transition condition in the sequential function chart.

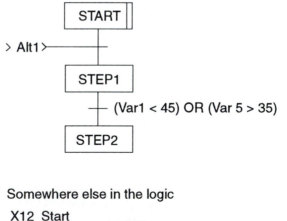

Somewhere else in the logic

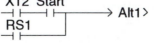

Figure 9-26. An example of a ladder diagram rung that ends in a transition connector named Alt1.

Structured text expressions can be used as transition conditions. In Figure 9-27 an ST expression is used to control the transition between STEP1 and STEP2. The ST expression must result in a true or false. In this case if Var1 is less than 45 or Var5 is greater than 35 the expression will result in a true (1) and the process will move from STEP1 to STEP2.

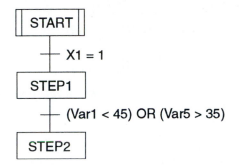

Figure 9-27. Use of a structured text expression to control a transition.

Transitions can also be defined by using instruction list (IL) programming. Figure 9-28 shows an example of the use of an IL program to control a transition. In this case the transition in the sequential function chart is called Cond1. Note that in the IL program the keyword TRANSITION is used to declare Cond1 as the name of the transition that is being defined. The conditions in the IL program are then evaluated when the program runs. If X12 or X15 is true this IL program will result in a 1 and Cond1 will be true and cause the program to change from the Fill step to the Heat step.

```
TRANSITION Cond1 :
    := X1 AND X15 OR X5;
END_TRANSITION
```

Figure 9-28. An example of the use of an IL program to control a transition.

Instruction list programming can also be used to program transition conditions. Figure 9-29 shows an example. The IL program starts with the keyword TRANSITION followed by the name that the programmer assigns to the transition condition. In this case the programmer chose Cond1 for the name. In this case the logic would be that if X12 OR X18 is true then Cond1 will be set to a 1, the transition condition between two steps will be met, and the program will move from one step to the next.

```
TRANSITION Cond1 :
    LD X12
    OR X18
END_TRANSITION
```

Figure 9-29. An example of instruction list programming to control a transition condition.

Ladder logic can also be used to control transitions. The logic shown in Figure 9-30 is an example. Note that the transition name appears at the right of the controlling logic.

Figure 9-30. Example of ladder logic to create a transition element.

Function block diagram language can also be used to define transitions. Figure 9-31 shows an example of an AND block ANDed with an OR block to define the transition logic.

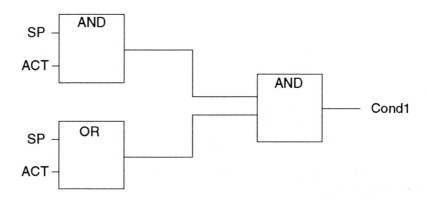

Figure 9-31. Example of an AND block ANDed with an OR block to define the transition logic

Step Actions

Steps are used to control actions that occur during that step of the sequence. Steps can control multiple actions. Figure 9-32 shows the general format for an action. The action block is shown to the right of the step. The first part of the action block can contain a qualifier. The qualifier controls how the action is performed. The qualifier in this case is an N. An N means that the action will be executed while the associated step is active. Additional qualifiers are shown in Figure 9-33. The second part of an action block is the action to be performed. In this case the action is StartSeq. The third part of the action block is optional. The user may use a variable in the third part of the action block. The variable is called an indicator variable. The variable will indicate when the action has completed its execution. In this example the variable is called Stat. In this example when StartSeq has finished its execution Stat would be set to a 1.

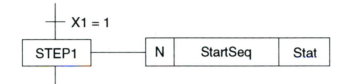

Figure 9-32. General format for an action.

Qualifier	Use
N	Not stored, executes while associated step is active
None	Default, not stored, executes while associated step is active
R	Resets a stored action
S	Sets an action active
L	Terminates after a given time period
D	Starts after a given time period
P	Pulse action that occurs only once when the step is activated and once when the step is deactivated
SD	Time delay and stored. The action is set active after a given period, even if the associated step has been deactivated before the time period elapsed
DS	Action is stored and time delayed. If the associated step is deactivated before the time period elapses, the action is not stored
SL	Time limited and stored. Action is started and executes for a given time period

Figure 9-33. Qualifiers that can be used with actions.

Figure 9-34 shows an example of the use of an action. The action is associated with STEP1. The qualifier is N so the action will be executed while STEP1 is active. There is no indicator variable. The action name is StartMotor. StartMotor is a very simple action. It is used to turn Motor1 on and Fan off. This has been specified in the block under the StartMotor action name. This example is quite simple. There are only two outputs associated with this action so the user programmed them right with the action block. The action could have been defined on another page if it was more complex. Another valuable asset of action blocks is that once they are defined they can be used repeatedly by the use of the qualifier and the action name. So StartMotor could be used anywhere in our program now without redefining the actual I/O to be performed.

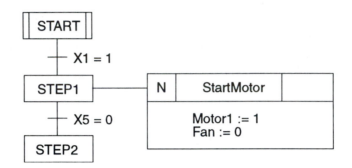

Figure 9-34. Example of the use of an action.

Actions can also be defined by using the function block diagram language. Figure 9-35 shows an example of the use of function block diagram language to define an action. The FBD logic that the user creates is enclosed in a box with the name of the action that it defines at the top.

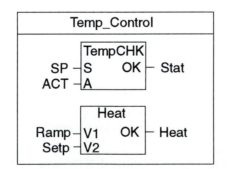

Figure 9-35. Example of the use of function block diagram language to define an action.

Ladder logic can be used to define what the action does. Figure 9-36 shows an example of the use of ladder logic for this purpose. The ladder logic is enclosed in a rectangle with the name of the action at the top.

Chapter 9: IEC 1131-3 Programming

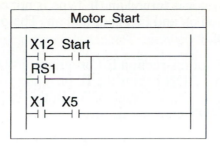

Figure 9-36. Example of the use of ladder logic to define an action.

Sequential function charts can also be used to describe the behavior of an action. The example shown in Figure 9-37 shows how complex action behavior could be defined simply with SFC. In this case there are three basic steps in the SFC, Fill, Mix and Pour. Also note that there is an action associated with each step.

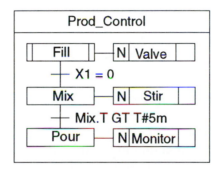

Figure 9-37. Example of how complex action behavior can be defined with SFC programming.

```
ACTION Stir :
IF Type = 1 THEN
    Speed1 = 75;
    Motor = ON;
ELSE
    SPEED1 = 25;
    Motor = ON;
    FAN = ON;
END_ACTION
```

Figure 9-38. Structured text program used to define an action.

Figure 9-38 shows an example of the use of structured text to define the behavior of an action. Note that the keyword ACTION starts the structured text program. Then the name of the action that is being defined is shown. The name of the action is Stir. This action decides if Type is equal to 1 or not. If the Type is 1 then

Speed1 is set to 75 and Motor is turned on. If Type is not equal to 1 then Speed1 is set to 25, Motor is turned on and Fan is turned on. The program is ended with the keyword END_ACTION.

Figure 9-39 shows the use of instruction list programming to define an action's behavior. The keyword ACTION begins the IL program followed by the name of the action. The IL program loads the state of X1 in this case and stores it to Y5. The keyword END_ACTION ends the IL program.

```
ACTION Monitor :
    LD X1
    ST X5
END_ACTION
```

Figure 9-39. Instruction list program used to define an action.

Use of Action Blocks in Graphical Languages

Action blocks can be used in ladder logic. Figure 9-40 shows the use of an action block in ladder logic. When there is power flow into the action block it is active. The indicator variable (Y5) can be used to indicate when the action is complete.

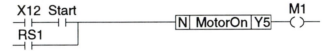

Figure 9-40. Example of the use of an action block in ladder logic.

Figure 9-41 shows the use of an action in the function block diagram language. Note that the indicator variable (Y5) can be used elsewhere to indicate when the action is complete.

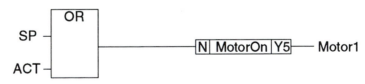

Figure 9-41. Use of an action in function block diagram language.

This chapter has been a very quick introduction to IEC 1131-3 programing. It is by no means complete. A book could be written on each of the programming methods. For those who want a more in-depth look at IEC 1131-3 programming, I would suggest a book by R.W. Lewis, *Programming Industrial Control Systems Using IEC 1131-3*.

Questions

1. Explain why the IEC 1131-3 standard was developed.

2. Explain what structured text programing is.

3. Explain function block diagram programming.

4. Explain IEC 1131-3 ladder logic programming.

5. Explain instruction list programming.

6. Explain sequential function chart programming.

7. Explain how these languages can be combined in control programs. Examples might be helpful.

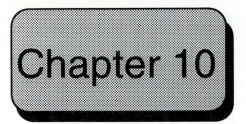

Chapter 10

Overview of Plant Floor Communication

If enterprises are to become more productive, they need to improve processes. This requires accurate data. Communications are vital. Production devices hold very valuable data about their processes. In this chapter we examine how these data can be acquired.

OBJECTIVES

Upon completion of this chapter, the student will be able to:

Describe the four levels of plant communications and characteristics at each level.

Compare and contrast human communications and machine communications.

Define such terms as serial, synchronous, RS-232, RS-422, device, cell, area, host, and SCADA.

Describe how computers can communicate with PLCs.

Explain the concept of MAP/TOP.

Describe the opportunity available through factory communications.

INTRODUCTION

The programmable controller has revolutionized manufacturing. It has made automation flexible and affordable. PLCs control processes across the plant floor. In addition to producing product, PLCs also produce data. The data can be more profitable than the product. This may not seem obvious; however, most processes are very inefficient. If we can use the data to improve processes we can drastically improve profitability. The inefficiencies are not normally addressed because people are busy. There are other, more pressing problems. (A good friend of mine in a small manufacturing facility said it best: "It's hard to think about fire prevention when you're in the middle of a forest fire.") Manufacturing people are usually amazed to find that most of the data they would like to have about processes is already being produced in the PLC. With very few changes the data can be gathered and used to improve quality, productivity, and uptime. There are huge gains possible if these data are used.

To be used the data must first be acquired. Many managers today will say that they are already collecting much of the data. Why should they invest in electronic communications when they are already gathering data from the plant floor manually? The reasons are many. Often, data gathered manually are very inaccurate. The data are not real-time either. The information must be written down by an operator, gathered by a foreman, taken "upstairs," entered by a data processing person, printed into a report, and distributed.

This all takes time. It can often make data many days or weeks late. If mistakes in entry were made on the floor, they are often too late to correct. The reports that are produced are often mazes of meaningless information: too much extraneous information to be useful to anyone. The lateness and inaccuracy of the data gathering makes it almost counterproductive.

The other communication that is required is real-time information to the operator: accurate orders, accurate instructions, current specifications, and so on. This is often lacking in industrial and service enterprises today. This communication is quite easily achieved with the use of electronic communications. Many of the data required already exist in the smart devices on the factory floor. Much of the data that people write on forms in daily production already exist in the PLC.

The improvements in computer hardware and software have made communication much easier. There are many software communications packages that make it easy to communicate with PLCs. Communications will increase in importance as American manufacturing tries to improve its competitive position.

LEVELS OF PLANT COMMUNICATION

Plant floor communications can be broken down into levels. Some authors break them into four levels, some into five. The basic concepts are the same. In this book we use a five-level model. The five levels are: device, machine, cell, area, and host (see Figure 10-1). Each level is a vital link. The device and machine levels are the production level. As we move up the pyramid, management of production becomes the task.

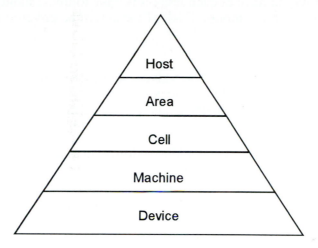

Figure 10-1. Five-level model of plant communication.

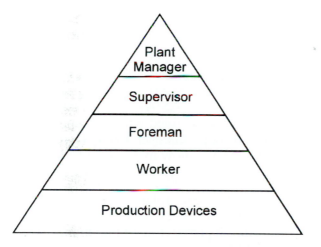

Figure 10-2. Comparison of the plant communication model and the typical industrial organization of people. It is necessary to communicate information down to the workers and also to send information back up to the top of the organization.

The easiest way to understand this model is to relate it to people. Figure 10-2 shows the human model of factory communications. The workers are at the machine level. The workers monitor and control the production devices that they are responsible for. Theses devices might be air valves and cylinders, gages, and so on. The foreman is at the cell level. The factory supervisors are at the area level, and the plant manager is at the host level. If we think of the typical duties that each person would have at each level it is easy to understand the function of each level in the electronic model. Each of these will be covered in more detail.

Device Level

This level consists of devices such as valves, sensors, motor drives, temperature controllers, and various other I/O devices. These are called field devices. The vast majority of field devices in industry today are very simple devices that have no ability to communicate other than an on or off signal to a PLC or computer.

Recently manufacturers have been adding additional communications capability to field devices. Inexpensive communications chips have been developed for consumer products such as cars. This has made it possible to add communications capability to even the simplest industrial field device. These field devices can be attached to a communications bus. There are a couple of advantages to this. First, it reduces the wiring expenses and complexity. In the past each device had to be connected to the controller. This often made for long wiring runs. It also often led to a maze of wires. A communications bus allows devices to just be "plugged in" to the bus. It also allows devices from any manufacturer that meets the LAN standard to be interchanged. This will be covered in greater detail in Chapter 11.

Machine Level

Think of machines as pieces of equipment that produce or handle product (see Figure 10-3). Some examples of machines are robots, conveyors, computer-controlled machine tools (CNC machines), hard automation, and automated storage/automated retrieval systems (AS/AR). These devices have several things in common. They are all directly involved with the product. Some move the product, others add value during production. Each machine produces valuable data during production. They know how many have been produced, how long it takes to produce the product (cycle time), uptime and downtime, and so on.

In the human model the machine level relates to production personnel. Production personnel add value to the product. They handle the product and add value in some manner. They also fill out paperwork so that management can monitor quality, productivity, and so on.

In the electronic model the machine might be a CNC machine. The CNC produces machined parts very efficiently. In addition to producing parts, the CNC is also producing data. The CNC can track cycle times, piece counts, downtime, etc. If it were used, the data could drastically improve productivity.

Unfortunately, very few manufacturers use the data. In the rest of the cases the data invisibly and continually spill out the back of the machine controller onto the floor or into the infamous bit bucket. The bit bucket is a fictitious object. The fact is that data generated in devices are not used. This is a huge lost opportunity.

One of the other things that machine controllers have in common is a need for data. Each machine needs a program to tell it what, when, and how many to produce. The programs for each machine will look different, but serve the same function. In the human model, people need the same information. What should I work on? How many should I make? What should I do next? Is there enough raw material? Etc. This information is crucial to efficient production. Unfortunately, inaccurate, untimely information is more the rule than the exception in enterprises. In most factories, people are used to coordinate the devices. People start and stop the machines, count pieces, monitor quality, monitor performance, and watch for problems. The importance of accurate, timely information is just as vital when people are involved.

Figure 10-3. Machine level. Machines are production-oriented equipment that add value to a product.

When you think of the machine level think of task-specific equipment that is adding value to the product. Think of production-type tasks. The machine level is where value is added and thus where the enterprise makes its money. It is this level that creates the wealth to support the rest of the organization. Some may be uncomfortable with that thought. The Japanese view the production level as the most important. They do anything they can to make that level more efficient. They use the concept of *kaizen*, or continuous improvement. Accurate, real-time data can be crucial in real improvement.

Cell Level

The cell is a logical grouping of machines used to add value to one or more products. A cell will typically work on a family of similar parts. A cell consists of various dissimilar machines (see Figure 10-4). Each machine typically has its own unique type of program and communication protocol. Machines do not want to communicate with each other. A CNC machine, for example, does not want to communicate with a robot.

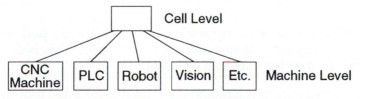

Figure 10-4. Typical cell control scheme. One cell controller is controlling several devices.

Compare this to the human model. The foreman is the cell controller. The foreman's job is to show up early for work and find out what needs to be produced that day. The foreman must then choose the appropriate production person to perform each task that needs to be accomplished. Note that all people have different personalities. A good foreman can communicate well with each individual. The foreman then monitors each employee's performance and by coordinating efforts produces that day's product. The foreman makes sure that people cooperate to get the overall job accomplished.

The purpose of the cell controller is to integrate the various machines into a cooperative work cell. The cell controller must then be able to communicate with each machine in the cell. Even if the devices are not able to communicate with each other, they must be able to communicate with the cell controller. The cell controller must be able to upload/download programs, exchange variable information, start/stop the device, and monitor the performance of each device.

There can be many foremen in a plant, and there can be many cell controllers in a plant. Each can control a group of machines. Each cell controller can talk to other cell controllers and also up to the next level. There are two main types of communication: primitive and complex (typically serial).

Primitive Communications
Some devices do not have the capability to communicate. Some simple PLCs, for example, cannot communicate serially with other devices. In this case primitive methods are used. In the primitive mode the devices essentially just handshake with a few digital inputs and outputs. For example, a robot is programmed to wait until input 7 comes true before executing program number 13. It is also programmed to turn on output 1 after it completes the program. A PLC output can then be connected to input 1 of the robot and output 1 from the robot can be connected to an input of the PLC. Now we have a simple one-device cell with primitive communications. The PLC can command the robot to execute. When the program is complete, the robot will notify the PLC. Note that it is very simple yes or no (binary) information.

Serial Communications
Many devices offer more communications capability. For example, we may need to upload/download programs or update variables. We cannot do that with primitive communications. Most machines offer serial communications capability using the asynchronous communications mode and have an RS-232 serial port available. One would think that any machine with an RS-232 port would easily communicate with any other device with an RS-232 port. This is definitely not true. Each machine may have its own protocol.

The RS-232 standard specifies a function for each of 25 pins. It does not say that any of the pins must be used, however. Some manufacturers use only three, as in Figure 10-5. Some device manufacturers use more than three pins, so some electrical handshaking can take place. In Figure 10-5 no handshaking is taking place. The first computer sends a message whether or not there is another machine there. The computer cable could be unplugged or the computer turned off; the sender would not know. Handshaking implies a cooperative operation. The first computer tells the second that it has a message it would like to transmit. It does this by setting pin 4 (the request to send pin) high. The second machine sees the request to send pin high, and if it is ready to receive, it sets the clear to send pin high. The first computer then knows that the cable is connected, the computer is on, and it is ready to receive. Some devices can be set up to handshake; others cannot.

Fortunately when a machine is purchased it is generally capable of communicating with an IBM personal computer. The user usually just has to open the device manual to the section on communication to find a pinout for the proper cable. It is still difficult and expensive to communicate when a wide variety of devices are involved.

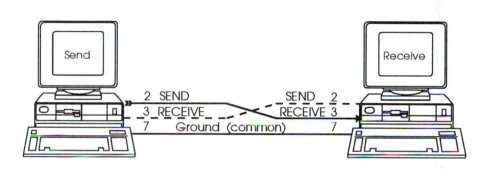

Figure 10-5. Simple RS-232 wiring scheme showing the simplest of RS-232 connections.

When a message is sent using asynchronous communications the message is broken into individual characters and transmitted one bit at a time. The ASCII system is used normally. In ASCII every letter, number, and some special characters have a binary-coded equivalent. There is 7-bit ASCII and 8-bit extended ASCII. In 7-bit ASCII there are 128 possible different letters, numbers, and special characters. In 8-bit ASCII, 256 are possible.

Each character is sent as its ASCII equivalent. For example, the letter A would be 1000001 in 7-bit ASCII (see Figure 10-6). It takes more 7 bits to send a character in the asynchronous model, however. There are other bits that are used to make sure the receiving device knows a message is coming, that the message was not corrupted during transmission, and even bits to let the receiver know that the character has been sent. The first bit sent is the start bit (see Figure 10-7). This lets the receiver know that a message is coming. The next 7 bits (8 if 8-bit is used) are the ASCII equivalent of the character. Then there is a bit reserved for parity. Parity is used for error checking. The parity of most devices can be set up for odd or even, mark or space, or none.

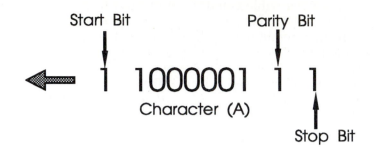

Figure 10-6. How the letter A would be transmitted in the asynchronous serial mode of communications. This example assumes odd parity. There are an even number of ones in the character A, so the parity bit is a one to make the total odd. If the character to be sent has an odd number of ones, the parity bit would be a zero. The receiving device counts the number of ones in the character and checks the parity bit. If they agree, the receiver assumes that the message was received accurately. This is rather crude error checking. Note that two or more bits could change state and the parity bit could still be correct but the message wrong.

Start Bit	Data Bits	Parity Bits	Stop Bit
1 Bit	7 or 8	1 Bit (Odd, Even, Mark, Space, or None)	1, 1.5, or 2 Bits

Figure 10-7. How a typical ASCII character is transmitted.

There are some new standards that will help integrate devices more easily. RS-422 and RS-423 were developed to overcome some of the weaknesses of RS-232. The distance and speed of communications are drastically higher in RS-422 and RS-423. These two standards were developed in 1977.

RS-422 is called balanced serial. RS-232 has only one common. The transmit and receive line use the same common. This can lead to noise problems. RS-422 solves this problem by having separate commons for the transmit and receive lines. This makes each line very noise immune. The balanced mode of communications exhibits lower crosstalk between signals and is less susceptible to external interference. Crosstalk is the bleeding of one signal over onto another. This reduces the potential speed and distance of communications. This is one reason that the distance and speed for RS-422 is much higher. RS-422 can be used at speeds of 10 megabits for distances of over 4000 feet, compared to 9600 baud and 50 feet for RS-232.

RS-423 is similar to RS-422 except that it is unbalanced. RS-423 has only one common, which the transmit and receive lines must share. RS-423 allows cable lengths exceeding 4000 feet. It is capable of speeds up to about 100,000 bps.

RS-449 is the standard that was developed to specify the mechanical and electrical characteristics of the RS-422 and RS-423 specifications. The standard addressed some of the weaknesses of the RS-232 specification. The RS-449 specification specifies a 37-pin connector for the main and a 9-pin connector for the secondary. Remember that the RS-232 specification does not specify what type of connectors or how many pins must be used.

These standards are intended to replace RS-232 eventually. There are so many RS-232 devices that it will take a long time. It is already occurring rapidly in industrial devices. Many PLCs' standard communications are done with RS-422.

Adapters are cheap and readily available to convert RS-232 to RS-422 or vice versa (see Figure 10-8). This can be used to advantage if a long cable length is needed for an RS-232 device (see Figure 10-9).

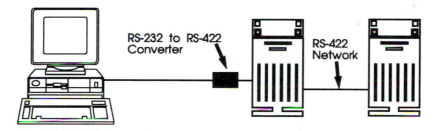

Figure 10-8. Use of a converter to change RS-232 communications to RS-422 communications. Note that the computer is then able to communicate with other devices on its network. The devices on the right are on an RS-422 network.

RS-485 is a derivation of the RS-422 standard. The main difference is that it is a multidrop protocol. This means that many devices can be on the same line. This requires that the devices have some intelligence, however, because the devices must each have a name so that each knows when it is being talked to. Many PLCs and other smart devices now utilize RS-485 protocol. The standard specifies the electrical characteristics of receivers and transmitters connected to the network. The standard specifies a differential signal between -7 to +12 volts. The standard limits the number of stations to 32. This would allow for up to 32 stations with transmission and reception capability, or 1 transmitter and up to 31 receiving stations.

Types of Cell Controllers

PLCs and computers can both be used for cell control applications. In fact the line between PLCs and computers for control is blurring. There are definite advantages and disadvantages for each type of controller.

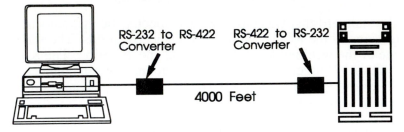

Figure 10-9. Use of two converters. In this case it is being done to extend the cable length. Remember that RS-232 is only reliable to about 50 feet. The use of two converters allows 4000 feet to be covered by the RS-422 and then converted to RS-232 on each end. Note that the speed will be limited by the RS-232.

PLCs as Cell Controllers

The PLC offers some unique advantages as a cell controller. It is easily understood by plant electricians and technicians. If the devices in the cell need to communicate in primitive mode, it is very easy to do with a PLC. If there are other PLCs of the same brand in the cell, it is easy for the PLC to communicate. The data highway of that brand PLC would then be used.

The PLC is not very applicable as a cell controller when there is more than one brand of device in the cell. Typical PLCs do not offer as much flexibility in operator information as computers do, although this is changing rapidly. Graphic terminals and displays are becoming very common for PLCs.

Computers as Cell Controllers

The use of computers as cell controllers is growing very rapidly. The computer offers more flexibility and capability than the PLC. The computer has much more communications capability than that of any PLC. In fact, all device manufacturers want their devices to communicate with a microcomputer. It is difficult to sell a device that does not communicate with an IBM microcomputer. Most device manufacturers do not care to communicate with other brands of PLCs, robots, and so on. They all want to talk to an IBM compatible, however. This makes it much easier for the computer to act as cell controller. It can communicate with each device in the cell.

This communication is usually accomplished by the use of SCADA (supervisory control and data acquisition) software. The concept is that software is run in a common microcomputer to enable communications to a wide variety of devices. The software is typically like a generic building block (see Figure 10-10).

The programmer writes the control application from menus or in some cases graphic icons. The programmer then loads drivers for the specific devices in the application. Drivers are software. A driver is a specific package that was written to handle the communications with a specific brand and type of device. They are available for most common devices and are relatively inexpensive.

The main task of the software is to communicate easily with a wide range of brands of devices. Most enterprises do not have the expertise required to write software drivers to communicate with devices. SCADA packages simplify the task. In addition to handling the communications, SCADA software makes it possible for applications people to write the control programs instead of programmers. This allows the people who best know the application to write it without learning complex programming languages. Drivers are available for all major brands of PLCs and other common manufacturing devices.

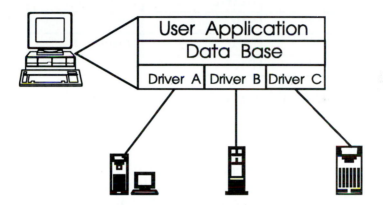

Figure 10-10. How a typical SCADA software package works. Note that the user application defines which variables from the devices must be communicated. These are collected through the drivers and stored in a database that is available to the application. Once the computer has the desired data, it is a relatively easy task to make the data available to other devices.

In general, an applications person would write the specific application using menu-driven software. The software is easy to use. Some are like spreadsheets and some use icons for programming. Instead of specific I/O numbers that the PLC uses, the programmer uses tagnames.

For example, the application might involve temperature control. The actual temperature might be stored in register S20 in the PLC. The applications programmer would use a tagname instead of the actual number. The tagname might be Temp_1 (see Figure 10-11).

This makes the programming transparent. Transparent means that the application programmer does not have to worry very much about what brands of devices are in the application. A table is set up that assigns specific PLC addresses to the tagnames (see Figure 10-11). In theory, if a different brand PLC were installed in the application, the only change required would be a change to the tagname table and the driver. Fortunately, there is more software available daily to make the task of communications easier. The software is more friendly, faster, more flexible, and more graphics-oriented. The data gathered by SCADA packages can be used for statistical analysis, historical data collection, adjustment of the process, or graphical interface for the operator.

Device	Actual Number	Tagname
PLC 12	REG20	Temp_1
PLC 12	REG12	Cycletime_1
PLC 10	S19	Temp_2
PLC 07	N7:0	Quantity_1
Robot 1	R100	Quantity_2

Figure 10-11. What a tagname table might look like.

Area Control

Area controllers are the supervisors (see Figure 10-12). They look at the larger picture. They receive orders from the host and then assign work to cells to accomplish the tasks. They also communicate with other area controllers to synchronize production. Area controllers use synchronous communications methods. Area controllers are attached via local area networks (LANs).

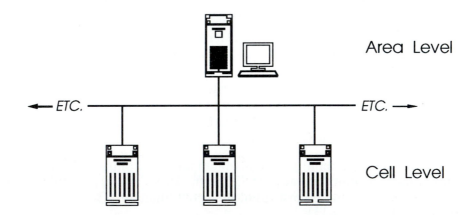

Figure 10-12. Typical area control diagram.

Local Area Networks (LANs)

Local area networks are the backbone of communications networks. The topic of

LANs can be broken down into various methods of classification. We examine three: topology, cable type, and access method.

Topology

Topology refers to the physical layout of LANs. There are three main types of topology: star, bus and ring.

Star Topology

The star style uses a hub to control all communications. All nodes are connected directly to the hub node (see Figure 10-13). All transmissions must be sent to the hub, which then sends them on to the correct node. One problem with the star topology is that if the hub goes down, the entire LAN is down.

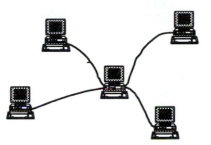

Figure 10-13. Star topology.

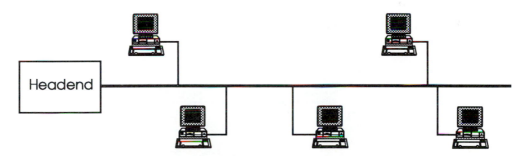

Figure 10-14. Typical bus topology. Each node (communication device) can speak on the bus. The message travels to the head end and is converted to a different frequency. It is then sent back out and every device receives the message. Only the device that the message was intended for pays attention to the message.

Bus Topology

The bus topology is a length of wire into which nodes can be tapped. At one end of the wire is the head end (see Figure 10-14). The head end is an electronic box that performs several functions. The head end receives all communications. The head end then remodulates the signal and sends it out to all nodes on another frequency. Remodulate means that the head end changes the received signal frequency to another and sends it out for all nodes to hear. Only the nodes that are addressed pay attention to the message. The other end of the wire (bus) dissipates the signal. Many industrial networks are based on bus technology.

Ring Topology

The ring topology looks like it sounds. It has the appearance of a circle (see Figure 10-15). The output line (transmit line) from one computer goes to the input line (receive line) of the next computer, and so on. It is a very straightforward topology. If a node wants to send a message, it just sends it out on the transmit line. The message travels to the next node. If the message is addressed to it, the node writes it down; if not, it passes it on until the correct node receives it.

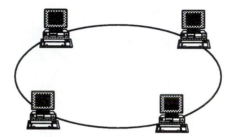

Figure 10-15. Ring topology.

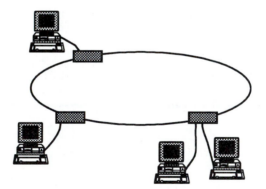

Figure 10-16. Typical ring topology. Note that it does not look like a ring. It looks more like a star topology. It actually is a ring, however. The multiple station access units (MSAUs) allow multiple nodes to be connected to the ring. The MSAU looks like a set of electrical outlets. The computers are just plugged into the MSAU and are then attached to the ring. The MSAU has relays for each port. This allows devices to be plugged in and removed without disrupting the ring.

The more likely configuration of a ring is as shown in Figure 10-16. This style is still a ring; it just does not appear to be. This is the convenient way to wire a ring topology. The main ring (backbone) is run around the facility and interface boxes are placed in line at convenient places around the building. These boxes are often placed in "wiring closets" close to where a group of computers will be attached. These interface boxes are called multiple station access units (MSAUs). They are just like electrical outlets. If we need to attach a computer, we just plug it into an outlet on the MSAU. The big advantage of the MSAU is that devices can be attached/detached without disrupting the ring. Communications are not disrupted at all.

Cable Types

There are four main types of transmission media: twisted pair, coaxial, fiber-optic, and radio frequency. Each has distinct advantages and disadvantages. The capabilities of the cable types are expanding continuously.

Twisted Pair

Twisted-pair wiring is, as its name implies, pairs of conductors (wires) twisted around each other along their entire length. The twisting of the wires helps make them more noise immune. The telephone wires that enterprises have throughout their buildings are twisted pair. There are two types of twisted-pair wiring: shielded and unshielded. The shielded type has a shield around the outside of the twisted pair. This helps further to make the wiring noise immune. The wiring that is used for telephone wiring is typically unshielded. The newer types of un-shielded cable are more noise immune than in the past. Higher speeds are being attained.

Unshielded cable is very cheap and easy to install. The hope is that companies may be able to use spare telephone twisted pairs to run the LAN wiring instead of running new cable. It must be remembered, though, that much of the twisted-pair wire in buildings today is the older, less noise-immune type.

The shielded twisted pair is now used commonly for speeds up to 16 megabits. There are companies that supply LAN cards for unshielded twisted pair that can also run at 16 megabits. There are committees studying the feasibility of up to 100 megabits for shielded twisted-pair wiring. There should soon be a standard for 100 megabits on shielded twisted-pair wiring.

Coaxial Cable

Coaxial or coax is a very common communication medium. Cable TV uses coaxial cable. Coax is broadband, which means that many channels can be trans-mitted simultaneously. Coax has excellent noise immunity because it is shielded (see Figure 10-17).

Figure 10-17. Coaxial cable. Note the shielding around the conductor.

Broadband technology is more complex than *baseband* (single channel). With broadband technology there are two ends to the wire. One of the ends is called the head end. The head end receives all signals from devices that use the line. The head end then remodulates (changes to a different frequency) the signal and sends it back out on the line. All devices hear the transmission but only pay attention if it is intended for their address.

Frequency-division multiplexing is used in broadband technology. The transmission medium is divided into channels. Each channel has its own unique frequency. There are also buffer frequencies between each channel to help with noise immunity. Some channels are for transmission and some are for reception.

Time-division multiplexing is used in the baseband transmission method. This method is also called *time slicing*. There are several devices that may wish to talk on the line (see Figure 10-18). We cannot wait for one device to finish its transmission completely before another begins. They must share the line. One device takes a slice of time, then the next does, and so on. There are several methods of dividing the time on the line.

Figure 10-18. How time-division multiplexing works. Each device must share time on the line. Device 1 sends part of its message and then gives up the line so that another device can send, and so on.

Fiber-Optic Cable

Fiber-optic technology is also changing very rapidly. (See Figure 10-19 for the appearance of the cable.) The major arguments against fiber are its complexity of installation and high cost; however, the installation has become much easier and the cost has fallen dramatically, to the point that when total cost is considered, fiber is not much different for some installations than shielded twisted pair.

Figure 10-19. Fiber-optic cable. Note the multiple fibers through one cable.

The advantages of fiber are its perfect noise immunity, high security, low attenuation, and high data transmission rates possible. Fiber transmits with light, so that it is unaffected by electrical noise. The security is good because fiber does not create electrical fields that can be tapped like twisted pair or coaxial cable. The fiber must be physically cut to steal the signal. This makes it a much more secure system.

All transmission media attenuate signals. That means that the signal gets progressively weaker the farther it travels. Fiber exhibits far less attenuation than other media. Fiber can also handle far higher data transmission speeds than can other media. The FDDI (fiber distributed data interchange) standard was developed for fiber cable. It calls for speeds of 100 megabits. This seemed very fast for a short period of time, but it is thought by many that it may be possible to get 100 megabits with twisted pair, so the speed standard for fiber may be raised.

Plastic fiber cable is also gaining ground. It is cheap and easy to install. The speed of the plastic cable is much less but is constantly being improved.

Radio Frequency

Radio frequency (RF) transmission has recently become very popular. The use of RF has exploded in the factory environment. The major makers of PLCs have RF modules available for their products. These modules use radio waves to transmit the data. The systems are very noise immune and perform well in industrial environments. RF is especially attractive because no wiring needs to be run. Wire is expensive to install. Wire is also susceptible to picking up electrical noise. Wire makes it harder and more expensive to move devices once they are installed. Wireless local area networks (WLANs) are finding a home everywhere from the factory floor to the office.

Wireless technology is becoming very transparent. Wireless technology has become virtually a "black box." A wireless network operates exactly the same as a hardwired network. Devices are simply attached to a transceiver (combination transmitter/receiver). The transceiver does all of the translation and communication necessary to convert the electrical signals from the device to radio signals and then sends them via radio waves. The radio waves are received by another transceiver, which translates them back to network signals and puts them back on the network.

There are two parts to a wireless LAN: an access transceiver and remote client transceivers. The access transceiver is typically a stationary transceiver that attaches to the main hardwired LAN. The remote client transceivers link the remote parts of a LAN to the main LAN via radio waves.

These transceivers can be used as *bridges* or *gateways*. A bridge is used to link two networks that utilize the same or very similar protocols. A gateway is more of a translator. It is used to connect dissimilar network protocols so that two different kinds of networks can communicate.

Spread spectrum technology is used for radio transmissions. Spread spectrum was developed by the US military during World War II to prevent jamming of radio signals and also to make them hard to intercept. Spread spectrum technology uses a wide frequency range.

In spread spectrum technology the transmitted signal bandwidth is much wider than the information bandwidth. Radio stations utilize a narrow bandwidth for transmission. The transmitted signals (music and voice) utilize virtually all of the bandwidth. In spread spectrum the data being translated is modulated across the wider bandwidth. The transmission looks and sounds like noise to unauthorized receivers. It is very noise immune.

A special pattern or code determines the actual transmitted bandwidth. Authorized receivers use the codes to pick out the data from the signal.

The FCC has dedicated three frequency bands for commercial use: 900 MHz, 2.4 GHz, and 5.7 GHz. There is very little industrial electrical noise in these frequencies.

Access Methods

Token Passing

In the token-passing method, only one device can talk at a time. The device must have the token to be able to use the line. The token circulates among the devices until one of them wants to use the line (see Figure 10-20). The device then grabs the token and uses the line.

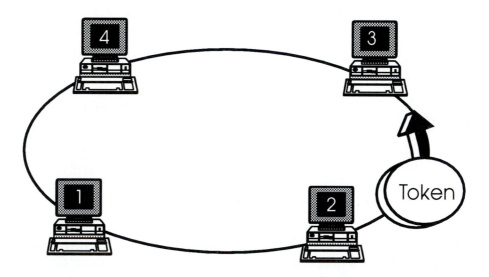

Figure 10-20. Token-passing method.

The device that would like to talk waits for a free token.

The sending station sets the token busy bit, adds an information field, adds the message it would like to send, and adds a trailer packet. The header packet contains the address of the station for which the message was intended. The entire message is then sent out on the line.

     Chapter 10: Overview of Plant Floor Communication

Every station examines the header and checks the address to see if it is being talked to. If not, it ignores the message.

The message arrives at the intended station and is copied. The receiving station sets bits in the trailer field to indicate that the message was received. It then regenerates the message and sends it back out on the line. The original station receives the message back and sees that the message was received. It then frees the token and sends it out for other stations to use.

Token passing offers very reliable performance. It also offers predictable access times, which can be very important in manufacturing. Predictable access is often called deterministic because actual access times can be calculated based on the actual bus and nodes.

Industrial Buses

There are many industrial buses that utilize token passing access schemes.

The first type of token-based industrial LAN we will look at is GE Genius I/O. GE Genius I/O employs a single network master that manages the token passing. Output nodes on the network are capable of logic. Output nodes listen for data from input nodes. Input nodes can send data when the token is available. Genius is considered to be a full peer-to-peer token-passing bus.

ArcNet is another type of token access bus. ArcNet also employs one network master to manage the token passing. All messages have source and destination addresses. In this bus scheme all nodes can have logic capability. ArcNet used to be extensively used in office networks. ArcNet is not very appropriate for single bit information exchange, but it does an excellent job of transmission for medium to long messages (128 to 4096 bits).

Interbus-S, Beckoff Light Bus, Sercos, and Seriplex mode 2 are all buses that are implemented with a single bus master to manage token passing. They also have a host that performs all logic. The host writes data to outputs when the token is free. Input nodes send their information when the token is free. In other words, input nodes send their data to the host when the token is available and the host performs logic based on the data. When the token is available it sends the data to the output nodes.

Seriplex mode 1 employs a single network master to manage the token passing. Input nodes are allowed to send messages when the token is available. Input nodes have exclusive corresponding output nodes. This could be called a limited peer-to-peer token bus.

CSMA/CD

This principle is called CSMA/CD (carrier sense multiple access/collision detection). Carrier sense multiple access means that many nodes can have access to the bus but each must listen to the line (carrier sense) and only talk when it is not busy.

It is somewhat analogous to access to a highway. The more cars there are on the highway, the more difficult it is to get on the highway. The driver who wants to merge onto the highway will have to stop if there is a car present.

Collision detection is a method by which any devices that wish to speak must listen to the carrier signal on the transmission line. If the line is not busy, the device may use the line to communicate. If two or more devices try to use the line at the same time, there is a collision. The collision is detected and the devices back off for awhile and try again. The device just retransmits after a short wait. The wait time for each node is determined by an algorithm that assures that two nodes will almost never have the same wait time.

Just as on the highway there are more collisions as the traffic increases. Typical industrial communications involve low line-utilization percentages. CSMA/CD performs well at these levels.

CSMA/CD could be described as unsolicited messaging. One cannot accurately determine access times for nodes. Access times will vary depending on the amount of traffic and collisions that occur and also on wait times.

CSMA/BA

This access method is called carrier sense multiple access/bitwise arbitration (CSMA/BA). The CSMA is the same as the previous method. The bitwise arbitration means that priorities have been established for individual nodes. In this case if two nodes begin to talk at the same time and a collision occurs the node with the lower address will win the right to talk. The higher addressed node must wait until the line is clear again.

One of the advantages of CSMA/BA is that priorities can be established. We can assure that critical devices will get a higher priority to access than less critical devices. Access times are still not deterministic, however, and are determined by the amount of traffic and the number of collisions.

Industrial buses that utilize CSMA/BA include DeviceNET, SDS and Lonworks. DeviceNET and SDS utilize a single network master that manages the startup operations. During this phase the master takes attendance of the devices. After initial startup the nodes can access the bus when there is a change that they need to communicate. Lonworks uses a single network master that handles startup tasks. After startup nodes can access the bus and freely exchange data. Logic can be performed by multiple masters. Input nodes can also write to output nodes. Input nodes can report changes.

CSMA/xx

This is still a CSMA system. The difference is that a node may not speak now unless it is told to. This is also called solicited messaging. This access method is deterministic. This means that access times are very predictable.

Master/Slave

In this bus access method both the master and slave are nodes. In this scheme the master is in charge. If the master calls out the slave's address the slave responds. The slave does not even have to look at the bus. The bus cannot be busy because the master controls the bus. It cannot be busy if the master has called the slave's address. In this type of access method slave nodes do not pay attention to other slave nodes' messages. Master/slave is very deterministic. Access times can be very accurately determined.

Examples of industrial buses that utilize master-controlled communication access include Sensoplex, ASI, Profibus DP, and Bitbus. They implement a traditional master/slave bus.

Manufacturing Automation Protocol

Imagine what it would be like if there were no standards for consumer goods and appliances. If you decided to switch your brand of refrigerator, you might have to add a transformer to supply the correct voltage, change outlets so that you could plug it in, and so on. Think what your house wiring would be like if every device manufacturer chose its own standard for voltage levels and connectors. It would be a nightmare! That may give you a small idea of what the situation is like in factory communications between devices. MAP (manufacturing automation protocol) was intended to solve many of the problems of connecting devices.

MAP was the idea of General Motors. GM had tens of thousands of devices in the early 1980s, very few of which were able to communicate outside their environment. Every brand and model of device had its own protocol. It would be almost impossible economically to write software drivers for each of the devices, so that all could communicate.

The idea of a standard for communications was born. The thought was that if a standard communications protocol existed and customers required it, manufacturers would produce devices that met the standard. The MAP standard is based on a standard called the open systems interconnect (OSI) model. It is a seven-layer method of standardizing communications among various devices (see Figures 10-21 and 10-22).

MAP had some early problems. The first standard, MAP 2.0, did not specify every layer. There also were not many devices available that were MAP-compatible. The early devices were expensive. Most plants already had substantial investments in other protocols also.

In 1987, Map 3.0 was released. MAP 3.0 addressed many of the complaints and weaknesses of MAP 2.0. MAP 3.0 addresses all seven layers of the OSI model. (See Figures 10-21 and 10-22.) It only allows compatible extensions for six years. This allows manufacturers to plan ahead and not worry about an updated standard that would make their equipment obsolete. In 1992 there were approximately 20 MAP installations in the United States. Some estimates show that Japan has a similar number of installations. MAP never really caught on.

The layers of the OSI model perform the following functions.

Physical layer: encodes and physically transfers the message to another network device.

Link layer: maintains links between devices and performs some error checking.

Network layer: establishes connections between devices attached to the network.

Transport layer: provides for transparent, reliable data transfer between devices.

Session layer: translates names and addresses, provides access security, and also synchronizes and manages the data.

Presentation layer: restores data to and from the standardized format that is used in the network.

Application layer: provides services to the user's application program in a format that it can understand.

OSI Seven-Layer Stack

| Application |
| Presentation |
| Session |
| Transport |
| Network |
| Data Link |
| Physical |

Figure 10-21. OSI model of communications.

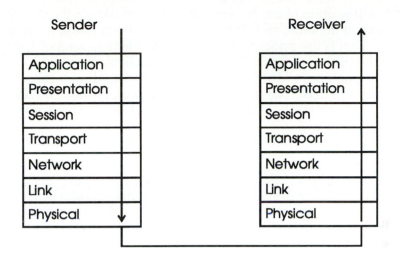

Figure 10-22. Message being sent from one device to a different device with a different protocol. The sender sends the message down through the seven layers. As it moves through the layers it is made generic, broken into transmission blocks, put on the network, and sent to the receiving device. The message moves through the stack at the receiver and the message is changed from generic to the protocol that the receiving device understands. The layers also provide error checking, security, choose the best path for the message, and so on.

Technical Office Protocol

The communication needs of the business portion of a business are seen to be different from those of the factory floor. Communication at this level is usually larger files, such as business data, engineering data, and so on. The TOP (technical office protocol) model was based on the contention access scheme. Boeing Corporation was instrumental in developing the standard.

Manufacturing Message Specification

The concept of MMS (manufacturing message specification) is that every machine does not have to be exactly the same to communicate. Imagine a convention of people from various countries with no common language and no interpreters. The convention would be less than successful. But what if each representative knew a core set of words that had the same meaning for each? They could get by quite well with a reduced vocabulary.

The same is true of machines. There are not too many commands we need to send to a machines. We need to upload/download, start/stop, monitor, update variables, and perform a few more tasks. If every device could handle the commands for these few tasks in the same way they could function very well in an integrated environment. That is the principle behind MMS. There is a set of functions that has been established as a standard. As of this time there are over 80 standard functions. If a machine is MMS-compatible, it can use all or a subset of these functions. This will make communications a much easier task. MMS will reside above the seven-layer stack. Like MAP and TOP, MMS has never been widely implemented.

Host Level

The host-level controller is generally one or more mainframes or minicomputers. The line between mainframe, minicomputer, and personal computer is rapidly blurring. The trend is definitely toward distributed processing. This level is responsible for the business software, engineering software, office communications software, and so on.

The business software is generally an MRP package. MRP stands for manufacturing resource planning. This software is used to enter orders, bills of materials, check customer credit, inventory, and so on. The software can then be used to generate work orders for manufacturing, orders for raw materials and component parts, schedules, and even customer billing. They are being used more and more for planning and forecasting. (They are typically called MRPII now because of the increased emphasis on planning and forecasting.)

The host level is going to be used more and more to optimize operation of the enterprise. Data from the factory floor will be gathered automatically from devices or from operator interface terminals. The host level's task will be to analyze the data to help improve the productivity of the overall system. The host level must also schedule and monitor the daily operations.

The real key to the future will be this data collection. Better business decisions can be made if accurate real-time data are available in a format that people can understand and use.

Figure 10-23 shows the levels of enterprise communications. The task of integrating these devices is rapidly becoming easier. Software and hardware are making remarkable advances in ease of use.

The rapid advancements in computers and networking are creating vast changes in industrial communication hierarchies. Figure 10-24 shows an example of what is occurring in networking. There are two important changes that Figure 10-24 shows. First you will note that the area level has disappeared. In reality the area level and host level have combined because of the power of micro and mini-computers. The host level now connects to a network. Cell controllers also connect to the network. The second important change is at the machine and device level. Note that the availability of device networks will radically change communication at this level. More and more devices will be connected to LANs. Note that some of the devices are still connected directly to the machine controller. There will certainly be a mix of device communications capabilities for the foreseeable future. But as the communications capabilities of devices rise and prices fall LANs will become more and more prevalent.

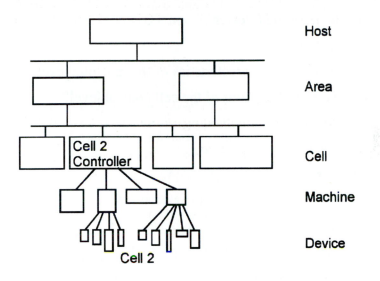

Figure 10-23. Typical host-level control. It is also called the enterprise level.

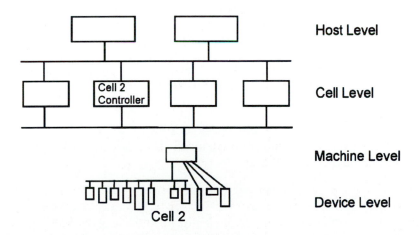

Figure 10-24. An example of networking.

Questions

1. In what ways are hardware communication hierarchies like human organizations?

2. What are the primary characteristics of the device level?

3. Describe the term *field device*.

4. Describe the field bus.

5. What are the primary characteristics of the machine level?

6. What are the main functions of the cell control level?

7. Describe the term *serial communications*.

8. Describe the term *asynchronous*.

9. Thoroughly explain the term *SCADA*.

10. What are the main functions of the area level?

11. What are the main functions of the host level?

12. Why collect so much data from the plant floor?

13. Complete the following table.

	Speed	Length Limit	Balanced or Unbalanced	Special Characteristics
RS-232				
RS-422				
RS-423				
RS-485				

Chapter 11

Industrial Networks

There are several types of industrial communication networks available. These networks are designed to allow industrial devices such as valves, sensors, motor starters and so on to be networked.

OBJECTIVES

Upon completion of this chapter, the student will be able to:

> *Describe at least two types of industrial communication networks.*

> *Describe the advantages of industrial communication networks.*

> *Define such terms as CAN, Fieldbus, DeviceNET, PROFIBUS.*

> *Describe the various levels of industrial networks available.*

> *Describe applications that would be appropriate for a network.*

OVERVIEW OF INDUSTRIAL NETWORKS

Industrial networks are becoming very prevalent in industrial control. They are sometimes called field buses or industrial buses. These networks have many similarities to the conventional office computer network. Office networks allow many computers to communicate without being directly hooked to each other. Each computer has only one connection to the network. Imagine if every computer had to be physically connected to every other computer. So networks minimize the amount of wiring that needs to be done. Networks also allow devices such as printers to be shared. Industrial networks share some of the same advantages.

Imagine a complex automated machine that has hundreds of I/O devices. Now imagine the time and expense of connecting each and every I/O device back to the controller. Conduit has to be bent and mounted as well as the hundreds of wires that need to be run. This can require hundreds of wires (or more) in a complex system and can involve long runs of wire. Typical devices such as sensors and actuators now require that two or more wires are connected at the point of use and then run to an I/O card at the control device. Industrial networks eliminate this need so only a single twisted-pair wire bus needs to be run (see Figure 11-1). All devices can then connect directly to the bus. This could also be called distributed I/O.

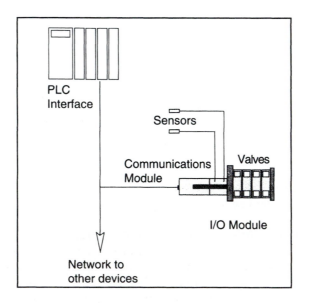

Figure 11-1. An example of a simple bus. Note the communications module and the I/O module. Courtesy Parker Hannefin Corporation.

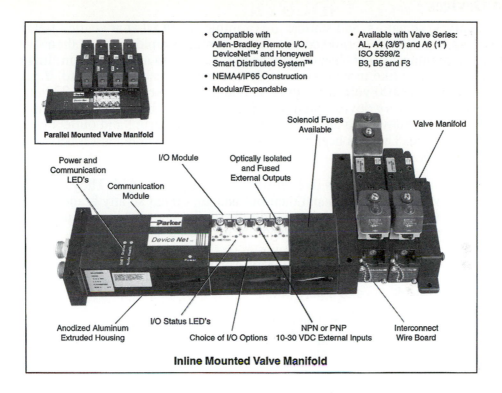

- Compatible with Allen-Bradley Remote I/O, DeviceNet™ and Honeywell Smart Distributed System™
- NEMA4/IP65 Construction
- Modular/Expandable

- Available with Valve Series: AL, A4 (3/8") and A6 (1") ISO 5599/2 B3, B5 and F3

Parallel Mounted Valve Manifold

Solenoid Fuses Available

Valve Manifold

Power and Communication LED's

I/O Module

Optically Isolated and Fused External Outputs

Communication Module

Anodized Aluminum Extruded Housing

I/O Status LED's

Choice of I/O Options

NPN or PNP 10-30 VDC External Inputs

Interconnect Wire Board

Inline Mounted Valve Manifold

Figure 11-2. A DeviceNET valve manifold. Courtesy Parker Hannefin Corporation.

Multiple devices can even share one connection with the bus in some cases. Several manufacturers have developed I/O blocks that allow multiple I/O points to share one connection to the bus. One example would be valve manifolds (see Figure 11-2). This particular unit could have up to 32 valves and up to 32 I/O with only two connections. Wiring and installation costs are drastically reduced when industrial networks are used.

The cost of field bus devices is higher than that of conventional devices. The field bus devices gain a cost advantage when one considers the labor cost of installation, maintenance and troubleshooting. The cost benefit received from using distributed I/O is really in the labor saved during installation and startup. There is also a material savings when one considers all the wire that does not have to be run. There is a tremendous savings in labor when one considers that only a fraction of the number of connections need to be made and only a fraction of the wire needs to be run. A field bus system is also much easier to troubleshoot than a conventional system. If a problem exists only one twisted-pair cable needs to be checked. A conventional system might require the technician to "sort out" hundreds of wires.

Field Devices

Sensors, valves, actuators, and starters are examples of I/O that are called field devices. The capabilities of field devices have increased rapidly as has their ability to communicate. This has led to a need for a way to network them. Imagine what it would be like in your home if there was no electrical standard. Your television might use 100 volts at 50 cycles. Your refrigerator might use 220 at 50 cycles, and so on. Every device might use different plugs. It would be a nightmare to change any device to another brand because the wiring and outlets would need to be changed. This has been the case in terms of communication between industrial devices.

Simple industrial digital devices like sensors can be interchanged. They typically operate on 24 volts or 110 volts and their outputs are digital so any brand can be used. In other words, we could replace a simple digital photo sensor from one manufacturer with a different brand because it is only an on/off signal.

Figure 11-3 shows a simple industrial network with several field devices attached to it. Note the I/O block (I/O concentrator) and the valve manifold. Both are used to connect multiple devices to a network.

Industrial applications often need analog information such as temperature, pressure, and so on. These devices typically convert the analog signal to a digital form and then need to communicate that to a controller. They may also be able to pass other information such as piece counts, cycle times, error codes and so on. Devices such as drives have also gained capability. Parameters such as acceleration and other drive parameters can be sent by a computer or PLC to alter the way a drive operates. Every device manufacturer has had different communication protocols until now for that communication. This required that a separate network had to be made for each brand of device.

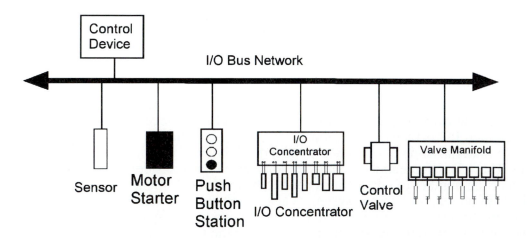

Figure 11-3. What a simple industrial network might look like.

TYPES OF INDUSTRIAL BUSES

There are two basic categories that we could divide industrial bus networks into: device and process. Device-type buses are intended to handle the transmission of short messages, typically a few bytes in length. Most devices in a device bus are discrete. They would be devices such as sensors, push buttons, limit switches, etc. Many discrete buses can also utilize some analog devices as well. They would typically be devices that only require a few bytes of information transmission. Examples would be some temperature controllers, some motor drives, thermo-couples, etc. Since the transmission packets are small, device buses can transmit data packets from many devices in the same amount of time it would take to transmit one large packet of data on a process bus.

Industrial buses range form simple systems that can control discrete I/O (device type bus) to buses that could be used to control a whole plant (process bus). Figure 11-4 shows a graph of the capabilities of several of the more common industrial buses. The left of the graph begins with simple discrete I/O control and goes up in capability to plant control. Block I/O would be devices like manifolds that would have several valves on one block and only two to four wires would need to be connected to control all of the I/O. By using an industrial bus only 2 to 4 wires would have to be run instead of a few dozen.

An example of a smart device would be a digital motor drive to which we could send parameters and commands serially. The peer level would be the capability for peer controllers to communicate with each other. The next level would be cell control; next would be control of a line of cells; and the top level would be overall plant control.

Device Buses

Device buses can be broken into two categories: bit-wide and byte-wide buses. Byte-wide buses can transfer 50 or more bytes of data at a time. Bit-wide buses typically transfer 1 to 8 bits of information to/from simple discrete devices. Byte-type systems are excellent for higher level communication, and bit-type systems are best for simple, physical level I/O devices such as sensors and actuators.

Process Buses

Process buses are capable of communicating several hundred bytes of data per transmission. Process buses are slower because of their large data packet size. Most analog control devices do not require fast response times. Process control-lers typically are smart devices. They are typically controlling analog types of variables such as flow, concentration, temperature, etc. These processes are typically slow to respond. Process buses are used to transmit process parameters to process controllers. Most devices in a process bus network are analog.

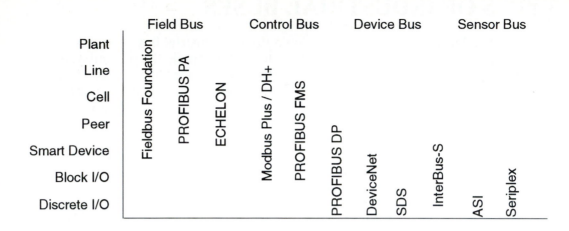

Figure 11-4. A comparison of some of the more common industrial buses.

PROCESS BUS STANDARDS

There are two main organizations working on establishing process bus standards: the Fieldbus Foundation and the PROFIBUS (Process Field Bus) Trade Organization.

WorldFIP and ISP joined forces to become Fieldbus Foundation (FF) in September of 1994, to speed up the process of completing the fieldbus standard. The Fieldbus Foundation is a group of control and process engineering suppliers, manufacturers and end-users from around the world. The organization is dedicated to accelerating the development of a single interoperable fieldbus.

DEVICE BUSES

Actuator Sensor Interface (ASI) Bus

AS Interface is a versatile, low-cost "smart" cabling solution that has been designed specifically for use in low-level automation systems. It is designed to be easy to install, operate and maintain and simple to reconfigure. It was developed by a consortium of European companies and the technology is now owned by AS International, an independent organization. It was specifically designed to ease installation and startup, and have a low cost per node. Cost savings of up to 40% are possible in typical automation situations. AS Interface technology has been submitted for approval under the proposed IEC 947 standard. There are user groups in eight European countries. Figure 11-5 shows a few of the characteristics of ASI.

Network Size	31 slaves
Network Length	100 meters, 300 with a repeater
Topology	Bus, multidrop, star
Physical Media	2-wire cable
Data Transfer	Master and slaves have 4 bits
Bus Access	Master/slave with cyclic polling

Figure 11-5. ASI bus characteristics table.

In an AS Interface network, data and power are carried over a single cable that links up to 31 slave devices to each master and up to 124 digital input and output nodes to a PLC or computer controller. Devices such as LEDs and indicators can be powered directly because the cable can handle up to 8 amps. Higher power devices such as actuators are powered separately from a second cable. Worst case cycle times for a full network are 5 ms per slave; faster times are possible when fewer devices are connected.

The cable is mechanically polarized and cannot be connected wrongly (foolproof). The cable is also self-healing, enabling I/O nodes to be disconnected and reconnected easily.

The network utilizes cyclic polling of every network participant (sensor or actuator). With 31 slaves the cycle time is 5 ms. Error detection is used to initiate a message repeat signal.

BITBUS

BITBUS is a serial communication system for industrial use. BITBUS is based on an RS-485 party line (several communication stations on the same twisted pair of wires). BITBUS is optimized for the transmission of short, real-time messages. A BITBUS communication network utilizes a master/slave access method. Each slave has its own network address which makes it uniquely identifiable in the network. The master operates the network by polling the slaves. The slaves may only respond when polled by the master. Repeaters must be used if long cable runs are necessary or more than 28 network stations are used.

BITBUS was officially accepted as international standard IEEE-1118 in 1991. The IEEE-1118 standard incorporates additional capabilities including the following new functions:

broadcast and multicast (group addressing)

additional network management and monitoring

master-transfer (master-passing)

new communications media (coax, twinax, fiber-optics)

A command will always be sent from the master to a specific slave or a group of slaves. The reply is a data frame coming from a slave as response to a command. If a reply is not ready as the slave receives a command, a sequence of control frames are sent between the master and the slave. BITBUS is a registered trade mark of Intel Corporation.

InterBus-S

The InterBus-S bus standard is designed to network devices such as actuators and sensors. InterBus-S is optimized for the throughput demands of sensor/actuator networks. It can be used for discrete and analog devices. The InterBus-S can have 256 nodes. Nodes can have built-in I/O interfaces. The network can handle up to 4096 field devices at a speed of 500 kbaud. I/O data is transmitted in frames that provide simultaneously and predictable updates to all devices on the network. Figure 11-6 shows a few of the characteristics of InterBus-S.

This bus uses a master/slave access method. A PLC, computer or other control device typically acts as the master. The master continuously scans the I/O devices. All inputs are read each scan after which output data is written. I/O addresses do not have to be set manually.

InterBus-S scan cycle times are deterministic. An equation, which takes all network variables into account, can be used to calculate scan time to the microsecond. The majority of the scan time is dependent on the number of process data words. Figure 11-7 approximates scan time based solely on the number of process data words.

I/O addresses are automatically determined based on the device's physical location. The InterBus-S uses ISO layers 1, 2, and 7. The length of this network can be up to 42,000 feet.

Network Size	256 nodes
Network Length	400 m per segment, 12.8 Km total
Topology	Segmented with T drops
Physical Media	Twisted pair, fiber, and slip-ring
Data Transfer Size	512 bytes
Method	Master/slave with total frame transfer

Figure 11-6. InterBus-S characteristics.

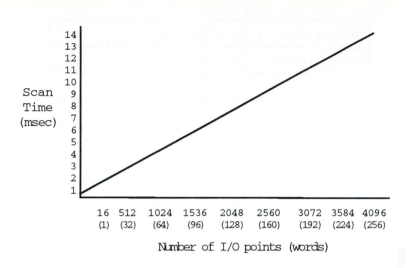

Figure 11-7. Scan times for InterBus-S.

InterBus-S is intended to be an open systems approach to a high performance, ring-based device network. Secure transmissions are ensured by the protocol's CRC error checking. Diagnostics capability exists to pinpoint the cause and location of errors. This helps provide maximum network uptime.

LonWorks

LonWorks has been used for a wide variety of applications. It has been used for very simple control tasks to very complex applications requiring thousands of devices. LonWorks networks can range in size from two to 32,000 devices. Figure 11-8 shows a few of the characteristics of LonWorks. LonWorks has been extensively used for building control. It has been successfully used for slot machines, home control, petroleum refining plants, aircraft, skyscraper building control, and so on.

LonWorks networks do not have a central control or master-slave arrangement. LonWorks uses a peer-to-peer architecture. Intelligent control devices called nodes have their own intelligence and can communicate with each other using a common protocol. Each contains intelligence that is able to implement the protocol and perform control functions.

Nodes typically perform simple tasks. Devices such as switches, sensors, relays, drives, motion detectors and other home security devices, and so on may all be nodes on the network. The overall network performs a complex control application, such as controlling a manufacturing line, securing your home, or automating a building.

Network Size	32,000/domain
Network Length	2000 m at 78 Kbps
Topology	Bus, loop, star
Physical Media	Twisted pair, fiber
Data Transfer Size	228 bytes
Method	Master/slave peer-to-peer

Figure 11-8. Characteristics of a LonWorks bus.

Controller Area Network (CAN)

CAN was originally developed by Bosch for the European automobile network. It was designed to replace expensive wire harnesses with a low-cost network. There are many safety concerns with automobiles such as antilock brakes and airbags, so CAN was designed to be high-speed and very dependable. CAN is also found in commercial products. All of the major European car manufacturers are developing models using CAN to connect the various electronic components.

The SAE (Society of Automotive Engineers) developed a standard called SAE J1939. This standard is designed to give "plug and play" capabilities to car owners. The standard allocates CAN identifiers to different purposes. The standard uses extended (29-bit) identifiers.

CAN was designed to be a high-integrity serial data communications bus for real-time applications. Electronic chips were designed and built to implement the CAN standard. Demand for CAN chips has exploded. In 1994 more than 4 million CAN chips were shipped. Over 10 million chips were expected to be shipped in 1996. This demand has led to the wide availability of reasonably priced CAN chips. CAN chips are typically 80 to 90% cheaper than chips for other networks. The CAN protocol is now being used in many other industrial automation and control applications

CAN is a broadcast bus-type system. It can operate up to 1 Mbit/sec. There are many controllers available for CAN systems. Messages are sent across the bus. The messages (or frames) can be variable in length. They can be between 0 and 8 bytes in length. Each frame has an identifier. Each identifier must be unique. This means that two nodes may not send messages with the same identifier. Data messages transmitted from a node on a CAN bus do not contain addresses of either the transmitting node or of any intended receiving node. All nodes on the network receive each message and perform an acceptance test on the identifier to determine if the message is relevant to that particular node. If the message is relevant, it will be processed; otherwise it is ignored. This is known as multicast.

The node identifier also determines the priority of the message. The lower the numerical value of the identifier, the higher the priority. This allows arbitration if two (or more) nodes compete for access to the bus at the same time.

The higher priority message is guaranteed to gain bus access. Lower priority messages are retransmitted in the next bus cycle, or in a later bus cycle if there are other, higher priority messages waiting to be sent.

The two-wire bus is usually a shielded or unshielded twisted pair.

The ISO 11898 standard recommends that bus interface chips be designed so that communication can still continue (but with reduced signal to noise ratio) even if:

Either of the wires in the bus is broken

Either wire is shorted to power

Either wire is shorted to ground

DeviceNET

DeviceNET is intended to be a low-cost method to connect devices such as sensors, switches, valves, bar-code readers, drives, operator display panels, and so on to a simple network. This can help drastically reduce wiring expenses. By adhering to a standard it is also possible to interchange devices from different manufacturers (see Figures 11-9 and 11-10). This means that one could replace a device from one manufacturer with a similar device from a different manufacturer. DeviceNET allows simple devices to be interchanged and makes interconnectivity of more complex devices possible.

In addition to reading the state of discrete devices, DeviceNET is able to report temperatures, to read the load current in a motor starter, to change the parameters of drives, and so on.

DeviceNET is an open network standard. The standard is not proprietary. It is open to any manufacturer. Any manufacturer can participate in the Open DeviceNET Vendor Association (ODVA), Inc. ODVA is an independent supplier organization that manages the DeviceNET specification and supports the worldwide growth of DeviceNET. Figure 11-11 shows a few of the characteristics of DeviceNET. ODVA works with vendors and provides assistance through developer tools, developer training, compliance testing and marketing activities. ODVA publishes a DeviceNET product catalog.

DeviceNET is based on the CAN standard. DeviceNET is a broadcast-based communications protocol. DeviceNET uses the CAN chip, which helps keep costs very low.

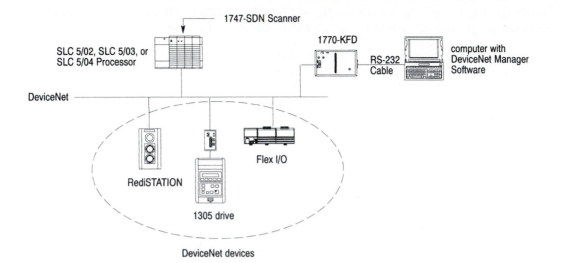

Figure 11-9. An example of a DeviceNET network. Courtesy Rockwell Automation/Allen-Bradley Company Inc.

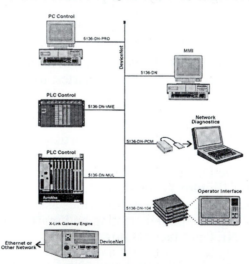

Figure 11-10. Example of a DeviceNET network. Courtesy S-S Technologies Inc.

Network Size	Up to 64 nodes
Network Length	500 meters at 125 Kbps 250 meters at 250 Kbps 100 meters at 500 Kbps
Topology	Linear (trunkline/dropline). Power and signal on the same cable.
Physical Media	Twisted pair for signal and power
Data Packets	0 to 8 Bytes
Method	Peer-to-peer with multicast (one-to-many): master/slave special case and multi-master: polled or change-of-state.
Bus Access	Carrier sense/multiple access

Figure 11-11. DeviceNET characteristics.

DeviceNET supports strobed, polled, cyclic, change-of-state and application-triggered data movement. The user can choose master/slave, multi-master and peer-to-peer or a combination configuration depending on device capability and application requirements.

Higher priority data gets the right-of-way. This provides inherent peer-to-peer capability. If two or more nodes try to access the network simultaneously, a bit-wise nondestructive arbitration mechanism resolves the potential conflict with no loss of data or bandwidth. By comparison, Ethernet uses collision detectors, which result in loss of data and bandwidth because both nodes have to backoff and resend their data.

Peer-to-peer data exchange means that any DeviceNET product can produce and consume messages. Master/slave operation is defined as a proper subset of Peer-to-Peer. A DeviceNET product may behave as a client or a server or both. A DeviceNET network may have up to 64 media access control identifiers or MAC IDs (node addresses). Each node can support an infinite number of I/O. Typical I/O counts for pneumatic valve actuators are 16 or 32.

Change-of-State and Cyclic Transmission

Change-of-state means that a device reports its data only when the data changes. To be sure the consuming device knows that the producer is still alive and active, DeviceNET provides an adjustable, background heartbeat rate. Devices send data whenever their data changes or the heartbeat timer expires. This keeps the connection alive and let the consumer know that the data source is still alive and active. The heartbeat timing prevents talkative nodes from dominating the network. The device generates the heartbeat, which frees the controller from having to send a nuisance request periodically just to make sure that the device is still there.

The cyclic transmission method can reduce unnecessary traffic on a network. For example, instead of a slow changing temperature being scanned many times every second, devices can be set up to report their data on a regular basis that is adequate to monitor their change.

Device Profiles

The DeviceNET specification promotes interoperability of devices by specifying standard device models. All devices of the same model must support common identity and communication status data. Device-specific data is contained in device profiles that are defined for various types of devices. This assures that devices from multiple manufacturers such as push buttons, valves, starters, and so on, that comply with the device type profile will be interchangeable. Manufacturers can offer extended capabilities for their devices but the base functionality must be the same.

Smart Distributed System (SDS)

The Smart Distributed System, developed by Honeywell's Micro Switch Division, is a bus system for intelligent sensors and actuators. SDS is a CAN-based network. Figure 11-12 shows a few of the characteristics of SDS.

The Smart Distributed System is based on the CAN protocol. The Smart Distributed System is uniquely and completely open. It works with any PLC or PC control device. The Smart Distributed System protocol will even accommodate peer-to-peer communication.

Device Diagnostics

Smart Distributed System devices can have diagnostics capability designed into them. For example, a photoelectric sensor could send warning messages if its lens is dirty or if it was out of alignment.

Device Functions

Smart Distributed System devices are intelligent and can be set up using PC-based software programs to perform high-level functions that non-SDS devices cannot do. Devices can perform simple control functions. This allows the host to concentrate on errors if they occur. Smart Distributed System device functions include:

> *On-delay, off-delay*
>
> *Normally open or normally closed (switches and sensor type devices)*
>
> *Light operate or dark operate (photoelectric controls)*
>
> *Number of operations count*
>
> *Batch counter*
>
> *Motion or jam detection*
>
> *Number of power cycles count*

Network Size	64 nodes, 126 addresses
Network Length	1600ft at 125Kbps, 800ft at 500 Kbps, 400ft at 1Mbps
Topology	Trunkline/dropline, peer-to-peer
Physical Media	Twisted pair for signal and power
Message Length	8-byte variable message
Method	Carrier sense/ multiple access

Figure 11-12. Characteristics of an SDS bus.

Seriplex

The Seriplex protocol is used by many manufacturers of actuators, sensors, and I/O blocks that devices can be connected to. Seriplex control buses can transmit digital and analog I/O signals in real time for control and data acquisition. Fieldbus type communication systems are best suited to transmitting large information data packets. Seriplex is intended for shorter data transmissions at the floor level. Seriplex resides primarily at the physical device level and provides deterministic, real-time I/O communication that is needed by most control systems. Seriplex leaves the higher level communication, which doesn't require fast throughput rates, to other higher level bus protocols.

The Seriplex bus utilizes a deterministic access method. Seriplex utilizes a serial multiplexed, intelligent, distributed network. Seriplex can use master/slave and peer-to-peer I/O control and logic. It features distributed and local I/O capability on the same bus.

A Seriplex network cable either connects directly to devices that contain an embedded Seriplex ASIC (application-specific integrated circuit) or through I/O blocks that contain an ASIC chip. Seriplex uses a low voltage, four-wire cable.

ASIC chips provide communication, logic, and addressing capability. ASIC chips can be put in almost any device, actuator, or sensor. ASIC chips can support communication between I/O devices and the host CPU over the control bus. The control bus can communicate using serial communications such as RS-232 and RS-485 or digital and analog device communication.

The Seriplex control bus can control over 7,000 digital I/O points, or 480 analog channels (240 input plus 240 output), or a combination of discrete and analog I/O over one four-wire bus. Figure 11-13 shows some of the characteristics of a Seriplex network.

The Seriplex control bus may be configured ring, star, multidrop, loop back or in any combination desired.

Network Size	510 non-multiplexed devices (255 input, 255 output), 7706 discrete multiplexed, or 480 analog multiplexed
Network Length	5000 feet
Topology	Tree, loop, multidrop, or any combination
Physical Media	4-wire shielded cable, 2 shielded for data and 2 unshielded for power
Data Transfer Size	1 to 255 bits
Method	Master/slave, peer-to-peer
Bus Access	Token passing

Figure 11-13. Characteristics of a Seriplex network.

Addressing, communications capability and logic functions are programmed directly into the EE (electrically erasable) elements. The communication capability is in the I/O block or device. It does not rely on a microprocessor for this function. This reduces the cost.

There is no host CPU in the peer-to-peer mode. The events occurring at each device can be communicated directly between I/O blocks or devices containing an ASIC. These events are acted upon at the device level and are based on the logic functions programmed into the ASICs.

The data signal utilizes large signal swings (12 volts peak-to-peak) and large hysteresis to maintain data integrity. There also is a provision for redundancy at the inputs to provide even further data protection, plus a bus fault detection scheme for additional data integrity.

I/O blocks are available from several manufacturers. I/O blocks allow sensors and outputs to be connected to a central point close to where they are used. AC and DC I/O termination blocks are available with from two inputs or two outputs per block, to as many as 8 inputs or 8 outputs per block. There are also combination blocks with two inputs and two outputs per unit. I/O termination blocks are typically used as smart terminal blocks for connecting dumb devices to the Seriplex bus.

Conditioning modules are also available. They can be plugged into I/O blocks to add analog to digital (ADC) and digital to analog (DAC) conversion capability. Analog (A/D) input blocks can convert an analog signal to a digital value for transmission over the control bus. Converters are available in 8, 12 and 16 bits resolution.

PROCESS BUSES

Fieldbus

The Fieldbus Foundation's Fieldbus supports high-speed applications in discrete manufacturing such as motor starters, actuators, cell control, and remote I/O. Fieldbus devices can transmit and receive multivariable data and can also communicate directly with each other over the bus. Performance of the system is enhanced due to the ability to communicate directly between two field devices rather than having to communicate through the control system.

Fieldbus is a new digital communications network that will be used in industry to replace the existing 4-20 mA analog signal. The network is a digital, bidirectional, multidrop, serial-bus communications network used to link isolated field devices, such as controllers, transducers, actuators and sensors.

Each field device has low cost computing power installed in it, making each device a "smart" device. Each device will be able to execute simple functions on its own such as diagnostic, control, and maintenance functions as well as providing bidirectional communication capabilities. With these devices not only will the engineer be able to access the field devices, but they are also able to communicate with other field devices. In essence Fieldbus will replace centralized control networks with distributed-control networks.

Foundation Fieldbus is an open bus technology that is compatible with Instrument Society of America (ISA) SP50 standards. It is an interoperable, bi-directional communications protocol based on the International Standards Organization's Open System Interconnect (OSI/ISO) seven-layer communications model. The Foundation protocol is not owned by any individual company, or controlled by any nation or regulatory body. Foundation Fieldbus allows logic and control functions to be moved from host applications to field devices. The Foundation Fieldbus protocol was developed for critical applications with hazardous locations, volatile processes and strict regulatory requirements. Each process cell requires only one wire to be run to the main cable, with a varying number of cells available. The Fieldbus protocol enables multiple devices to communicate over the same pair of wires. New devices can be added to the bus without disrupting control. Figure 11-14 shows a few of the characteristics of Fieldbus.

Fieldbus eases system debugging and maintenance by allowing on-line diagnostics to be carried out on individual field devices. Operators can monitor all of the devices included in the system and their interaction.

Measurement and device values are available to all field and control devices in engineering units. This eliminates the need to convert raw data into the required units. This also frees the control system to perform other tasks.

With Fieldbus technology, field instruments can be calibrated, initialized, and operated directly over the network. This reduces time for technicians, operators and maintenance personnel.

Network Size	240 per segment, 6500 segments
Network Length	1900 m at 31.25K or 500m at 2.5M
Topology	Multi-drop with bus powered devices
Physical Media	Twisted pair
Data Transfer Size	16.6 M objects/device
Method	Client/server, publisher/subscriber
Bus Access	Centralized scheduler

Figure 11-14. Characteristics of a Fieldbus network.

PROFIBUS

PROFIBUS specifies the functional and technical characteristics of a serial fieldbus. This bus interconnects digital field devices in the low (sensor/actuator level) up to the medium (cell level) performance range. The system contains master and slave devices. Masters are called active stations. Slave devices are simple devices such as valves, actuators, sensors, etc.

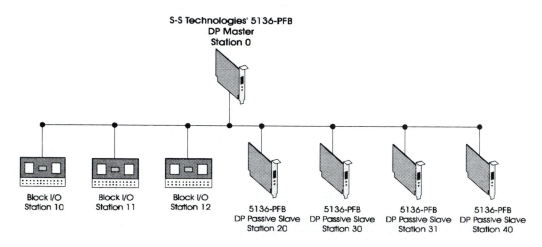

Figure 11-15. The use of an S-S Technologies PC card to act as a PROFIBUS DP master. Courtesy S-S Technologies Inc.

Chapter 11: Industrial Networks

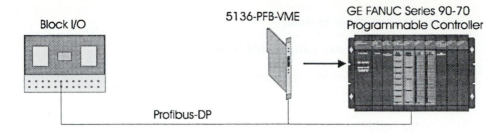

Block I/O   5136-PFB-VME   GE FANUC Series 90-70
                            Programmable Controller

Profibus-DP

Figure 11-16. The use of a S-S Technologies PC card to enable
a GE Fanuc PLC to communicate with and control PROFIBUS
DP I/O. Courtesy S-S Technologies Inc.

Slaves can only respond to received messages, or if requested by the master, send messages. Master devices have the capability to control the bus. When the master has the right to access the bus it may send messages without a remote request. Figure 11-15 shows an example of a PROFIBUS network with a PC card used as a PROFIBUS master. Figure 11-16 shows a GE Fanuc communicating with PROFIBUS I/O by using a PC card.

The PROFIBUS Access Protocol includes the token-passing method for communication between complex stations (masters) and the master-slave method for communication between complex stations and simple peripheral devices (slaves). It is circulated in a maximum token rotation time between all masters. This maximum time is user configurable. PROFIBUS uses the token-passing method only between masters. Figure 11-17 shows some of the characteristics of a PROFIBUS network.

The master-slave method allows the master that currently has the token to communicate with slave devices. Each master has the ability to transmit data to the slaves and receive data from the slaves.

Network Size	Up to 127 nodes
Network Length	24 Km
Topology	Line, star, ring
Physical Media	Twisted pair or fiber
Data Transfer	244 bytes
Method	Master/slave, peer-to-peer
Bus Access	Token passing

Figure 11-17. Characteristics of a PROFIBUS network.

PROFIBUS allows the user choices of access method:

a pure master-slave system,

a pure master-master system (token passing),

a system with a combination of both methods, also called a hybrid system.

CONTROLNET™

ControlNet™ is a high-speed, deterministic network developed by Allen-Bradley for the transmission of critical application information. Allen-Bradley has a 3-level model of communications (see Figure 11-18). The information layer is for plantwide data collection and maintenance.

The second level is the automation and control layer. This layer is used for real-time I/O control, messaging and interlocking. The lowest level is the device layer. The device layer is used to cost effectively integrate low-level devices into the overall enterprise. Figure 11-18 shows an Ethernet network at the information level. ControlNet™ is used at the second level for automation and control. The third level utilizes DeviceNet to communicate.

ControlNet™ is deterministic. ControlNet™ has very high throughput, 5 Mbit/sec for I/O, PLC interlocking, peer-to-peer messaging and programming. This network can perform multiple functions. Multiple PLCs, man/machine interfaces, network access by PC for programming and troubleshooting from any node can all be performed on the network. The capability of ControlNet™ to perform all these tasks can reduce the need for multiple networks for integration. Figure 11-19 shows an example of integrating various networks.

ControlNet™ is compatible with Allen-Bradley PLCs, I/O and software. ControlNet™ supports bus, star, or tree topologies. It utilizes RG6 cable, which is identical to cable television cable. This means that taps, cable and connectors are all easy to obtain and very reasonable in price. ControlNet™ also has a dual-media option (see Figure 11-20). This means that two separate cables can be installed to guard against failures such as cut cables, loose connectors, or noise. The figure also shows some of the types of computers and controllers that can be integrated as well as the wiring.

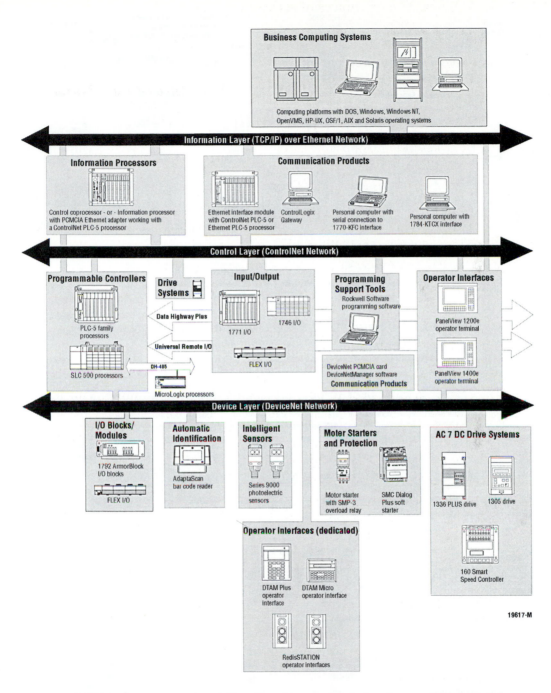

Figure 11-18. The Allen-Bradley communication architecture model. Courtesy Rockwell Automation/Allen-Bradley Company Inc.

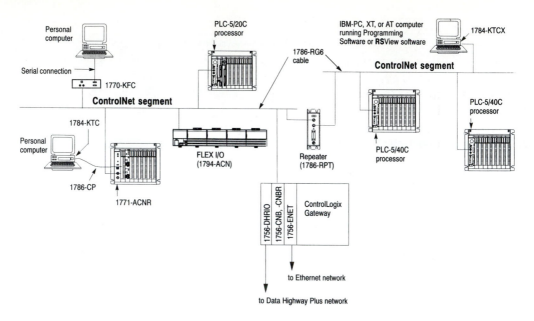

Figure 11-19. Example of integrating various networks. Courtesy Rockwell Automation/Allen-Bradley Company Inc.

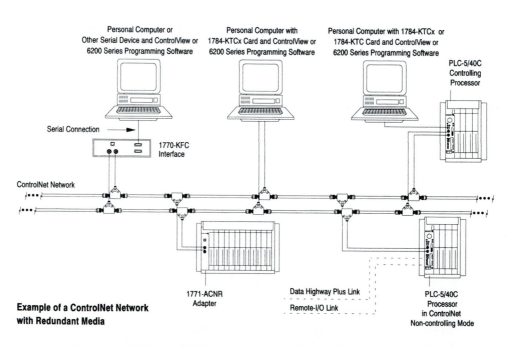

Figure 11-20. Redundant cabling. Courtesy Rockwell Automation/Allen-Bradley Company Inc.

INTEGRATING NETWORKS

One thing is sure, networking and integrating industrial devices will become increasingly easier. There will be a huge need to connect various floor-level devices and networks together. These devices and networks have different protocols. There will be software and hardware available to make these network differences transparent. Virtually all devices will be able to communicate with a network and all networks will be able to communicate with each other.

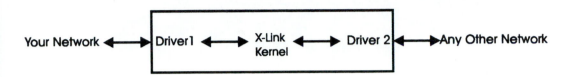

Figure 11-21. How two dissimilar networks can communicate. Courtesy S-S Technologies Inc.

The solution to the different protocols is a combination of hardware and software. There are already products like X-Link from S-S Technologies Inc. that can be used to interconnect the various network buses. Figure 11-21 shows an example of how this is accomplished.

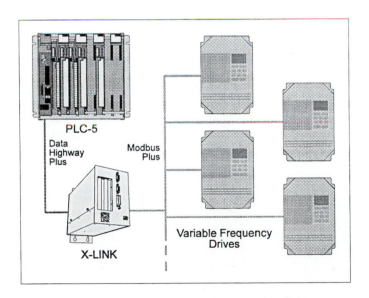

Figure 11-22. Example of connecting an AB PLC-5 to a Modbus Plus network. Courtesy S-S Technologies Inc.

Figure 11-23. What the X-Link hardware looks like. Courtesy S-S Technologies Inc.

Three software modules cooperate in an X-Link: the X-Link kernel, network driver 1 (communicates with the user's network) and network driver 2 (communicates with any other network). Each of the network drivers only interfaces with the X-Link kernel, meaning that any two drivers can be combined to meet the needs of a particular application. This means that the user does not have to worry about their interconnection. The hardware/software takes care of it. It appears as if there is only one network.

Figure 11-22 shows an example of connecting an AB PLC-5 to a Modbus Plus network that has several drives on it.

X-Link is a gateway that can be used to connect proprietary networks and devices that support serial communications to any industry standard network (see Figures 11-23 and 11-24). Think of it as a bridge to connect equipment to any network or one network to another. Hardware and software like X-Link and the move toward more standardization will make it easier to integrate enterprises.

CONTROL UNITS

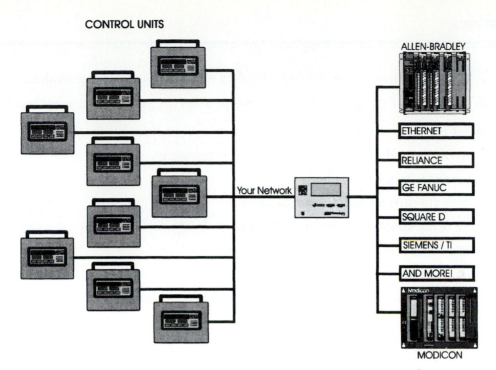

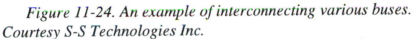

Figure 11-24. An example of interconnecting various buses. Courtesy S-S Technologies Inc.

Questions

1. Explain industrial networks and why they are becoming more prevalent.

2. What are some of the differences between a device bus and a process bus?

3. What is a field device?

4. Describe how the capabilities of field devices are changing.

5. Where are device buses appropriate?

6. Where are process buses appropriate?

7. Why is the CAN system so important?

8. Briefly describe the DeviceNET standard.

9. Briefly describe the Fieldbus standard.

10. Explain how will different networks be able to communicate with each other?

Chapter 12

Supervisory Control and Data Acquisition

Supervisory control and data acquisition software has become very prevalent in industrial automation. This chapter will examine the use of one of the more common supervisory control and data acquisition systems.

OBJECTIVES

Upon completion of this chapter, the student will be able to:

Describe a typical SCADA application.

Describe terms such as **driver, topic, item, recipe, historical log,** *and so on.*

Describe how typical SCADA software is programmed.

Describe how SCADA software can be used to improve the performance of an enterprise.

Describe how a SCADA system can be integrated with the business software in an enterprise.

OVERVIEW OF SUPERVISORY CONTROL AND DATA ACQUISITION (SCADA)

Many different technologies have come together in a relatively short time to revolutionize manufacturing. Almost all of these technologies are computer based. The rapid advancements in microcomputers in the last 20 years have given them the power that only mainframes had in the past. Networking technology has been developed to enable these microcomputers to become even more powerful through shared resources. The reduction in the price of memory and increase in the speed of microprocessors has allowed incredibly powerful software to be developed and used.

In the past software development was very proprietary. Software developers reaped large monetary rewards from developing, installing and servicing proprietary software. When an industry decided to automate a process they normally hired a system integrator who would develop the necessary software to operate the system. Software that did exist was too complex for most enterprises to develop their own automation applications.

Slowly software developers started to "partner" with other software developers to offer software packages that would work together.

While this was occurring a revolution was taking place in the offices of enterprises. The mainframe and "dumb" terminal had been the prominent technology. Microcomputers began to appear for many tasks. Speadsheets, databases, word processing packages, and computer-aided drafting (CAD) began to be used in offices. Productivity increased drastically. The old "dumb" ASCII terminals only displayed text. Microcomputers had the ability to display graphics and colors. They were also user-friendly and more intuitive.

There were two main microcomputer systems battling for supremacy: Apple and IBM. Both were leaders in some fields. Apple was particularly strong in graphics. IBM gained prominence in business and industrial use. Apple developed a very easy-to-use, intuitive operating system. Microsoft developed the Windows operating system which addressed the need for a user-friendly, intuitive operating system. Meanwhile industrial controllers were gaining capability also. Faster processors, more networking capability, more powerful communications modules were all being offered. This began to make it easier to communicate with industrial controllers. Until this time microcomputers were typically just used to program the PLCs or upload/download programs to them. Communication was still very proprietary, however. Each PLC had its own communication protocol.

Until the end of the 1980s very few software development companies had written Windows-based industrial software. Most still utilized the DOS environment. One of the first software packages developed for the industrial automation world was a package called Wonderware InTouch. It was designed to be a man-machine interface (MMI) package. It was one of the first to make good use of the Windows graphical user interface (GUI). This made it very user-friendly and intuitive to use. It also broke new ground in graphics capability in an industrial software package. MMI is also called supervisory control and data acquisition (SCADA) software. It was intended to be easy to use so that companies could easily develop and modify their own applications.

About the same time, Windows Dynamic Data Exchange (DDE) protocol began to be used for a new type of I/O driver. These DDE servers made it easy and convenient to acquire data from industrial control devices such as PLCs. DDE used report by exception techniques to create databases that could then be used to analyze processes. The data could be gathered and analyzed on-line. The data could be displayed graphically to show historical trends and to analyze the process. DDE also was designed to share data with other Windows applications. This allowed production data obtained from industrial controllers to be shared automatically with other Windows-based spreadsheets, word processors, databases and many other types of software. For example, a graph could be created in a spreadsheet program that would show production for the day compared with the past week. The data that is used in the graph is retrieved by the MMI software and shared via DDE with the spreadsheet. Data is updated whenever it changes in the system so the graph is always current. All of this occurs automatically. The same can be done with a production report created in word processing software. The report can be automatically updated with current production data and is available instantly.

Meanwhile, Windows-based software flourished. All Windows-based software had a common user interface. The software was also highly graphical in nature, making it very intuitive. This meant that once a person learned to use one Windows application it was very easy to learn a new Windows software package. Windows-based software drastically reduced the learning curve.

Industrial software developers also started to make their products user-friendly by utilizing the Windows environment. Wonderware's InTouch is a good example of user-friendly industrial software. InTouch was designed to make it easy for an applications person to develop industrial applications that could communicate with industrial controllers. The basis of InTouch is a graphical interface. By drawing graphics on the screen and then answering a few configuration questions specific to the particular industrial controller the application is developed and is able to communicate with the controller. If we decide to switch to a different brand of controller, the application is still usable. All we would have to do is reconfigure the driver and tagnames for the new controller. Notice that with InTouch developing an industrial application did not require a programmer. An applications person is able to easily develop applications. InTouch made it possible for manufacturers to develop their own applications.

SCADA was traditionally used for data collection from PLCs and plant floor controllers. SCADA systems were also used for monitoring and supervisory control of processes. The role of SCADA systems has expanded. SCADA systems are a vital part of many manufacturers' information systems. They provide manufacturing data to many other software systems in the typical manufacturing enterprise.

As companies automated and evolved to improve productivity, quality changes also occurred that affected the workforce. Workers are expected to have broader capabilities these days. Maintenance personnel are trained in many skills that were not traditionally a part of their job duties. This is sometimes called cross training. For example, electricians may be taught some mechanical skills while mechanical maintenance personnel are taught some electrical skills. This makes them more valuable in today's manufacturing environment. Production workers are also expected to be responsible for more than they were previously. In the past a worker was most likely only responsible for running one machine. Now that same worker is probably responsible for several computer-controlled machines or one or more automated systems. The result is that more is expected of both the skilled trades personnel and the production personnel. This has made SCADA systems more important also. When we broaden workers' responsibility they have less time and experience with each particular machine or technology. This makes operator information crucial. A SCADA system can be used to provide operator information, prompts for required action and/or input, alarms, detailed instructions, plans, and so on. This information can appear on the operator's screen, in the maintenance department, or with today's technology and the Internet anywhere in the world, if it is desired. The alarm can notify the maintenance department that a repair is needed. It can notify them which part probably needs repair and even give detailed instructions or video on how to repair or replace the part. Any of these capabilities can be designed into SCADA systems today.

SCADA systems have also been given recipe capability. Recipes for various products can be stored and downloaded to controllers as needed. SCADA systems have also gained the capability to share their information over the Internet. It is possible today for a manufacturer to watch graphics and have data from a process anywhere in the world.

SCADA systems are continually gaining capability and are able to handle many of the functions associated with MRP systems such as inventory tracking, scheduling, etc. Today's SCADA systems integrate very easily with a wide variety of other types of enterprise software up to and including manufacturing execution system (MES) software. MES software is designed to fill the gap between MRP system and the process control system. MES systems are concerned with monitoring, tracking, operator information and instruction, and archival of manufacturing data for historical and process improvement purposes. The MES system typically provides direction much like a routing that guides all activities that occur as a product moves through the enterprise. It defines what occurs in each step and keeps track of which operator was involved, when and where, materials, quality data, and other relevant information. A complete production record then exists.

The graphical user interface (GUI) is the key to manufacturing systems and in fact all software applications. As previously mentioned, Windows has given most software a common interface and "feel." To be successful software systems must have user-friendly interfaces (GUIs) and must be very easy to use to develop applications. The key to all technology will increasingly be ease of use.

SAMPLE APPLICATION

The best way to understand a SCADA system is to see how a simple industrial application would be developed. We will develop a simple temperature control application. Note that this is not intended to teach you every key to press. It is intended to give you a broad, overall understanding of how SCADA applications are developed.

Figure 12-1 shows a simple temperature control system. It consists of a conveyor to move product through two heat chambers. There are two temperature controllers in the system. The product moves along the conveyor line and though each furnace for a controlled time in each furnace. The first furnace acts as a preheat chamber and is controlled at a lower setpoint than the second heat chamber. Figure 12-2 shows a table that contains the temperature setpoints for various products this process produces. There are 5 values that need to be changed in each temperature controller for each different product run. There is also an Allen-Bradley SLC 500 to take input from sensors and to control the conveyor.

Figure 12-1. A simple temperature control system. A conveyor and two ovens are to be controlled.

At this point we need to use some specific hardware and software to develop our application. We will use an IBM-compatible computer running Windows 95, Wonderware InTouch, two Omron temperature controllers (see Figure 12-3), and an AB SLC 500. Figure 12-4 is a table that shows all of the controllers and I/O that are used in the system.

The Omron temperature controllers in this example have RS-485 communication modules installed.

Product	Setpoint Chamber 1	Setpoint Chamber 2	P	I	D
P134	180	225	90	10	30
P135	192	238	85	3	30
P136	163	207	97	12	32
P137	193	267	76	23	25
P138	215	237	85	12	24
P139	199	332	90	10	25
P140	137	183	87	8	22

Figure 12-2. The parameters for different products that are loaded into the temperature controllers.

Figure 12-3. Three models of Omron temperature controllers. Photo courtesy of Omron Electronics Inc.

Control Device	I/O Number or Name	Tagname	Use
Allen Bradley SLC 500	O:3/15	Conveyor	This is a PLC output that turns the conveyor on and off (Output 15, slot 3)
Allen Bradley SLC 500	I:7/8	Part-present	This is a sensor in the oven to check for part presence (Input 8, slot 7)
Omron Temperature Controller	temperature	Temp_1	This is the variable that holds the actual temperature value in the temperature controller
Omron Temperature Controller	setpoint	Set_1	Used to change the setpoint of the temperature controller
Omron Temperature Controller	proportional	Prop_1	Used to change the proportional gain of the temperature controller
Omron Temperature Controller	integral	Int_1	Used to change the integral gain of the temperature controller
Omron Temperature Controller	derivative	Der_1	Used to change the derivative gain of the temperature controller
Omron Temperature Controller	status	Stat_1	This is a bit in the temperature controller. It is a 1 if communications are normal and a 0 if there is a communications error

Figure 12-4. Information about the actual I/O and controllers used in the application. Tagnames have also been chosen for each I/O point.

We will use communications port one (serial port 1) from our microcomputer to talk to the Omron controllers. The Omron controllers have RS-485 communications modules so we will need to convert our RS-232 computer output to RS-485 output. This is done with an RS-232 to RS-485 converter (see Figure 12-5). Converters like these are very inexpensive. They typically cost about $60.

There are two temperature controllers so we need to give each a unique name or address. This is called a unit address by Omron. We will set the unit address of the first temperature controller to 1 and the second temperature controller to 2.

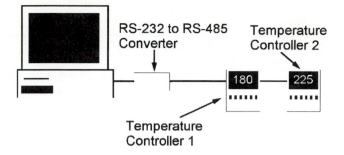

RS-232 to RS-485
Converter

Temperature
Controller 2

180 — 225

Temperature
Controller 1

*Figure 12-5. How the two temperature controllers are con-
nected to the computer.*

DDE is an acronym for dynamic data exchange. It is a communication protocol
that was designed by Microsoft to allow Windows applications to send/receive
data to/from each other. It can establish a client-server relationship between two
concurrently running applications. The server accepts requests for data from other
applications and provides requested data to them. Applications that request data
are called clients. This relationship can change. A client can become a server and
a server a client as they share data with each other. DDE is a standard feature for
most Windows applications that need data links to other applications. DDE-
compliant applications include Microsoft Excel, Lotus 1-2-3 for Windows,
InTouch, and many others. This means that if we are putting data into a database
program we can share that with another software program running in the computer. For example, we might want to put investment data into a database every
day and then graph the performance of our investments using a graph in a spreadsheet program. DDE could share the data between the applications so that we
would not have to enter it twice.

DDE can also share data with other computers. It has the capability to communicate data over computer networks or over modems. In fact, NetDDE extends the
DDE standard to make it possible to communicate over local area networks and
through serial ports. DDE can be used to collect and distribute factory data. For
example, we might want to get daily production data and put it into a spreadsheet
where it can be analyzed and graphed. It would be useful for a foreman or management to see a graph that could show production over the last 10 work days, for
example. This data can be accessed automatically and shared via DDE with a
spreadsheet. DDE can also be used to send production data to applications.
Imagine a temperature control system that has several variables that change with
the type of product that is manufactured. A spreadsheet could be set up that
would hold the variables needed for each product. When needed the data could be
sent via DDE to the controller.

Application Development

Developing an application with Wonderware is a very simple and straightforward
process. To develop an application the programmer just draws a picture of the

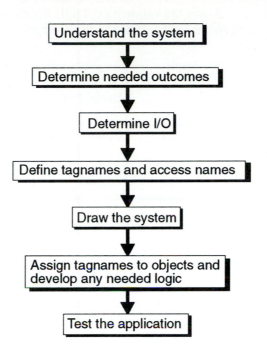

```
┌─────────────────────────┐
│ Understand the system   │
└─────────────────────────┘
            │
            ▼
┌─────────────────────────┐
│ Determine needed outcomes │
└─────────────────────────┘
            │
            ▼
┌─────────────────────────┐
│ Determine I/O           │
└─────────────────────────┘
            │
            ▼
┌─────────────────────────────┐
│ Define tagnames and access names │
└─────────────────────────────┘
            │
            ▼
┌─────────────────────────┐
│ Draw the system         │
└─────────────────────────┘
            │
            ▼
┌─────────────────────────────┐
│ Assign tagnames to objects and │
│ develop any needed logic       │
└─────────────────────────────┘
            │
            ▼
┌─────────────────────────┐
│ Test the application    │
└─────────────────────────┘
```

Figure 12-6. Steps in developing a SCADA application. Note that SCADA software is quite flexible and experienced developers will develop their own preferences for the order in which they develop an application.

objects that are needed to represent the application and then answers a few questions to describe the real-world I/O that each object represents.

Understanding the System

Developing a Wonderware SCADA system is a relatively easy and straightforward task. Figure 12-6 shows a simple flow diagram of system development steps. The key to success however, is understanding the manufacturing system. The application developer must know what types and brands of controllers are used in the system and what role they play.

First we must examine and understand our system. There are three controllers in our simple system. There are two Omron temperature controllers and one AB SLC 500 PLC. Let's give each of the controllers a name. Let's call the SLC 500 "AB_SLC." We have two temperature controllers so let's give each a descriptive name. Let's call the temperature controller for oven 1 Temp_Contrl_1. We could then call the second temperature controller Temp_Contrl_2. Note that almost any name can be used as long as naming conventions are used but simple descriptive names are best. The names we just created are called DDE access names. They will be used by the application we create to tell Wonderware which controller we are talking to. Figure 12-7 shows the controller, DDE access name, communications port used, and the purpose for each of the controllers.

Controller	DDE Access Name	Com Port	Use
AB SLC 500	AB_SLC	Serial Port 1	Sensor input, conveyor control and logic.
Omron Temperature Controller	Temp_Contrl_1	Serial Port 2	Controls the temperature of oven 1
Omron Temperature Controller	Temp_Contrl_2	Serial Port 2	Controls the temperature of oven 1

Figure 12-7. The controller, DDE access name, communications port used, and the purpose.

The application we will develop is shown in Figure 12-8. The figure shows what the application will look like in runtime. There are product selection buttons on the top of the application. These can be used by the operator to send parameters to the temperature controllers. For example, if the operator clicks on the first button, the parameters for product P134 are automatically sent to the temperature controllers. There are two indicator lights. One of the lights is a communications status light and the other is used to indicate when there is product present in the furnace.

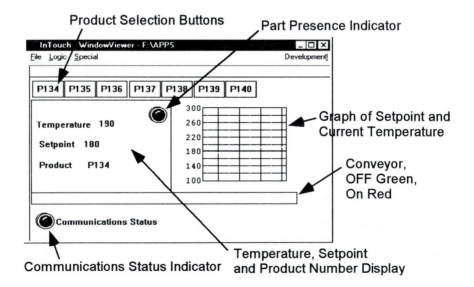

Figure 12-8. A graphic of the runtime screen of the application that will be developed.

The conveyor (long thin rectangle on the bottom) turns red if the conveyor is on and green when it is off. The graph is used to show the setpoint and the current temperature of furnace 1. There are also text displays that show the current temperature, the setpoint and the product number being manufactured.

It will be helpful to your understanding to keep the final outcome in mind as the application is developed.

Remember that this sample application development is not meant to teach you each specific step in application development. It is meant to show you that application development is a very straightforward and relatively simple task. It does not take a programmer to develop most SCADA applications. An applications technician can easily develop applications.

Communications Configuration for the Omron Controllers

The first task we will do is to create names for our controllers. Wonderware calls these DDE access names. These will be used in our application to determine which controller we are talking to. Think of the DDE access name as containing all of the specific information needed to communicate with a controller. We will create the names for the Omron temperature controllers first. Remember that we decided to call them Temp_Contrl_1 and Temp_Contrl_2.

DDE access names are created by running the driver software for each controller. Remember that driver software is like translator software for a particular controller (see Figure 12-9). Let's create the Temp_Contrl_1 DDE access name first. OMRONHL is the name of the driver that talks to Omron PLCs and Omron temperature controllers. Remember that a driver is software that we would purchase for each brand of controller we would need to talk to. Drivers are inexpensive translators that handle all communications protocol between the computer application we create and the controller we need to talk to. This makes communication with any controller a transparent, easy task.

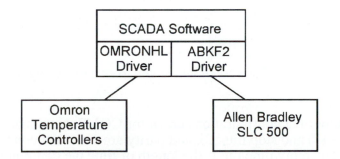

Figure 12-9. How SCADA software communicates with specific controllers. A software driver for each particular controller handles all communication with each controller. In this application we need two drivers, an Omron software driver (OMRONHL) and an Allen-Bradley driver (ABKF2).

The programmer started the OMRONHL driver software and the screen shown in Figure 12-10 appeared. The user then chose Configure from the menu with the left mouse button.

OMRONHL
Configure Help

Figure 12-10. Configure screen for the OMRONHL communication software driver.

The screen shown in Figure 12-11 then appears. This is the configuration screen for communications parameters. Note that there are three things that can be configured. The user chose COM port settings.

OMRONHL
Configure Help
Com Port Settings
Topic Definition
DDE Server Settings...

Figure 12-11. Configure screen for the OMRONHL communication software driver showing the three things that can be configured.

COM2 was chosen from the menu. Remember that serial port 2 (COM2) will be used to talk to temperature controllers 1 and 2 (see Figure 12-12). The programmer chose the modify button from the screen so that the parameters for COM2 could be set.

Select a Communications Port

| COM1: | COM2: | COM3: | COM4: |

Modify

Figure 12-12. The screen where the user chooses which port to configure.

Figure 12-13 shows the configuration screen for COM2. Note that baud rate (9600), data bits (7) and stop bits (2), and parity are set to the controller's protocol specifications. The reply time out is the length of time the computer should wait for a reply from the controller before generating an error message. That is all that is required to set up COM2 for communications. You may have a question at this point, "How can one COM port (COM2) be used to talk to 2 temperature controllers?" We will set that up next.

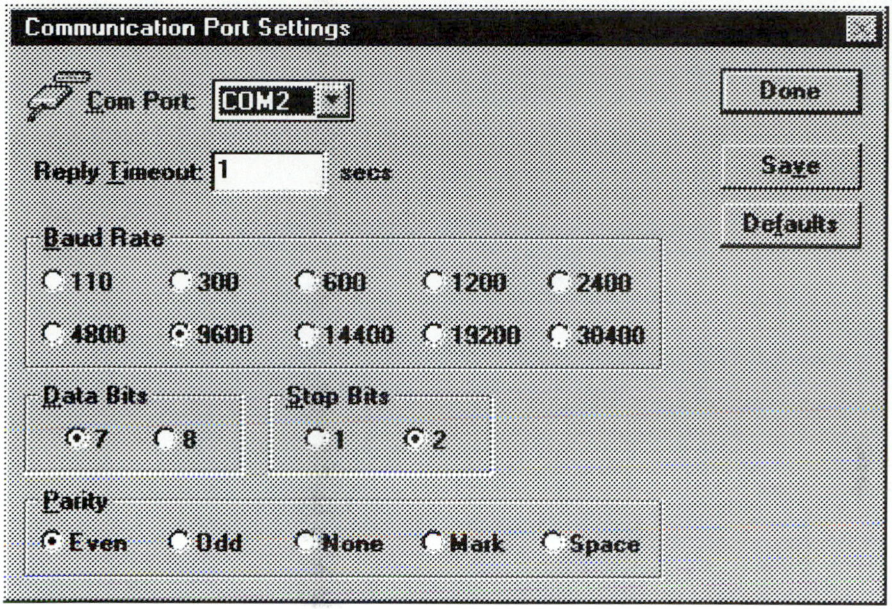

Figure 12-13. The actual port settings that were made.

The programmer then chose Done and was taken back to the configuration screen (see Figure 12-14) and chose Topic Definition. The topic definition screen appears (see Figure 12-15). The user then chose New to add a new topic name. For now, think of Topic and DDE access name as the same thing.

The programmer typed the topic name Temp_Contrl_1 in the Topic Name space (see Figure 12-16). Next COM2 was chosen. Then Temperature Controller was chosen. Note that other controllers are also shown. This driver could be used to talk to several different Omron products. It can be used with their PLCs and temperature controllers.

The correct model of temperature controller was then chosen from a drop-down menu. Next a unit address was entered. Note that 1 was entered. The unit number will identify which temperature controller we are talking to (see Figure 12-17). The programmer was defining Temp_Contrl_1 now so 1 was entered. A unit address of 2 would be entered for Temp_Contrl_2.

The actual Omron temperature controllers (hardware) need to be internally addressed. Each is given a unique address. The unit address for oven controller 1 was set to 1 and the unit address for oven controller 2 was set to be 2. Note that this was done in the actual controller. COM2 will be used to talk to both temperature controllers.

Figure 12-14. The OMRONHL configuration menu screen.

Topic Definition

Topics

Done

New...

Modify...

Delete

Figure 12-15. Topic definition screen. Note that any topics that had been created would appear as a list here. None have been created thus far.

OMRONHL Topic Definition

Topic Name:	Temp_Contrl_1	OK
Com Port:	COM1	Cancel
Device Type	● Temperature Controller	Model: E5AX
○ C-Series PLC	○ Digital Controller	
○ CV-Series PLC	○ Signal Processor	
Unit Address: 0	Update Interval: 1000	msecs

Memory Area Sizes

Automatically Switch to Monitor Mode on writes

Communication Type   Unit Number   0
● Local   ○ Network   Node Number   1

Figure 12-16. The actual port settings that were made.

An update interval can also be entered. The update interval determines how often values from this controller are to be updated. This completes the steps for configuring Temp_Contrl_1.

The same steps would be followed to configure Temp_Contrl_2 except that Temp_Contrl_2 would be substituted for Temp_Contrl_1 and the unit address would be set to 2. Next the programmer configured the driver for the AB SLC PLC.

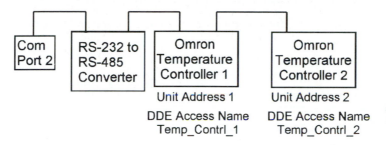

Figure 12-17. The connection of the two temperature controllers to COM2. The DDE access names and unit address are also shown for each. Remember that the unit addresses are set in the actual controller.

Communications Configuration for the Allen-Bradley SLC

The driver software for the AB SLC 500 is called ABKF2. The programmer started the ABKF2 driver software and the screen shown in Figure 12-18 appeared. Configure was chosen from the menu with the left mouse button.

Figure 12-18. Configuration screen for the AB driver.

The screen shown in Figure 12-19 then appeared. This is the screen that is used to configure COM port settings. Note that there are three things that can be configured. The user chose COM Port Settings.

COM1 was chosen from the menu (see Figure 12-20). Remember that serial port 1 (COM1) will be used to talk to the AB SLC 500 PLC. The user chose the modify button.

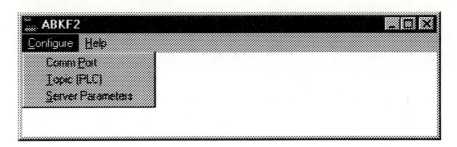

Figure 12-19. The screen where the user chooses what to configure.

Select a Communications Port

| COM1: | COM2: | COM3: | COM4: |

Modify

Figure 12-20. The screen where the user chooses which port to configure.

A-B Serial Comm Port Configuration

Com Port: COM1:   **Reply Timeout:**   3   **Secs**

Checksum
  ○ BCC   ⊙ CRC

Baud Rate
  ○ 110   ○ 600   ○ 2400   ⊙ 9600
  ○ 300   ○ 1200   ○ 4800   ○ 19200

Protocol Mode
  ⊙ Full Duplex (DF1, Point-to-point)
  ○ Half Duplex (Master-slave, multidrop)

  OK   Cancel

Figure 12-21. COM port configuration screen.

Figure 12-21 shows the configuration screen for COM2. Note that baud rate (9600), CRC (cyclic redundancy check) and full duplex were chosen. The reply time out is the length of time the computer should wait for a reply from the controller before generating an error message. That is all that is required to set up COM1 for communications.

Next the programmer chose Done and was taken back to the configuration screen and chose topic definition. The topic definition screen appeared (see Figure 12-22). The user typed in the topic name AB_SLC in the Topic Name space. Next COM1 was chosen. The PLC station number was set to 1. Note that we could have multiple controllers, each with a different station number. The programmer also chose SLC 500 for the model type.

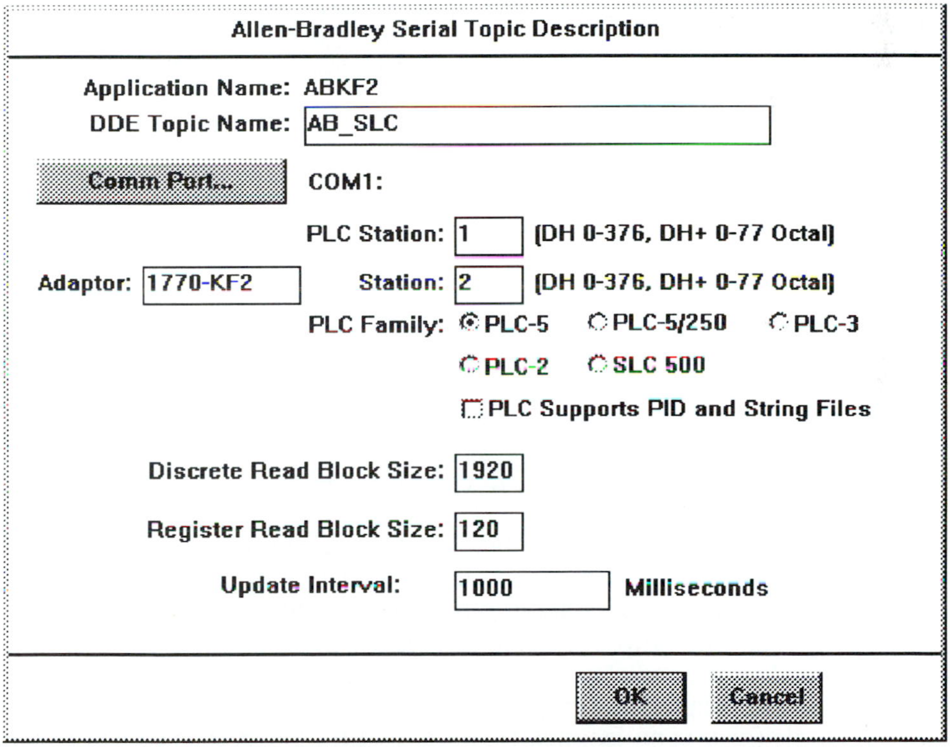

```
┌─────────────────────────────────────────────────────────────┐
│            Allen-Bradley Serial Topic Description            │
├─────────────────────────────────────────────────────────────┤
│     Application Name: ABKF2                                  │
│     DDE Topic Name: │AB_SLC                              │    │
│                                                             │
│   ░Comm Port..░      COM1:                                   │
│                                                             │
│              PLC Station: │1   │   [DH 0-376, DH+ 0-77 Octal]│
│                                                             │
│  Adaptor: │1770-KF2 │    Station: │2   │ [DH 0-376, DH+ 0-77 Octal]│
│              PLC Family: ⊙ PLC-5   ○ PLC-5/250   ○ PLC-3    │
│                         ○ PLC-2   ○ SLC 500                  │
│                         ☐ PLC Supports PID and String Files │
│                                                             │
│     Discrete Read Block Size: │1920 │                        │
│                                                             │
│     Register Read Block Size: │120  │                        │
│                                                             │
│          Update Interval:   │1000      │   Milliseconds      │
│                                                             │
│                              ░ OK ░   ░ Cancel ░             │
└─────────────────────────────────────────────────────────────┘
```

Figure 12-22. Configuration screen.

That is all there is to setting up DDE access names. The programmer created two DDE access names: one for temperature controller 1 (Temp_Contrl_1) and one for the AB SLC 500 (AB_SLC). Temp_Contrl_1 will be used to communicate with the Omron controller on oven 1. AB_SLC will be used to communicate with the AB SLC 500.

APPLICATION DEVELOPMENT

Next a simple temperature control application will be examined. We will examine the development of half of the whole temperature control system. We will develop the window (display) for temperature controller 1.

Wonderware refers to a screen of information as a window. Figure 12-23 shows a development screen with one window. We will develop the application in the development mode of InTouch. Development is very easy. It really just involves drawing a picture of what the operator screen should look like and linking objects on the screen to I/O in the controllers so that things will change on the screen as things change in the real world.

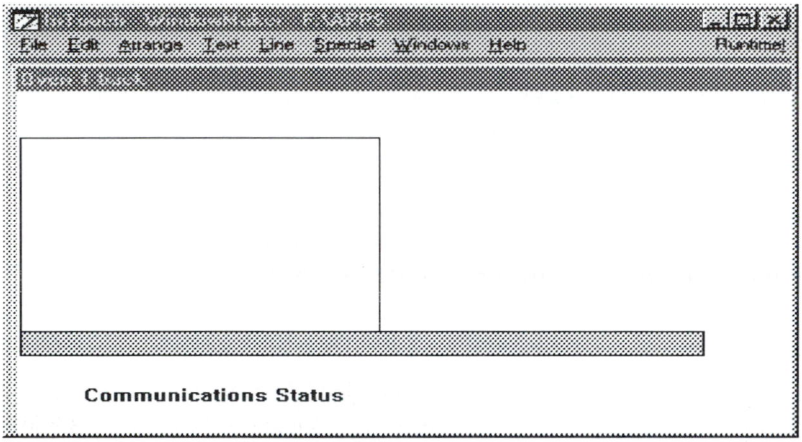

Figure 12-23. Wonderware development screen.

Study Figure 12-24. Note the tool palette. The tool palette is very similar to the tool palette that any drawing package would have. The tool palette is used to draw the application. The tools in the top row are: arrow (selection tool), rectangle, rounded rectangle, ellipse, line, horizontal/vertical line, polyline, polygon, and the text tool. The second row has a wizard tool, a bitmap tool, a real-time trend (graph), a historical trend (graph) tool, and the button tool. The third row has text-type tools for text styles, justification, and font size. The fourth row has tools for changing the colors of lines, fills, and window color. The fifth row's tools are used to align objects. The sixth row's tools are used to copy, space and group objects. The last row's tools are used to undo/redo (change last user action), reshape, rotate, flip, and to turn the snap to grid on/off. We will use some of these tools as we develop this application. Note that as a tool is selected the name of the tool appears at the bottom of the tool window. In this case the Snap to Grid tool is selected.

Study Figure 12-25. The rectangle tool was used to draw a rectangle that represents the oven. The rectangle tool was also used to draw a long thin rectangle under the oven rectangle. This rectangle represents the conveyor. The text tool was used to add the words Communications Status to the lower left portion of the window. The title bar above the working area shows the window name. The window's name is Oven 1 Temperature Controller.

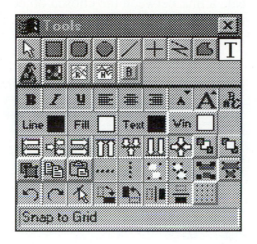

Figure 12-24. The Wonderware InTouch tool palette. These are the tools that are used to create an application.

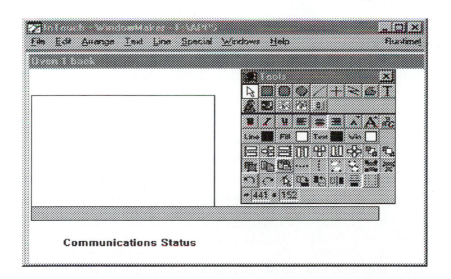

Figure 12-25. A typical development window in Wonderware InTouch. Note that 2 rectangles have been drawn and text (Communication Status) was added to the window. The tool palette is also shown in the window.

The conveyor is controlled by an Allen-Bradley SLC 500. The output number used is O:3/15 (output 15, slot 3). The programmer then linked the conveyor rectangle and the real-world I/O in the PLC. There are a couple of ways this can be done. We will just look at one. The programmer double-clicked the left mouse button on the desired object (conveyor rectangle in this case). A new screen appeared that helped create a link between the chosen object and the real-world I/O in the controller. Figure 12-26 shows this input screen.

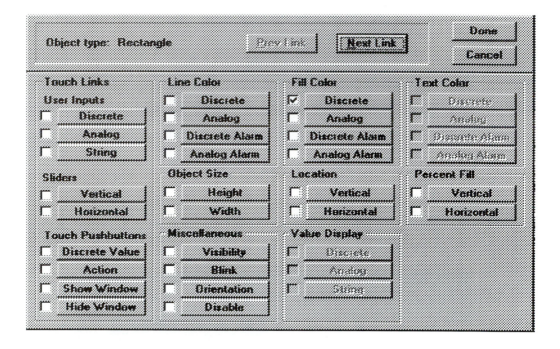

Figure 12-26. The screen that enables the user to decide how this object will behave. In this case it is the rectangle that represents the conveyor. Fill Color Discrete was chosen. Note the many other possible choices that could have been made. The choices are not exclusive. We could choose to have any combination of the object's attributes change depending on the real-world value if we wanted.

Note that this screen allowed the programmer to choose how this object relates to the real-world I/O. The rectangle could be used for user inputs or to change the line color and/or fill color based on real-world I/O. The rectangle could be used as a slider to change the value of real-world I/O, or we could change the rectangle's size, and/or location, and/or fill level based on the value of the real-world I/O. The rectangle could also be used as a touch push-button to input discrete values or to change values, perform calculations, or to show or hide windows. There are also some other miscellaneous choices that control the object's visibility, blinking, orientation and/or disable. The user chose the button for Fill Color Discrete, and a new input screen appeared (see Figure 12-27). Note that multiple choices can be made. We could, for example, change an object's fill color, size, location or other attributes if we needed to.

In this case the programmer wanted the rectangle's color to change to red when the conveyor is on and green when the conveyor is off.

In this screen the programmer entered the tagname for the object in the Expression entry area. The programmer entered "conveyor" for the tagname. Any name can be used for a tagname as long as naming rules are followed. The programmer clicked the mouse in the 0,FALSE,Off box. A color menu appeared and the programmer chose the color for the OFF condition. In this case green is chosen. Then the programmer clicks on the 1,TRUE,On box and chose the color red.

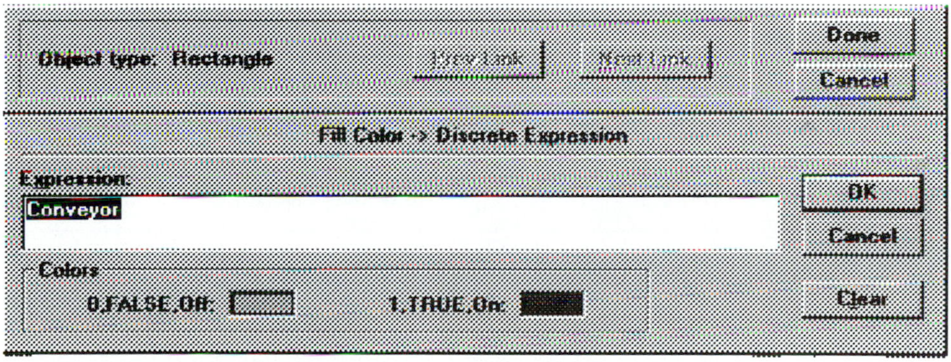

Figure 12-27. The configuration screen for the rectangle we called "conveyor." Fill Color Discrete was chosen for its type so this screen allows the off and on colors to be chosen for the rectangle (conveyor).

The programmer was done with data entry on the screen and chose the done button. The software realized at this point that it didn't know what the tagname conveyor was. A new screen appeared asking the programmer if he/she wished to define "conveyor" (see Figure 12-28). The programmer chose OK with the left mouse button.

A new input screen appeared which allowed the programmer to define the "conveyor" tagname (see Figure 12-29). The programmer chose the type from a drop-down list. In this case the type was DDE discrete. This means that this real-world I/O point can only have a value of 0 or 1.

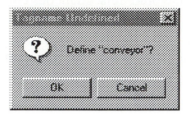

Figure 12-28. This screen appears when the programmer uses a tagname that has not been defined yet. This allows the programmer to define the tagname at this point.

| Dictionary - Tagname Definition | | • Details | ○ Alarms | ○ Both | ○ None |

| New | Restore | Delete | Save | << | Select | >> | Cancel | Done |

Tagname: conveyor   Type: ... DDE Discrete

Group: ... $System   ○ Read only • Read Write

Comment:

☐ Log Data   ☐ Log Events   ☐ Retentive Value

Initial Value   Input Conversion   On Msg:
○ On   • Off   • Direct   ○ Reverse   Off Msg:

DDE Access Name: ... Unassigned

Item:

☐ Use Tagname as Item Name

Figure 12-29. Tagname definition screen for the "conveyor" tagname.

Next the programmer chose a DDE access name. One called AB_SLC was already created so the programmer chose it (see Figure 12-30). Remember that the DDE access name tells Wonderware which controller the application needs to talk to. It associates the name of the controller and the I/O point (tagname) with each other so that when the application refers to the tagname "conveyor" Wonderware knows that this tagname refers to the AB PLC that we need to talk to.

The programmer then entered O:3/15 for the item name (see Figure 12-31).

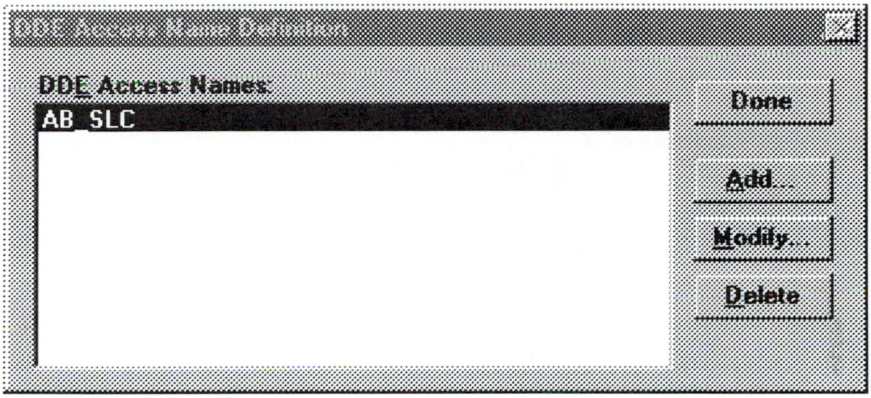

Figure 12-30. DDE access name screen.

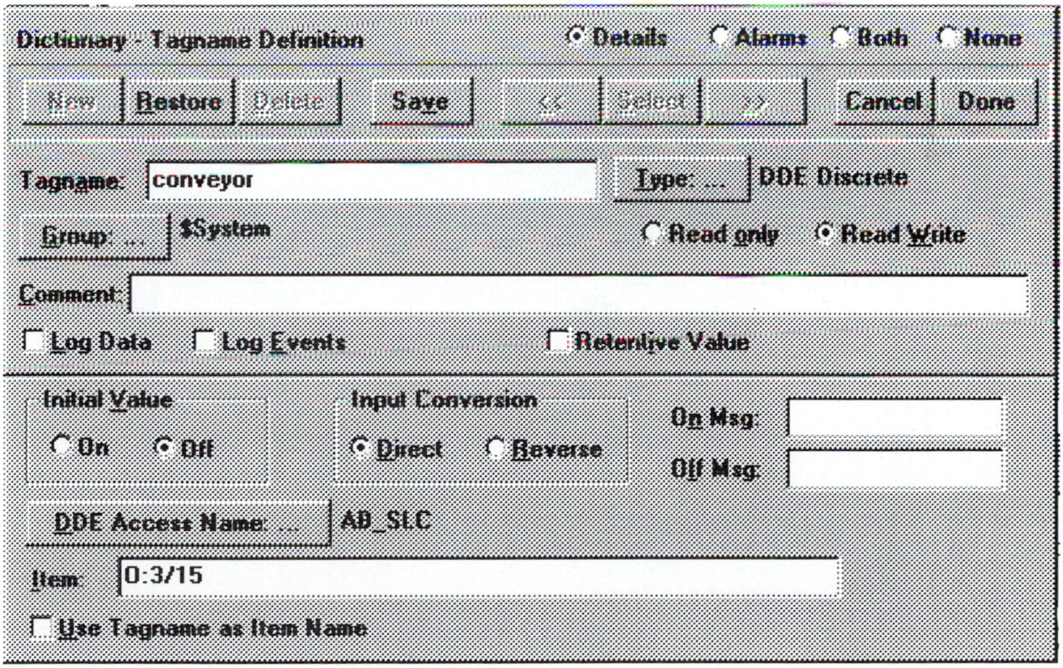

Figure 12-31. Tagname definition screen.

The programmer then added a status light to indicate whether or not the computer is communicating with the Omron temperature controller. The programmer also added a part presence light to show when there is product inside the oven. These were added using the Wizard tool.

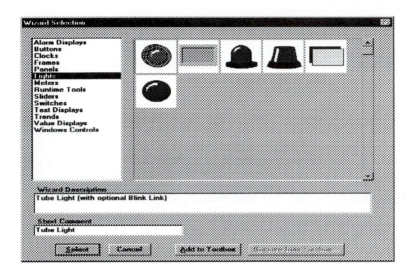

Figure 12-32. Wizard screen

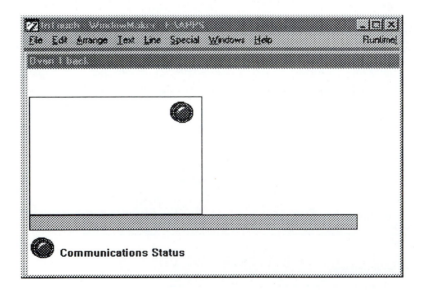

Figure 12-33. Application development screen showing the addition of a status light and product present light.

Figure 12-32 shows a Wizard screen. Note that there is a list of types of Wizards that are available on the left side of the page. Lights was chosen so the light Wizards are shown. The programmer chose the first light and placed it in the upper right corner of the oven (see Figure 12-33). The programmer then placed another light on the lower left for a communications status indicator for the first temperature controller.

The programmer double-clicked on the light that is used for communications status and an input window appeared (see Figure 12-34). The programmer entered a tagname (Status_1) for the status light that represents the temperature controller and then chose red for the OFF state and green for the ON state. ON will mean that the computer and Omron temperature controller are communicating and red will indicate that they are not communicating. Most control devices will have a status bit in memory that indicates whether or not communications are normal. The programmer chose OK and the screen in Figure 12-35 appeared asking the operator if he/she wanted to define Status_1. The programmer chose yes and the screen in Figure 12-36 appeared. The programmer defined Status_1 (see Figure 12-36). The programmer chose DDE discrete for the type. The programmer also chose "Read only" because this bit can only be read. It cannot be written to by the computer. The programmer chose Temp_Contrl_1 (remember that this is the DDE access name that we created earlier) for this controller's DDE access name.

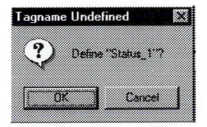

Figure 12-34. Tagname definition screen.

Figure 12-35. Tagname undefined screen.

Figure 12-36. Status_1 Definition screen.

The programmer then entered the actual I/O item that the Omron temperature controller will understand. The driver for the Omron temperature controller understands the word "status." Status is entered as the item name. Remember that our tagname is Status_1. This is because the application has two temperature controllers and tagnames should clearly describe their purpose.

Next the programmer added a "real-time" graph to the display. The programmer needed to display a graph that shows how the oven has performed during the previous hour. The programmer decided to display the temperature and the setpoint on the graph. The trend-graph tool is chosen from the tool palette and a graph is created just the way a rectangle would be drawn (see Figure 12-37).

Figure 12-37. Application screen showing graph.

Note that there is also a historical trend graph that can be used to look at logged data. The historical graph could be used to look at logged data from past performance. The programmer adjusted the size of the graph by moving the side edges or the corners.

The programmer then double-clicked on the graph and an input window appeared (see Figure 12-38). First the time span was chosen for the graph. It was decided that the total time that the graph would show at one time would be 1 hour; 1 minute was chosen as the sample interval. This means that the graph will show from the present to the past 60 minutes and it will sample and display new data every 1 minute.

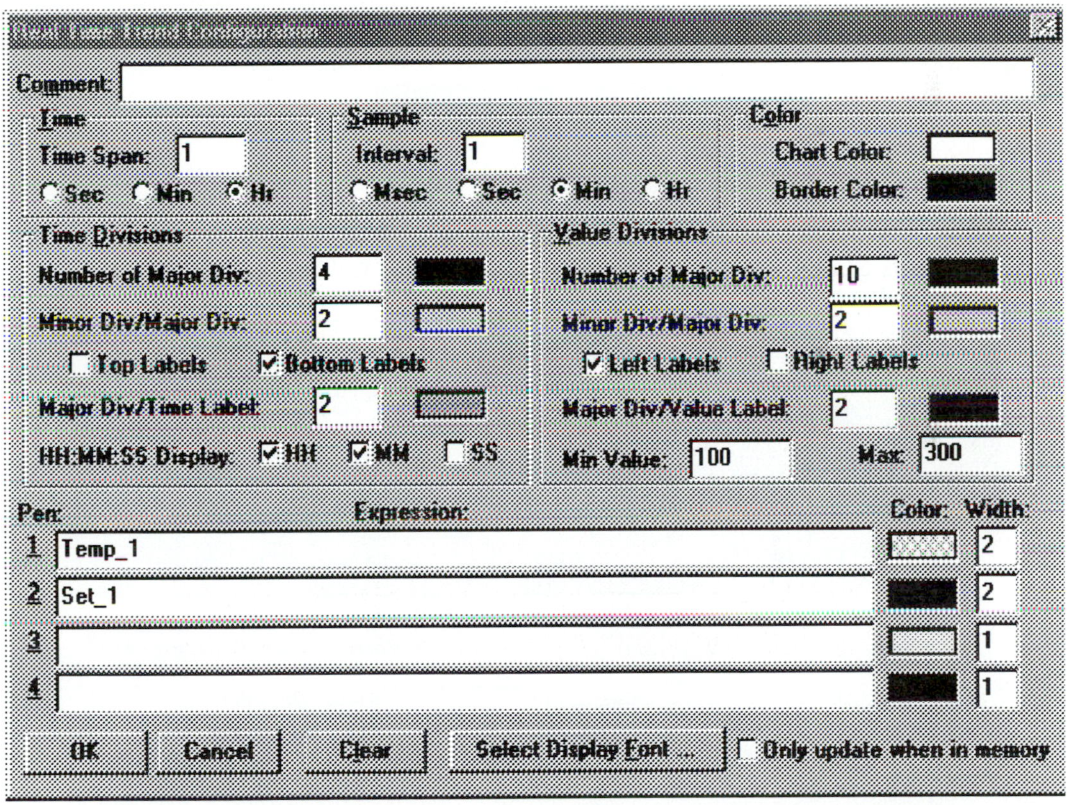

Figure 12-38. Graph configuration screen. Note that colors and time increments can be configured on this screen. The programmer can set the length of time that the graph will show on the screen. For example, the user might want to show the last hour's temperatures on the screen.

Next the programmer entered the tagnames of the items to be displayed on the graph. Temp_1 was chosen for the tagname of the temperature variable in the Omron temperature controller 1. Set_1 is chosen for the tagname of the setpoint variable in temperature controller 1. The programmer then changed the line width and color for each of the tagnames. The line width was set to 2 to make the lines more visible on the graph. Green was chosen for the temperature line (Temp_1) and blue was chosen for the setpoint line color (Set_1).

The programmer was done entering graph information at this point and clicked on the OK button. The programmer had not defined the Temp_1 tagname or the Set_1 tagname so a window appears as shown in Figure 12-39. The programmer chose yes and a new window appeared as shown in Figure 12-40.

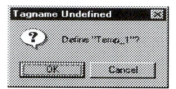

Figure 12-39. Tagname undefined screen.

| Dictionary - Tagname Definition | | | • Details | ⚬ Alarms | ⚬ Both | ⚬ None |

| New | Restore | Delete | Save | << | Select | >> | Cancel | Done |

Tagname: `Temp_1`   Type: ... DDE Integer

Group: `$System`   ⚬ Read only   • Read Write

Comment: `                    `

☐ Log Data   ☐ Log Events   ☐ Retentive Value ☐ Retentive Parameters

| Initial Value: | `0` | Min EU: | `0` | Max EU: | `9999` |
| Deadband: | `0` | Min Raw: | `0` | Max Raw: | `9999` |

Eng Units: `                    `   Conversion
                                           • Linear
DDE Access Name: ... `Temp_Contrlr_1`   ⚬ Square Root

Item: `temperature`

☐ Use Tagname as Item Name   Log Deadband: `0`

Figure 12-40. Tagname definition screen for Temp_1.

The programmer chose DDE integer for the type, Temp_Contrl_1 for the DDE access name and temperature for the item name. Temperature is the name that the Omron driver uses to get the current temperature value from the controller. The programmer then chose done. The same procedure was followed to define tagname Set_1. The item name used was "setpoint." The Omron driver uses the name "setpoint" to send the setpoint value to the controller.

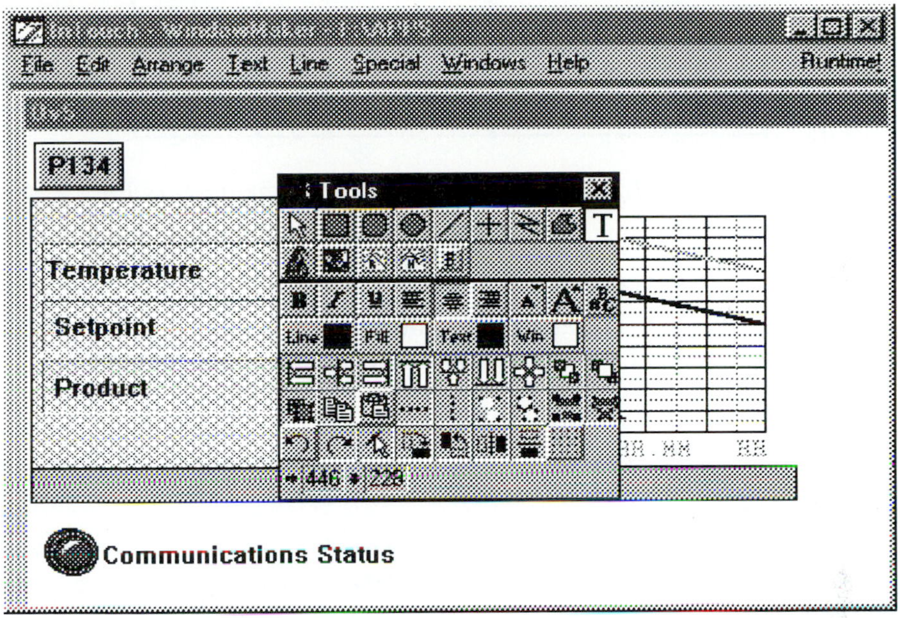

Figure 12-41. Application development screen showing graph, conveyor and furnace and a product button.

Next the programmer added a button so that the operator will be able to automatically change all of the variables in the temperature controller (proportional gain, integral gain, derivative gain, and the setpoint). The programmer used the button tool to create a button on the screen. Figure 12-41 shows the application with a button added for product P134. The programmer then substituted the name P134 for the button name. The programmer also used the text tool to add three labels: Temperature, Setpoint, and Product. The rectangle tool was used to draw a small rectangle to the right of the labels. The programmer will create links later so that the actual values can be displayed during runtime.

Examine Figure 12-42. This table shows the seven products that are manufactured in this process and the parameters for each product. Note that the setpoints are different for each controller.

Product	Setpoint Chamber 1	Setpoint Chamber 2	P	I	D
P134	180	225	90	10	30
P135	192	238	85	3	30
P136	163	207	97	12	32
P137	193	267	76	23	25
P138	215	237	85	12	24
P139	199	332	90	10	25
P140	137	183	87	8	22

Figure 12-42. Table showing temperature parameters for products.

The programmer then double-clicked on the button and a new input screen appeared (see Figure 12-43). The action type button was chosen for this object. The programmer chose OK and the screen shown in Figure 12-44 appeared.

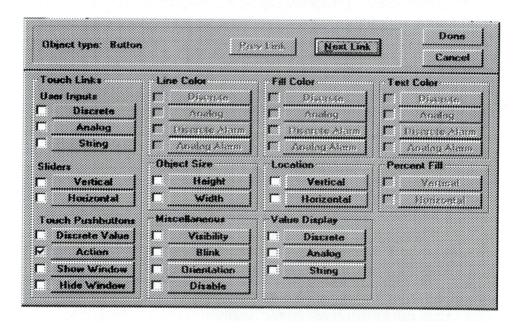

Figure 12-43. Link definition screen. Note that the action touch push-button was chosen.

Figure 12-44 shows the input screen for an action button. Note that 4 assignment statements were made in the input area. If the operator chooses this button while the actual application is running these value assignments will be made. The values for this product will be sent to temperature controller 1 to set its parameters. The first statement assigns the value 180 to Set_1. Remember that Set_1 is the tagname for the temperature variable in the Omron temperature controller. The values for Prop_1, Int_1, and Der_1 are also assigned in the same way. Note that Set_1, Prop_1, Int_1 and Der_1 would have to be defined in the same way that Temp_1 was defined.

The last assignment statement assigns the character string P134 to the variable named "product." This will be used to display which product is being produced on the screen while the application is running.

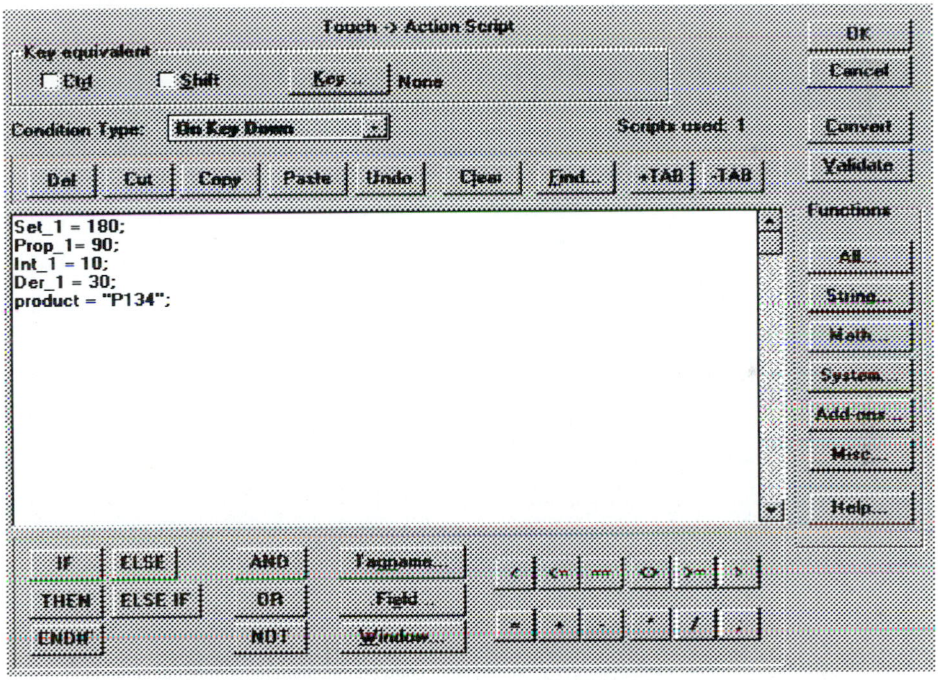

Figure 12-44. Script definition screen. This is the screen where the user writes simple scripts to define actions. In this case the parameter values for temperature controller 1, product 1 are being set. When the button linked to this action is pressed this script will execute and send the new values to the controller.

To create this display link for the product name, the programmer used the text tool and typed a space in the rectangle labeled product in Figure 12-45. The programmer then double-clicked on the space and chose a string display type. The programmer entered "product" for the tagname in Figure 12-46. Remember that "product" was entered as a variable name in the script for the product button. When the button is pressed the script will set variable product equal to P134. The link that we are now developing will display it on the screen.

Tagname product was never defined so the programmer had to define it. Figure 12-47 shows that the user chose memory message for the tagname type. This completes the display for product.

Next the programmer would have followed basically the same steps to create display links for the actual temperature and the setpoint. They have already been defined so their addition to the application would be very easy. They would use a DDE integer type because they both involve the actual controller. We will not add the links for temperature and setpoint.

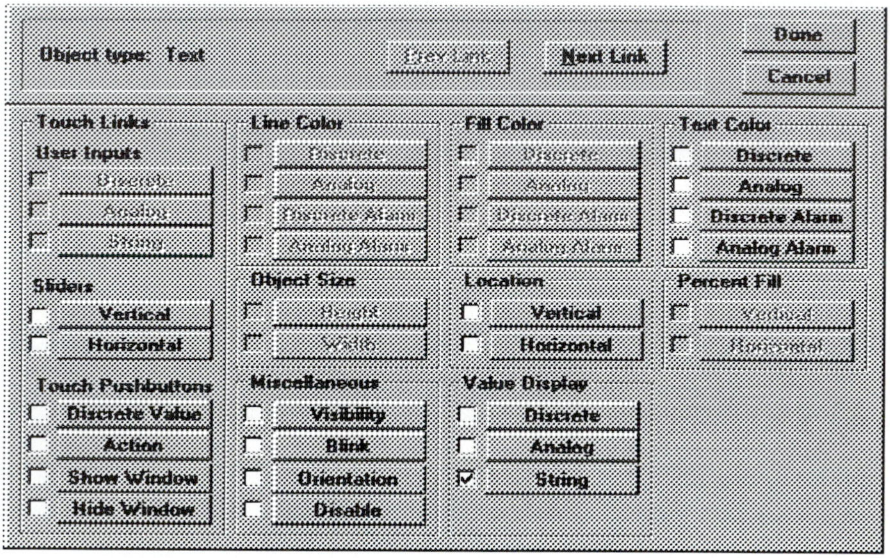

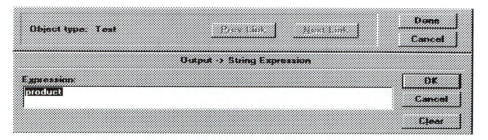

Figure 12-45. A string type value display was chosen.

Figure 12-46. Tagname definition screen.

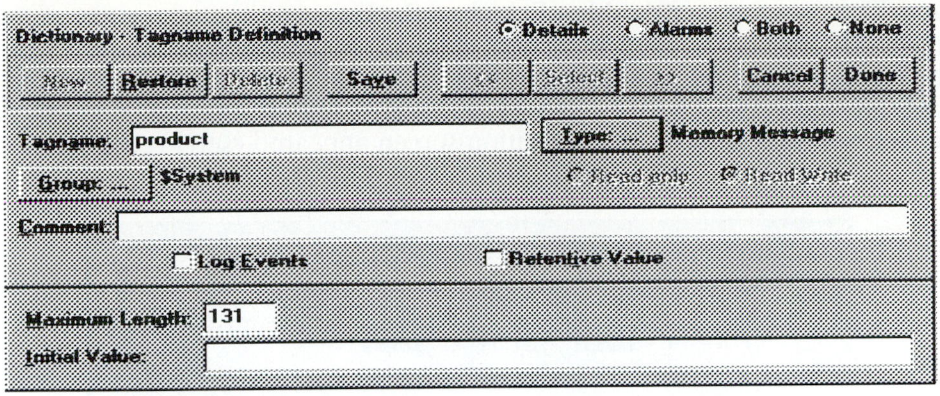

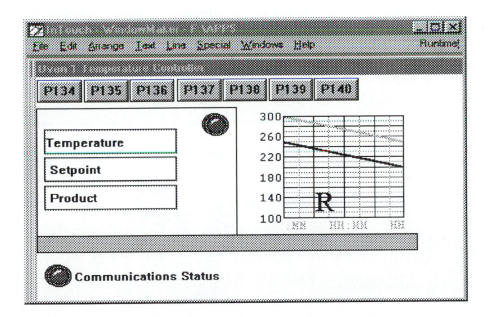

Figure 12-47. The programmer chose memory message for the type.

The same process was followed and buttons were created for each product (see Figure 12-48). The product variables and product name were changed for each product. Note that in a real application the buttons would be used to change the parameters in both temperature controllers at once.

Figure 12-48. Application development screen showing that all product buttons have been added.

The programmer then created a link to the part present indicator light on the furnace. The programmer double-clicked on the light on the upper right of the oven (see Figure 12-47). Part_present is entered for the tagname. Green is chosen for the ON color and red is chosen for the OFF color. The programmer then entered OK and was asked if he/she wanted to define Part_present. The programmer chose yes and the tagname definition screen appeared (see Figure 12-49).

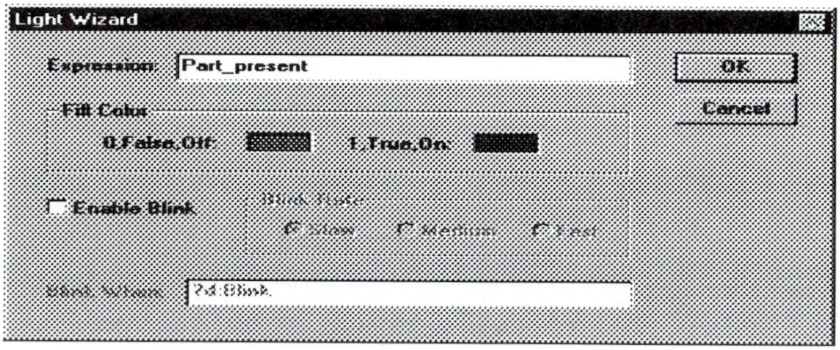

Figure 12-49. Tagname definition screen for Part_present. This is where the ON and OFF colors were chosen.

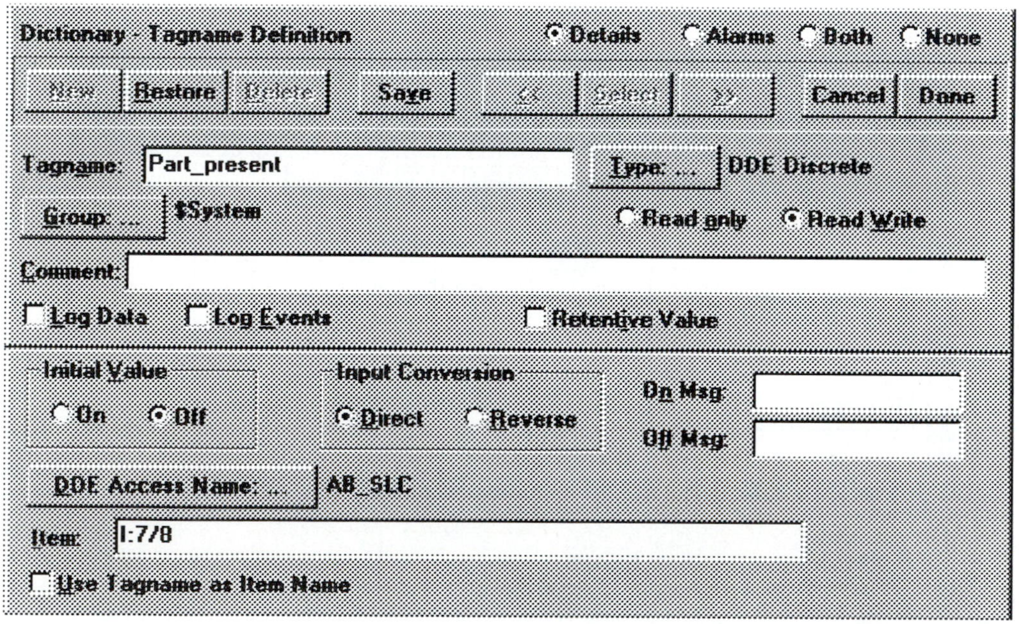

Figure 12-50. Tagname definition screen for Part_present indicator light.

The programmer chose DDE discrete for the I/O type (see Figure 12-50). AB_SLC was chosen for the DDE access name. Remember that this access name was created before to access the AB SLC 500 used in this application. I:7/8 was entered for the item name. This is the actual PLC input address for the presence sensor in the AB PLC.

The completed application is shown in Figure 12-51. This is what it would look like in the run mode. The operator could monitor the process and use the buttons to change to a different product. Note that the actual temperature appears on the screen as well as the setpoint and the product being run.

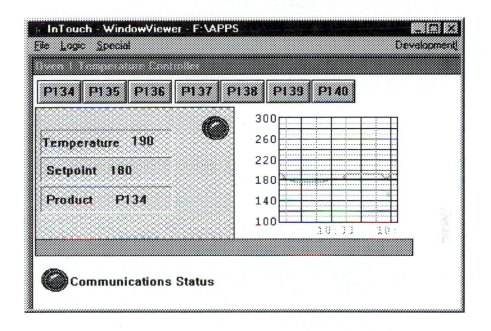

Figure 12-51. The complete application.

The effort that was used to create this temperature control window could be used to create a window for the second temperature controller. The window would be saved under a new name such as Temperature Controller 2. The programmer would change the tagnames and DDE access names so that the second controller would be used. A button could be added to each so that the operator could switch back and forth between screens. It would be very easy to finish the application.

This was a very simple application example to get a feel for how SCADA applications are developed. We could have made this application much more complex and useful. For example, we could have added animation effects so that we could have watched the product move down the conveyor line. Alarms could have been added in case something went wrong in the application. The alarm information could have been sent to another window, in the maintenance department, for example. We could have set up data logging so that process data could be saved for later use. In fact, any information in any of the controllers could now be available anywhere at any site worldwide. Internally the information can be exchanged over the internal computer network and the Internet can be used between worldwide sites. Figure 12-52 shows an example of a person monitoring a temperature-controlled system.

The data could be shared with any and all internal business systems. Figure 12-53 shows a graphic of the overall integration of an enterprise. There is tremendous capability in SCADA systems. We have taken only a quick look at maybe 5% of their power. Their biggest asset may be their ease of use and their power in easily integrating industrial devices and networks.

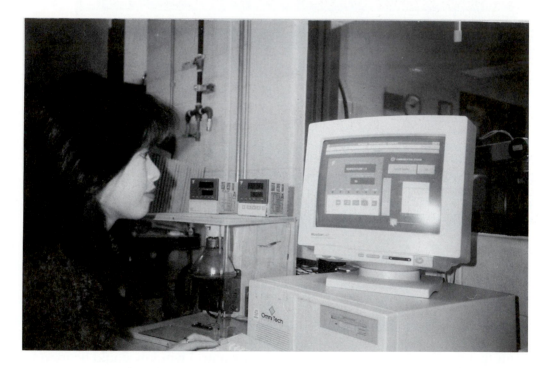

Figure 12-52. Photo courtesy Naho Takahashi.

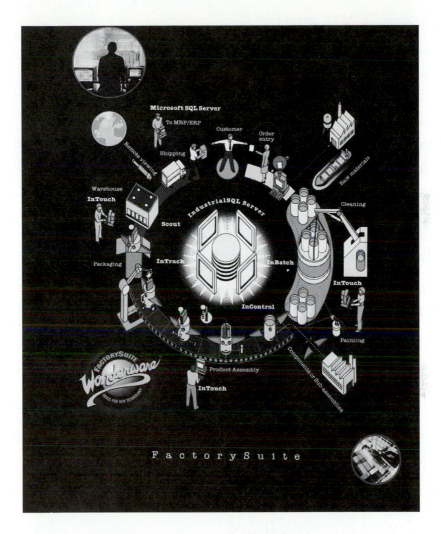

Figure 12-53. The overall integration of an enterprise and the Wonderware products that could be used for the integration. Graphic courtesy of Wonderware Corporation.

Questions

1. What does SCADA stand for?

2. What does the acronym GUI stand for?

3. What does the acronym HMI stand for?

4. What is an MES system?

5. What is a topic?

6. What is an item?

7. What is a driver?

8. What would historical logging be used for?

9. List and explain at least three different functional areas whose performance might be improved by the use of a SCADA system.

10. Draw a diagram that illustrates typical communication in a SCADA system. Make sure you include two computers and at least two different industrial controllers.

Chapter 13

PC-Based Control

The use of computers for industrial control is expanding at a very rapid rate. This chapter will examine the use of PCs for industrial control. Different approaches to PC-based control will also be examined.

OBJECTIVES

Upon completion of this chapter, the student will be able to:

>*Describe the types of applications that are most appropriate for industrial PC-based control.*

>*Describe the advantages of PC-based control.*

>*Compare and contrast the use of PC-based control vs. PLC-based control.*

>*Describe some of alternative approaches to PC-based control.*

INTRODUCTION TO PC-BASED CONTROL

PLCs were first sold as simple control devices that could be used to replace relays. Computers at this time were not user-friendly and users had to be software programmers in order to develop an application. There was almost no automation software available. Industrial personnel were unfamiliar and uncomfortable with computers. There were no fancy graphics or user-friendly help screens. Everything was text-based. This all made PLCs very attractive to manufacturers who wanted easier, quicker, and more flexible controllers than relay control methods.

PLCs were often called "relay replacers." They were intended to be easy for plant electricians to use and program. PLCs were very proprietary. Many manufacturers began producing and selling PLCs. They all used ladder logic for programming but even that varied substantially between PLC brands. It should be remembered that a PLC is really just a special-purpose microcomputer. The simple functions that the first PLCs could perform soon were not enough. People were demanding more capability. Users wanted math functions and analog capability. Users soon wanted to control more complex systems with PID functions. Manufacturers were more than happy to add and sell additional capability to users. There was heavy competition among PLC manufacturers.

Users also began to demand networking capability. PLC manufacturers offered proprietary solutions again. The networks, for the most part, were only intended to be used with the manufacturer's own brand of PLCs. They would not allow other brands of PLCs on their network. All used proprietary protocols. Some examples are AB Data Highway, Modbus, and many others. PLC networks did however have the capability of linking to IBM-compatible computers for programming and data exchange.

All PLCs are proprietary in software and hardware. You cannot buy I/O modules from one manufacturer and a CPU from another and expect them to work together. When you choose a PLC brand for an application you are forced to buy all related equipment from that manufacturer.

Personal computers started this way also. There were many companies that tried to sell their proprietary computers. Most have fallen by the wayside. The IBM PC became a standard. DOS and then Windows became the standard operating system. Today there are many manufacturers of PC microprocessor motherboards, memory, hard drives, monitors and so on. This has caused extreme price competition and real values for consumers. In fact, today almost anyone could start a computer business by buying standard components, assembling them into a computer, putting a name on them and then selling them. Apple was not so open with their system and has not flourished. Apple is very strong in niche markets like graphics and the printing industry.

This has not occurred with PLCs. PLC manufacturers have kept their products proprietary.

Ladder diagramming was usually done on a dedicated programming terminal purchased from the PLC manufacturer. These were very costly and cumbersome to program with. If you had multiple brands of PLCs you had to have multiple dedicated programming terminals, one for each brand of PLC.

Computers, meanwhile, were becoming more user-friendly and commonplace. People were becoming more competent at using computers. Third-party software companies began to sell ladder diagram programming software that could be used on a standard microcomputer. The software was more user-friendly than the manufacturers' dedicated systems and cheaper because the computer could be used for other purposes also. Instead of buying multiple programming terminals users could purchase programming software for each brand of PLC and utilize one computer to program all of their PLCs. This was a more economical solution. It also helped make microcomputers very popular in industry.

PLCs and industrial computers are just microprocessor-based systems. Both use a microprocessor for the logic and control functions.

Meanwhile computers were improving at an incredible pace. Software was becoming more and more user-friendly. The Windows environment standardized the look, feel and use of all software, vastly reducing the complexity of learning new software.

Hardware improvements were even more incredible. Processing power and speeds doubled every 6 to 12 months. Memory prices as well as the price of all related hardware steadily decreased. More and more people were able to have computers on their desks and in their homes. Computers have become an ever present and widely used tool in industry and our daily lives.

People are now realizing that the PLC is really just a special-purpose microcomputer. Users are now willing to consider the microcomputer as a control device in industry.

Advantages of Computers for Industrial Control

The interface between user and automated system is crucial. Computers have a distinct advantage in operator interface because of their graphics capability. A whole computer system is often cheaper than purchasing one high-quality color display panel for a PLC. The software that is available for the computer is often easier to use to develop these interface screens also. It is also possible to utilize the computer for other tasks, if the application allows. The user may be able to use other software in Windows to accomplish tasks. Data from the process might be used in spreadsheets, word processors or other analysis software. Microcomputers are excellent for data logging also. They can easily store vast amounts of historical process data.

The ease of use and understandability of the Microsoft Windows environment makes all software similar, and makes training easier and less extensive.

It is also easier to connect the computer to a standard LAN such as Ethernet and share manufacturing data throughout the organization. This ease of networking is a major advantage. The networking cost is also incredibly low. Network cards are available at a fraction of the cost of a PLC network card.

The network capability allows integration of other systems for data exchange, tracking, maintenance, production planning, quality control, recipe downloading, order tracking, etc.

Another advantage is that the same hardware can be used for programming and control. The hardware can be used to develop the program and then to operate and troubleshoot the system. Microcomputer hardware is standardized and can survive several generations of product change. This is not always the case with PLCs.

There is an endless array of peripheral devices available for microcomputers: printers, bar code scanners, bar code printers, sound cards, multimedia cards, and so on. The user can purchase hardware from any manufacturer. The prices of hardware are continually falling. Several years ago an observation was made that if luxury cars had made the same improvements in power and speed and decreases in price that computers had made a luxury car today would cost about 50 cents. The improvements in computers have been phenomenal.

The drive toward a common programming standard (IEC 1131-3) has also made computers more attractive. There are a variety of programming methods available for PLCs and microcomputers: ladder, flow diagram, function block, statement list, state logic, C, BASIC, etc.

Microcomputer Concerns

Anyone who has ever had a computer lock up for no apparent reason realizes that this cannot be allowed to happen when industrial systems are involved, especially when safety might be a concern. A manufacturer will not tolerate a control system that locks up.

PLCs are designed to be very rugged in an industrial environment. They are very noise immune. They can operate in extremes of temperature, shock and vibration, and in dirty environments.

Several companies have come up with microcomputer solutions to these concerns. There is a wide array of equipment and software to utilize computers for industrial control. Next we will consider a few of the alternative approaches to computer control.

There are two types of PC-based control systems, soft logic and hard real-time control (see Figures 13-1 and 13-2). Most soft-logic systems run the control as a high priority real-time task under Windows NT. Real-time tasks can be interrupted by deferred procedure calls which are used to service NT system functions such as disk access, network communications, and so on. Soft logic does not provide the same level of deterministic control as a PLC because higher priority functions can interrupt and delay real-time control systems. If the application requires high-speed control the lack of determinism may be unacceptable or worse yet, unsafe.

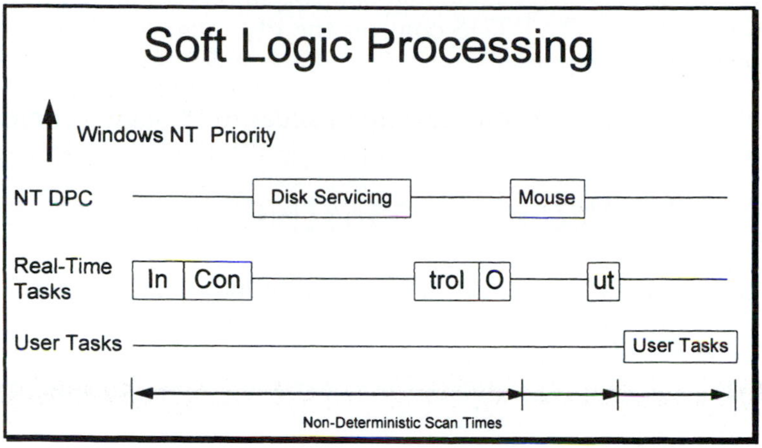

Figure 13-1. Soft logic processing. Courtesy Steeplechase Software, Inc.

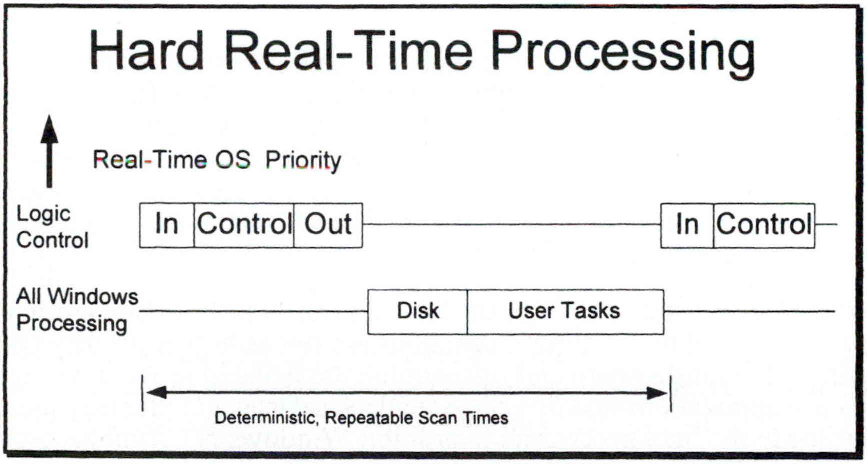

Figure 13-2. Hard real-time processing. Courtesy Steeplechase Software, Inc.

Hard-real time control is used by PLC systems. In a hard real-time system, the real-time operating system is loaded first. The control engine runs as the highest priority task. All Windows functions run as the lowest priority task within the operating system. The logic control engine always has priority and is completely protected from Windows. Windows is not allowed to preempt real-time control.

Remember that a PLC is really just a microprocessor with a real-time operating system (RTOS). The RTOS is the code that controls all tasks and operations that run on the microprocessor. The RTOS controls the PLC scan and logic and gives the PLC its deterministic response.

Next we will take a look at a few different approaches to PC-based control.

FLOWCHART PROGRAMMING

Flowchart programming is becoming very popular in industrial control. There are many reasons for its popularity as a programming language. Steeplechase Software, Inc. has a PC-based control system that utilizes flowchart programming as their fundamental programming method. Flowcharts allow systems to be developed as a simple, intuitive, graphical description of a process that everyone can understand. Flowcharting allows engineers, operators and plant floor technicians alike to easily interpret and understand the step-by-step process used to program, operate and troubleshoot machines. The same cannot be said about ladder logic. Steeplechase calls the product the Visual Logic Controller, or VLC. Their system was the first Windows-based system on the market to offer integrated flowchart programming, control, simulation and a MMI (man-machine interface) an a simple PC.

Steeplechase's approach marries the benefits of hard real-time control and Windows NT. The Windows NT environment provides many benefits over the traditional PLC. Windows-based systems can be used to provide network communications and graphical user interface, and can utilize Windows-based software to process information. There is also a substantial cost savings in some systems. In systems where communication and graphics are important a PC-based controller can be more cost effective (see Figure 13-3).

Steeplechase loads Windows NT as the lowest priority task in the hard real-time operating system. All of the control functions are run as higher priority tasks in the real-time operating system and furthermore are isolated in memory from Windows NT applications and drivers. Steeplechase uses the memory protection functions inside the Intel processors to prohibit Windows NT from accessing any of the memory or CPU cycles that are dedicated to the real-time engine.

This approach yield several benefits. Windows NT can crash without affecting system control. This allows the control program to continue to operate as normal or execute an orderly shutdown to put the machine into a safe state.

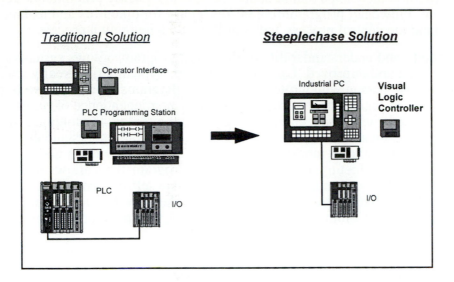

Figure 13-3. A comparison of a PLC system to a system con-trolled by a PC-based controller. Diagram courtesy Steeplechase Software, Inc.

The system can survive a hard disk crash. The hard real-time operating system is loaded into the PC, so the dependence on the hard disk is eliminated. The entire RTOS is loaded and active in memory, so a failure of the hard disk, regardless of its impact on Windows NT, will not affect the control activities.

Benefits of Flowchart Programming

Flowcharting, unlike ladder diagramming, allows programs to be easily broken down into logical steps. Control system design can be reduced 50 to 70% using flowchart programming compared to ladder logic. It should be remembered that a large part of every ladder diagram is logic to make sure outputs are not on when they should not be. Flowcharts break the process into logical steps that avoid that problem. Flowcharts allow the programmer to draw the desired machine control sequence as a simple set of flowchart steps. "Boxes" are actions, where outputs are turned on/off and math or logic functions are performed. "Diamonds" are decisions which direct the control sequencing based on input states or logic conditions. The operation of any process can be easily described by a set of flowcharts. Flowcharting is very similar to the way in which a system would be designed anyway. The engineers, technicians, and operators would get together to develop and understand the system. They would discuss and write down the process as a series of steps. Flowchart programming makes good use of that effort and the flowchart becomes the program. During runtime troubleshooting is made much easier using flowcharts. In fact, operators can often find the problem by watching the control program flowchart as it highlights every step of the process.

This information then helps maintenance technicians quickly pinpoint and repair the problem. Flowchart programming makes it easy for operators and technicians to understand the control process and contribute to debugging a process.

The simplicity and understandability of flowchart programs allows systems to be easily modified and improved. Ladder diagrams are typically only really understood by the one person who wrote them. Ladder diagrams typically have hundreds of rungs of logic. The interrelationship and interdependence of rungs make it difficult to make wholesale changes to logic after a system is running. Flowcharts, however, break even the complex system into small understandable blocks. This ability allows modifications to be easily made without undue worry about the change's effect upon other logic in the program. This allows process improvement to be easily made. Flowcharting is also self-documenting. A flowchart is easily understood by all plant personnel. The flowchart acts as program and documentation. Comments are easily added to explain the process even further.

APPLICATION DEVELOPMENT

There are three steps in application development.

> *Planning the control sequences and entering the flowcharts into the PC,*

> *Setting up the operator control panel screens (MMI), and*

> *Simulating the control system and when ready assigning the real-world I/O to tagnames and running the system.*

Planning Control Sequences and Entering the Flowcharts

The first step is to plan the machine sequences. Let's consider a simple tank fill application. The sequence is very simple. First the pump must be turned on, then the level must be monitored so that the pump can be turned off when the tank is full.

Programming

Each flowchart begins with a Start element (see Figure 13-4). The elements (Start, Action, Decision, Subprogram, and Stop) are simply chosen from the flow tool palette and placed on the screen. This is done by clicking the mouse on the desired element in the tool palette and then moving the cursor to the desired position on the screen and clicking the mouse button again. The programmer simply places all of the desired elements on the screen in the desired location. The programmer then uses the flow tool to connect the elements together.

In our example (see Figure 13-4) the first element was a Start element. The start was connected to an action element with a flow line. This action element is used to turn the pump motor on. Action elements can also be used to initiate functions such as turning outputs on/off, performing mathematical computations, logic or any combination of these. If the programmer double clicks on any element a new screen will appear to allow the programer to configure the element. Figure 13-5 shows a typical input screen for an action element. Flowchart Enhancements

Loop commands are also available, IF-THEN-ELSE and WHILE. In addition, multidimensional arrays have been incorporated to aid in the generation and processing of complex algorithms traditionally used in a variety of material handling and high-speed sorting applications.

▭	Action	Control outputs, perform math, other functions; each Action element can include up to any number of statements (up to 64K of text).
◇	Decision	Decides to take the "Yes" or "No" branch based on a logical result
▱	Comments	Each comment contains up to 1024 characters to document your program
↓	Flows	Connects flow chart elements
⬭	Start/Stop	Defines where programs begin and end
○	Connectors	Indicates connections to flows at other points in the flow chart
▯▯▯	Sub-programs	Calls a program as a sub-program; parameters can be passed with the current tag value or by a tag name
⬭	I/O Specific	Calls an I/O function specific to an I/O family, such as diagnostics
═══	Parallel	Parallel branch and merge allow separates flow to execute simultaneously, starting at a branch and ending at a merge; up to 255 separate flows can operate in parallel

Figure 13-4. The elements that are used to create a flow diagram program. Diagram courtesy of Steeplechase Software, Inc.

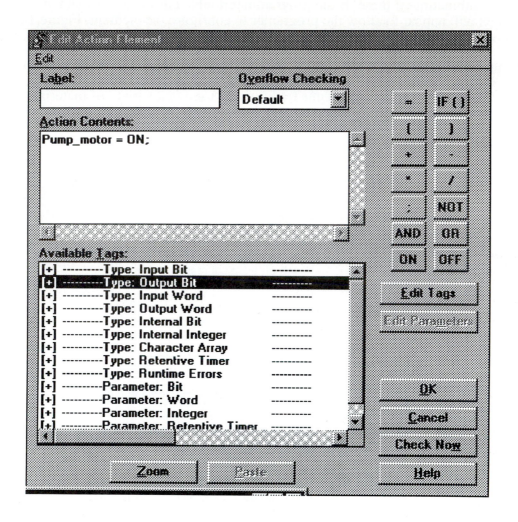

Figure 13-5. A configuration screen for a tag.

If the programmer chooses edit tag from the screen shown in Figure 13-5 the screen shown in Figure 13-6 appears. The programmer then enters the specific information for this tag. The name of the tag will be Pump_motor. The device is then chosen from a drop-down list. The specific point is then entered. The point is the actual address of the I/O point in the device that was chosen. This is an important point. With this type of control (industrial PC) there can be multiple devices and/or I/O devices attached to the computer. They will all respond as if they are one system. The programmer just chooses the device from the list for each particular tag.

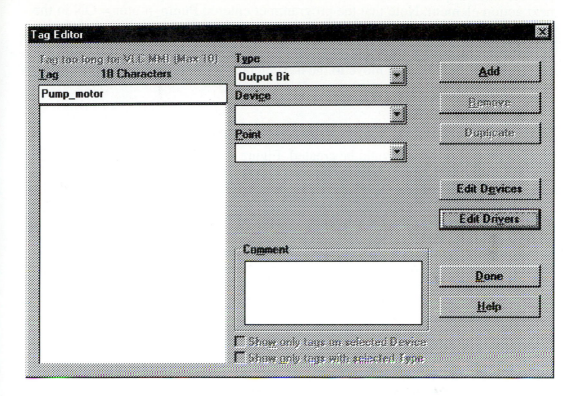

Figure 13-6. A tag definition screen.

This is quite an advantage over a PLC, in that devices and I/O from different manufacturers can be easily integrated. The software supports I/O modules from Allen Bradley, GE Fanuc, APC Seriplex, DeviceNet, Modicon Remote I/O, Interbus-S, PID Controller, RS-232/RS-422 Drivers, Smart Distributed System (SDS) and many others.

The tag configuration can be done at the time the element is placed on the screen or later after all graphic elements have been placed. This same process would be followed for all elements.

The next element is a Decision element. This element is used to decide if the tank is full. Note that there are two paths that flow out from a Decision block, a YES and a NO path. Each of the paths can connect to other elements. In our system the YES flow line is connected to the next Action element. The No flow line is connected back to the input flow line to the decision block. This Decision element will be used to determine if the tank is full. If the tank is full, the YES flow line will be followed to the next element. If the tank is not full the NO flow line will be followed back to the Decision element again. It will continue to repeat this process until the tank is full. When the tank is full, processing then moves to the next action element. Note that the programmer entered Pump_motor = ON in the Action contents box. The programmer also chose the output type bit for the tag.

This Action element will be used to turn the pump motor off. After the pump is turned off processing will move to the next element. The next (and last) element is the Stop element.

Setting Up Operator Control Panel Screens

Next the operator panel would be created. Figure 13-7 shows what the completed operator panel looks like. There is a meter shown in the upper right of the panel. The meter will be used to show the pump speed. The slider control will be used by the operator to set the pump speed. The operator can use the mouse to move the slider. The pump speed can also be controlled by using the up and down arrow keys. The meter, slider and up/down arrows are all linked to the actual analog register in the controller that controls the pump speed. All of the controls on this screen were created by selecting them from a library (see Figure 13-8) and placing them on the screen. Figure 13-9 shows an example of the types of sliders available. The programmer then double-clicks on each and assigns a tagname, device and the actual I/O point address.

The graph is used to show tank level versus time. Note that multiple variables can be graphed at one time so the operator can get instant real-time feedback on actual system operation.

The programmer can also draw or import text objects and animate them. By double clicking on any drawn object it is possible to link the graphic to a tag and control its color, position, size and fill. Figure 13-10 shows an example of the screen during development.

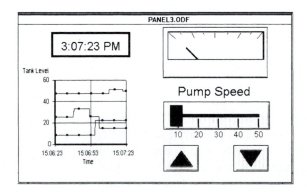

Figure 13-7. The completed operator panel. Diagram courtesy Steeplechase Software, Inc.

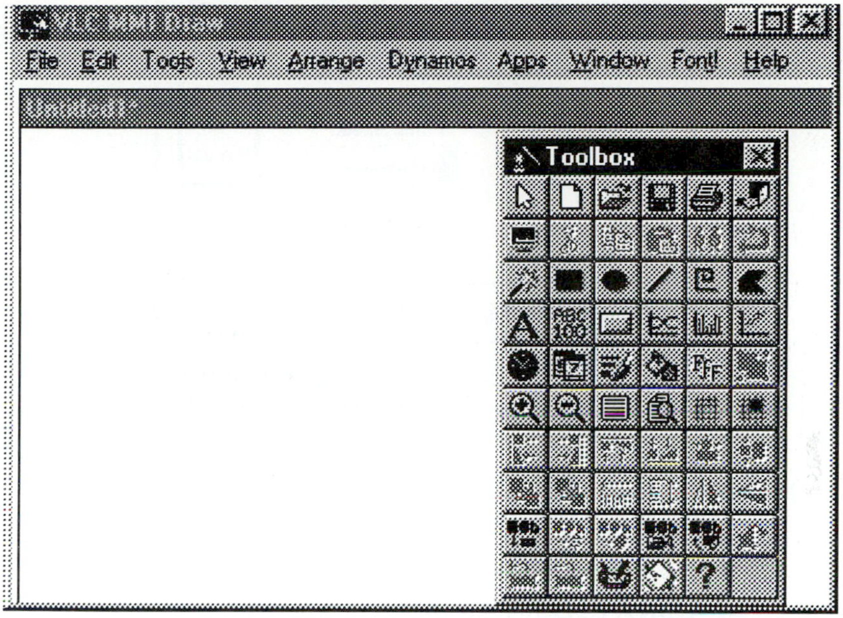

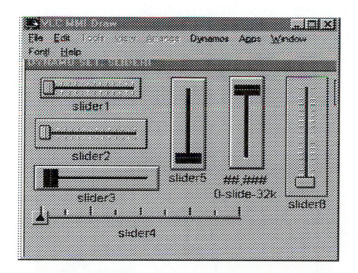

Figure 13-8. The MMI development screen. Note the tool palette.

Figure 13-9. Some of the slider controls that are available in the library.

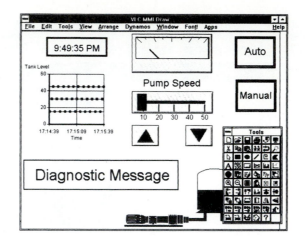

Figure 13-10. The operator panel for the system during its development. Diagram courtesy of Steeplechase Software, Inc.

Simulating the Control System and Assigning I/O

After the flowchart and operator panel have been developed the system can be simulated to be sure that everything works correctly. Figure 13-11 shows the way the screen might appear for our system during simulation. Note that the active step is highlighted. Note also that the operator panel shows the tank about half full. The level will change as the real-world level changes. The tank level display was linked to an analog register in the controller. The programmer can utilize any of the controls on the screen for test purposes.

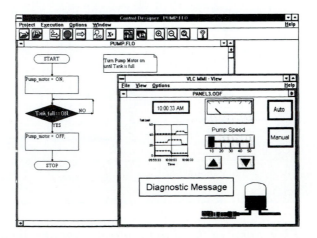

Figure 13-11. The system in simulation mode. Diagram courtesy of Steeplechase Software, Inc.

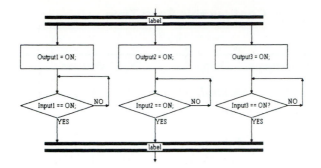

Figure 13-12. An example of parallel branch elements. Diagram courtesy of Steeplechase Software, Inc.

At this point, if it had not been done already, the programmer could assign real-world I/O to the tags and run the machine. It should be noted that this was a very simple example and does not show the real power or ease of use of this software.

Concurrent processing can also take place. Figure 13-12 shows an example. In this case three different flows were started from one point. These are called parallel branches. The double lines at the top and bottom can be used to create a parallel branch or parallel merge.

Another type of very powerful element is a subprogram element. These can be used to simplify the view of the process. A subprogram element is actually used to call another flowchart. This can be used to hide portions of programs that have already been tested or to organize programs in hierarchies so that each view remains simple and compact. The number and nesting of subprograms is almost unlimited.

Ladder Logic Editor

Ladder logic is still the most commonly used control language on the factory floor. Steeplechase software allows users to write programs in traditional flowcharts or a ladder logic editor based on the IEC-1131 specification.

To increase the power and usability of ladder logic, powerful flowchart elements can be added, allowing users to create more powerful program routines than with pure ladder logic. The incorporation of self-documenting flowchart elements in a ladder logic program reduces the time required to develop a control program, and greatly simplifies troubleshooting.

Scan Time

The programmer can set the desired scan time. The scan rate can be set in 5 millisecond increments from 5 to 500 milliseconds. During each scan the VLC reads inputs, executes the program, and writes outputs. In the time remaining, it executes DOS and Windows applications (see Figure 13-13).

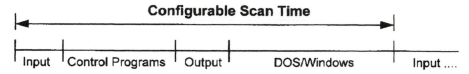

Configurable Scan Time

| Input | Control Programs | Output | DOS/Windows | Input .... |

Figure 13-13. An example of scan time. Diagram courtesy of Steeplechase Software, Inc.

If the control programs exceed the target scan time during any cycle, the DOS and Windows applications do not execute in that cycle. The control program can also detect the scan overrun error and report it.

As you can see, flowcharting is a very powerful and yet easy to understand programming language.

GELLO

GELLO is a product that was developed by Event Technologies, Inc. It is an object-oriented system. GELLO can be used in a number of configurations to control systems.

A PC would be used to create the program, simulate the application, and also debug the application. The program can then be downloaded to a target controller to run the application. The target controller could be:

> *A regular PC (see Figure 13-14). This figure shows an off-the-shelf PC as the controller. Special-purpose cards in the PC can directly control I/O or rack-based, device, or distributed I/O.*

> *An embedded co-processor. The embedded controller would be a separate card with its own CPU. The co-processor card would be plugged into a host PC's bus. The card would run the GELLO engine. The co-processor card can also communicate with the host controller.*

> *PC-based rack I/O control. In this configuration the target controller is a separate, dedicated, rack-mount PC controller operating with custom, integrated I/O modules.*

> *Multitasking system control. The target in this case would be the development PC. The GELLO engine in this case runs as the highest priority task. It is also preemptive, which ensures that control execution is not compromised by other tasks.*

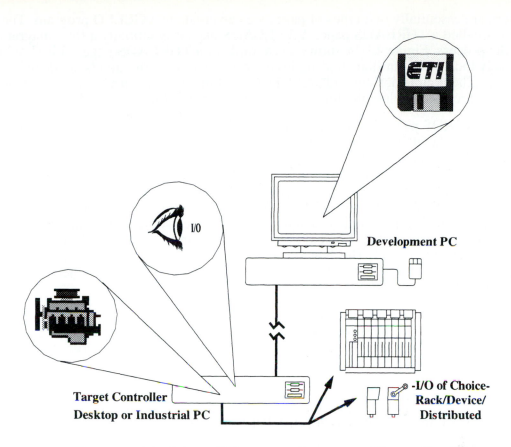

Figure 13-14. Using a PC to control a system. The application was developed in the development PC. The GELLO software is represented by the ETI disk (Event Technologies Inc.). The "eye" represents the system I/O. The engine represents the GELLO runtime engine running in the controller. Courtesy Event Technologies Inc.

Boiler Application

Let's look at a typical industrial application to see how GELLO would be used to control the application. Figure 13-15 shows a boiler system. The temperature of the boiler is controlled by varying the fuel input into the heater. We will use PID to perform the closed-loop control.

Next we will look at how the actual application would be developed. We will create a PID controller. The controller will have adjustable setpoints and several variables to monitor and control the tuning of the PID loop. We will also add a strip chart recorder to record the input and output values and to monitor loop performance as tuning variables are adjusted.

There are essentially two types of pages we can create in a GELLO program. The first is called a THREADS page. A THREADS page is essentially a flow diagram of the process. Figure 13-16 shows an example of a THREADS page. This should already look quite familiar. Rectangles are process blocks, diamonds are decision blocks. The double horizontal lines indicate that there is concurrent processing of the two processes below the lines. Arrows indicate the flow between blocks.

The THREADS page is created by choosing the appropriate objects from a tool palette and placing them on the page. Figure 13-17 shows the object palette for the THREADS page. The leftmost tool is the *decision* object. It could be used to make decisions as to time, counts and so on. The second tool is the *processs* object. The process object can be used to create a whole application or a piece of the application. We will use a process object later to create our boiler PID controller. The third object is the *fork* object. This can be used to create alternatives for concurrent processing. Look at Figure 13-16 again to see its use. The fourth object is the *sync* object. The sync object is used in conjunction with the fork (see Figure 13-16). The last object is the *I/O disable/enable* object. This object can be used to enable/disable I/O.

So remember that the THREADS page is the logical flow of the application. Also note that it is very easy to document and comment the objects on the page (see Figure 13-16). The next step would be to define how each object works. To do this let's see how the boiler application would be developed.

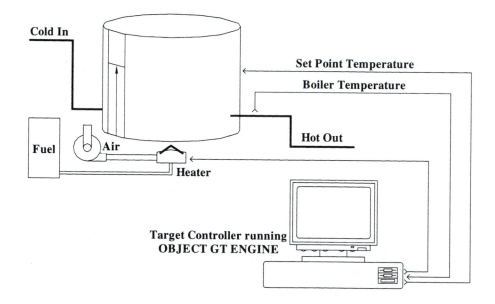

Figure 13-15. PID control for boiler temperature control.
Courtesy Event Technologies Inc.

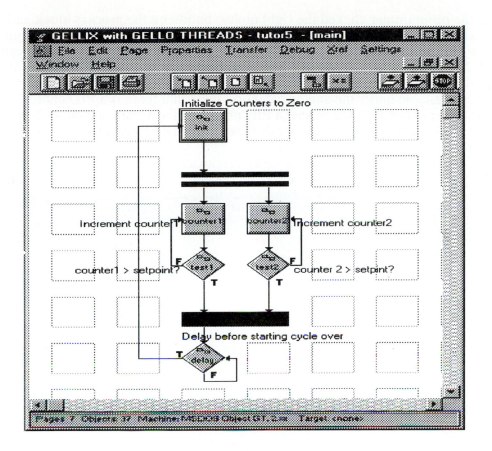

Figure 13-16. A GELLO THREADS page.

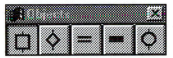

Figure 13-17. Object palette for the THREADS page.

The first step would be to create our THREADS page. Figure 13-18 shows the THREADS page for our PID application. The programmer chose a process object and used the mouse to draw a flow line from the bottom of the box (process object) to the top of the box. This means that the process will operate continuously. The programmer also added a label for the name of the process block: Boiler PID. The programmer also created a page underneath the process block and named it Page 1. This page will be used to define the actual application. The THREADS page is now complete. As you can see, it is very simple for our application. It only has one process block that will execute continuously.

Next the programmer clicked the right mouse button on the process block and was taken to the next level (Page 1). This level will be used to define what our Boiler PID process block does. Figure 13-19 shows the type of objects that can be used. Note that there are analog, binary, bit, conversions, float, float_math, page, and table. Each of these types has more alternatives. Figure 13-20 shows the choices available under the analog category. Note that there is a PID choice. The programmer chose PID and placed a PID object on the screen, shown in Figure 13-21. Note that all the programmer did was place the object on the screen. There was no programming involved.

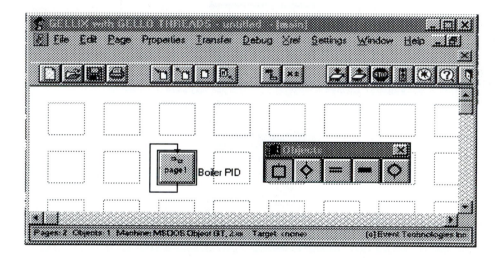

Figure 13-18. Development screen showing a GELLO THREADS page with a process object for the boiler PID application.

Figure 13-19. Drop-down menu of object types.

*Figure 13-20. Drop-down submenu showing just the analog
type object that are available.*

The programmer then placed the rest of the objects need for the application.
Figure 13-22 shows the completed application. The programmer added two
constant blocks to the upper left of the page. These were objects that were placed
just like the PID object. The uppermost constant is the high setpoint limit for our
sensor. The lower object is the low setpoint for our system. The programmer
clicked on the objects to add a label and clicked in the lower right of the object to
enter a value. The high setpoint limit was set to 200 and the low was set to 40.

The programmer then added a setpoint wave generator object and labeled it SP

Wave Generator. This object will be used to simulate a change in setpoints so that we can test our PID system's ability to control the temperature. This object will vary the setpoint between 40 and 200: the high and low limits. Note that the programmer used the mouse to connect the output from the high and low setpoint objects to the high and low input of the waveform generator. The wave generator chosen was a sine wave type. The programmer then clicked in the lower right of the waveform generator and entered a period of 10 seconds. The range of the period could be between .07 and 327 seconds.

Next the programmer added five variable objects. These will be used for the setpoint, proportional gain, integral gain, derivative gain, and for a loop output variable. The programmer clicked on each variable and entered a label (name) for each. The programmer also clicked on the lower right of the proportional, integral and derivative variable objects and entered a value for each. Next the programmer added a filter type object to the right of the PID object. The programmer then connected the objects with flow lines. Study the figure closely to understand the relationship between the inputs and outputs. For example, note that the PID block output is fed to the filter and the output from the filter is sent to the input of the loop variable. The output of the loop variable is an input to the PV input on the PID object. PV is the present value of the system: the current temperature in our case. Note the other inputs to the PID object. The system is now functional and can run in simulation. The programmer added one more thing to the application screen, a strip chart (real-time graph) so that the operation would be easier to see in simulation mode. Figure 13-23 shows the system in simulation mode.

Figure 13-21. The analog objects drop-down menu. Note that PID was chosen and the programmer placed a PID object on the screen.

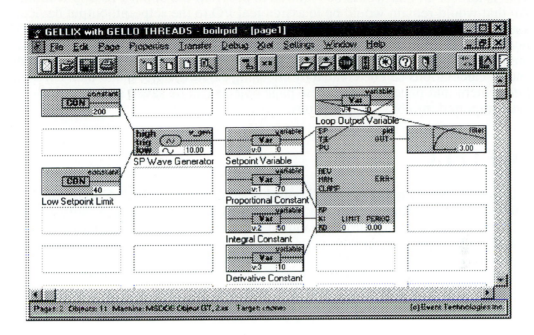

Figure 13-22. The completed application.

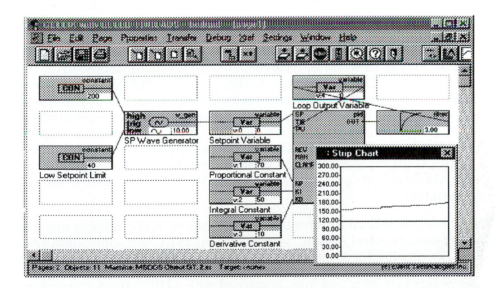

Figure 13-23. The system in simulation mode with the strip chart showing system performance during simulation.

That is almost all that would be required to program the boiler PID application. All that would remain would be to assign real I/O to the setpoint, output device, and temperature input device. You will note that there was no programming involved. The mouse was used to choose objects and then each object was configured. GELLO also allows the user to create custom objects. Custom control logic and/or calculations can be created by choosing existing objects and then *encapsulating* them into page objects. These page objects can be saved and reused and even imported into other applications.

SOFTPLC

A different approach to PC-based control was taken by SoftPLC Corporation. They decided to utilize ladder logic as their programming language. SoftPLC is software technology that turns a standard industrial computer (PC) into a full-function PLC-like process controller. SoftPLC combines PID, discrete and analog I/O control with the data handling, computational and networking capabilities of computers. A multitasking control kernel, SoftPLC provides a powerful instruction set, fast and deterministic scan time, reliable operation, and an open architecture for connection to a wide variety of I/O systems, other devices and networks.

SoftPLC, when added to supporting computer hardware components, creates a control system with throughput, performance and programming capabilities exceeding those of conventional programmable controllers.

SoftPLC is not a DOS application; it is a real-time, multitasking operating system, or kernel. Once SoftPLC is loaded into memory, it is in control of the computer CPU at all times. SoftPLC loads from DOS, then runs in computer RAM memory. SoftPLC can be loaded from any defined drive available to the computer, including EPROM or flash memory, a local hard/floppy drive, a network drive, etc. DOS is never called. This gives SoftPLC the reliability and characteristics of proprietary dedicated hardware PLCs, such as deterministic scan time, protection from bugs in the operating system or other software applications, etc. In general, SoftPLC's instruction execution times are 2 to 10 times faster than traditional PLCs on a 486 system. On-line run mode program changes may be made. SoftPLC logic can be changed while the machine(s) are running. There is no need to recompile the application or to stop the machine control to download the new program.

SoftPLC provides several features that ensure consistent and reliable operation of a computer-based control system. SoftPLC manages its own application program and data files directly for automatic backup/restore (upload/download) purposes.

Advance power loss detection logic combined with a UPS (uninterruptible power supply) or battery-backed RAM may be used to enable SoftPLC to save the ladder program and data table to a disk file (which may reside on a network file server).

SoftPLC's status file includes fault bits to enable easy detection of the reason for a runtime error and to clear/correct the fault. (Examples of these faults include divide by 0 and jump to nonexistent label.)

The keyboard lockout feature can be used to protect the operation of the system. Even [Ctrl]-[Alt]-[Del] won't shut down the system.

SoftPLC Features

Runs as a 32-bit real-time program on 386, 486, or Pentium platforms. There are four user-configurable communication channels (network, data paths) for data or program logic access from other computer applications or PLCs programmed with TOPDOC. It runs imported/converted Allen-Bradley PLC, PLC-2, PLC-5 or SLC-500 programs.

Instruction Set Programming

SoftPLC's ladder logic execution capabilities are very similar to those of an Allen-Bradley PLC-5. In fact, they are similar enough that an existing PLC-2, PLC-5 or SLC-500 program can be easily converted, then loaded into SoftPLC. Standard ladder instructions include: contacts, coils, timers/counters, data comparisons/moves, math and logical operations, shift registers/sequencers, branches, jumps and subroutines, and specials (PID, message, diagnostics). SoftPLC's standard instruction set also includes a number of loadable functions such as COMGENIE, which is used for general-purpose ASCII communications to/from RS-232, RS-422 or RS-485 devices through COM ports. TOPDOC also provides a PID loop display and a PID auto-tuning utility as well as advanced math instructions such as statistics, trigonometric functions, and others. Users can create their own instructions with C, C++, or Java.

A-B Conversion Utilities

Software utilities are available to import and convert PLC, PLC-2, PLC-5 or SLC-500 programs and documentation into SoftPLC/TOPDOC format, thus providing an easy upgrade path from existing Allen-Bradley PLC systems to SoftPLC.

I/O Capabilities

SoftPLC connects to I/O systems through the use of loadable software I/O drivers. The user can select whichever I/O vendor hardware is best for the specific application. The drivers for some of the more popular I/O systems are included with SoftPLC. Other drivers have been developed and are available from I/O vendors or third parties. SoftPLC I/O drivers are called loadable modules (or TLMs).

SoftPLC works with a variety of I/O types. The hardware is available from several I/O vendors. SoftPLC Corporation also distributes some I/O interface cards. I/O can reside on the backplane of the computer or be remote. Some I/O cards fit directly into the computer, in which case the I/O driver talks to the I/O over a PC bus or backplane. Remote or fieldbus I/O systems typically utilize an I/O scanner or interface card that fits into the PC. In these cases, the SoftPLC I/O driver communicates to RAM memory on the card, which stores the I/O status information from inputs or sends it to outputs.

SoftPLC's maximum I/O capacity is 8192 I/O, and includes support for analog, digital and special I/O. I/O forcing is independently controllable for inputs and outputs. Motion applications can also be controlled. The user just adds a motion control card to the PC. The combination of SoftPLC and a computer resident motion control card makes a powerful, tightly integrated system. Tuning, scaling, trajectory, and velocity data can come from SoftPLC, using simple ladder logic commands. SoftPLC can then send data to each axis on an event-driven or timed basis.

HARDWARE

There are major changes coming in control hardware. The line between PLCs and PC-based controllers will certainly blur. One of the approaches will be to get dual use from the equipment. Allen-Bradley has introduced a product called the 1747 Open Controller. This controller is their first entry in a family of PC-based controllers.

The heart of this controller is the 1747-OC CPU module. This is the processor module. It can be used in the standard SLC PLC chassis. The CPU is capable of using any existing Allen-Bradley hardware, I/O chassis, and power supply. The controller module has dual CPUs. One is used as an I/O scanner and the other for the operating system (see Figure 13-24). This assures rapid deterministic response times and reliable operation.

Users can choose the operating system and software they would like to use to control the system (see Figure 13-25). Any PC-based commercial software control package can be used with the controller and standard Allen-Bradley I/O. The system can be expanded to utilize other I/O also.

The CPU utilizes SRAM, DRAM, and Flash Drive™ technology so that it is not dependent on a hard drive. There are many communications options because it is a PC-based system. Ethernet interfaces, Allen-Bradley communications cards, SCSI adapters, modems and so on are all available to expand the capability of the system. Figure 13-26 shows examples of the use of modules to connect to networks.

The system is capable of booting without a monitor or keyboard. It can also be used with keyboard and monitor for operator interface.

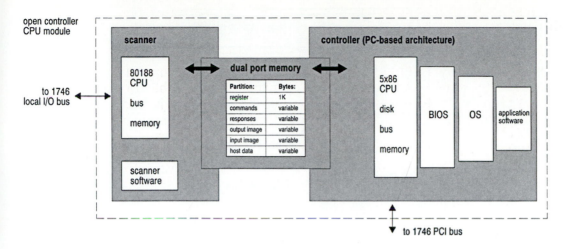

Figure 13-24. Diagram of the Open Controller architecture.
Courtesy of Rockwell Automation/Allen-Bradley Company Inc.

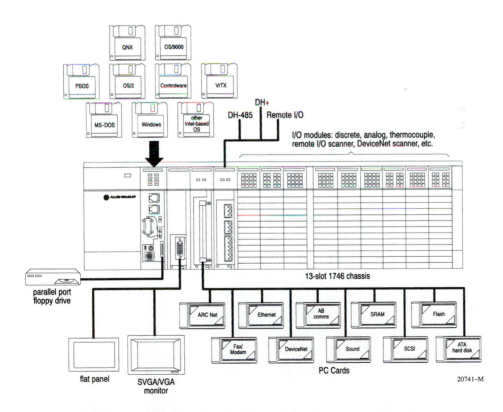

Figure 13-25. 1747 Open Controller System and options.
Courtesy of Rockwell Automation/Allen-Bradley Company Inc.

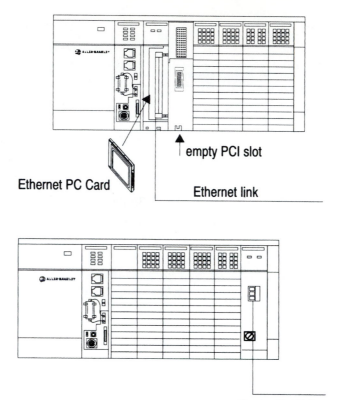

Figure 13-26. Examples of connecting to communications networks. Courtesy of Rockwell Automation/Allen-Bradley Company Inc.

Questions

1. List at least five advantages of PC-based control systems.

2. List at least three potential disadvantages of PC-based control.

3. What is flowchart programming?

4. Describe at least 3 advantages of flowchart programming.

5. Describe at least 3 advantages of ladder logic programming for PC-based control.

6. Explain what object-oriented programming is.

7. Compare and contrast the methods and advantages of each of the four methods of PC-based control in this chapter.

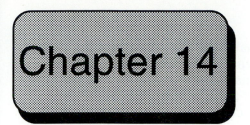

Chapter 14

Industrial Automation Controllers

There is a new type of controller that is becoming more popular in industry. It is not a PLC and it is not easily classified as an industrial computer either.

OBJECTIVES

Upon completion of this chapter, the student will be able to:

>*Describe what a industrial automation controller is.*

>*Describe applications that are appropriate for industrial automation controllers.*

>*Describe how a typical industrial automation controller is programmed.*

>*Develop a flow chart for a typical program.*

OVERVIEW OF INDUSTRIAL AUTOMATION CONTROLLERS

There is a new breed of industrial controllers that do not fit neatly into the PLC or personal computer classifications. They are often used for special application controls such as motion and process control. This chapter will focus on one brand of controller. Control Technology Corporation's (CTC) controller will be used to explain industrial automation controllers. Other companies such as Giddings & Lewis also offer industrial automation controllers.

Industrial automation controllers have been designed to fill a need. Traditional PLCs and industrial computers were very adequate for the vast majority of control applications. Applications have become increasingly more complex and have required more speed. This has led to more and more servo control. Closed-loop servo control has become very prevalent in industry. Traditional controllers have not addressed this need adequately.

Industrial automation controllers offer the traditional discrete and analog control capabilities and also extended capabilities. The CTC controller uses a state logic-type programming language called Quickstep State Language. Quickstep State Language is very English-like, which makes it very user friendly. The language is also designed to help break complex tasks into logical stages or steps. Next we will take a look at a simple application that will explain Quickstep Language though an application that involves discrete I/O and a motor.

Automated Bottle Capping Application

The bottle capping operation involves moving each bottle into position and then putting a cap on it. Figure 14-1 shows the I/O that is used in this application. There are 2 digital outputs and three digital inputs. There is also one servo motor to advance the conveyor and move the bottles into position. Study the I/O chart and Figure 14-2 to understand the application.

Bottle Capping Application I/O		
Tagname (register)	Controller Resource	I/O Type
Capper_Up/_Down	Output 1	Digital Output
Alarm_Horn_On/_Down	Output 2	Digital Output
Bottle_Ready/_Not ready	Input 1	Digital Input
Up_Confirm	Input 2	Digital Input
Down_Confirm	Input 3	Digital Input
Conveyor	Servo 1	Servo Motor

Figure 14-1. I/O chart for the bottle capping application.

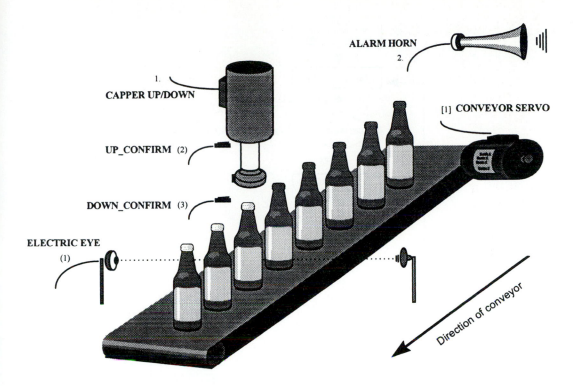

ALARM HORN
2.

1.
CAPPER UP/DOWN

[1] CONVEYOR SERVO

UP_CONFIRM (2)

DOWN_CONFIRM (3)

ELECTRIC EYE
(1)

Direction of conveyor

Figure 14-2. The actual bottle capping application.

Figure 14-3 shows a flowchart of the application. The flowchart breaks the application into logical operation steps. The first step is called INITIALIZE. This step will be used to shut the alarm off and profile the servo motor to the correct parameters. The next step is the RETRACT CAPPER step. This is done to make sure the capper is retracted before the conveyor is moved (indexed).

The next step is a decision. Decisions can be used to control program flow (see Figure 14-5). The controller must verify that the capper has retracted. It uses input 2 to confirm that the capper is in the up position. If the capper is up, the program moves to the next step. If not, the program moves to the next decision step to see if 5 seconds has elapsed. If not, the program goes back to the UP CONFIRM step. If 5 seconds has elapsed the program goes to the ERROR step. The ERROR step sounds the alarm and halts program execution. If the capper was confirmed as being in the up position, the program moves to the NEXT BOTTLE step.

The NEXT BOTTLE step is used to index the conveyor. The conveyor is driven by a servo motor. This step will move the conveyor the proper amount. The SERVO STOPPED decision step is performed next. This step assures that the move is complete before the program moves to the next step. When the move is

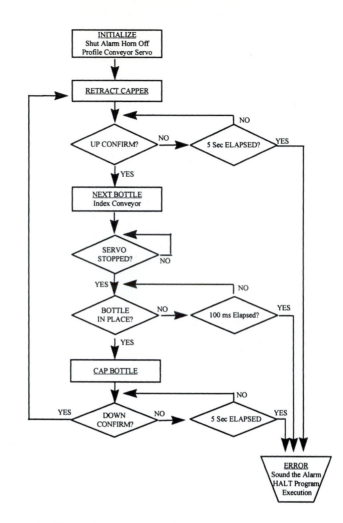

Figure 14-3. Flow diagram for the bottle capping application.

complete, the program goes to the BOTTLE IN PLACE decision step. This step waits up to 100 milliseconds for a bottle. If no bottle is sensed in 100 milliseconds, the program goes to the ERROR step.

If a bottle is in place within 100 milliseconds, the program moves to the CAP_BOTTLE step. This is the step in which the bottle is actually capped. The next step is the DOWN CONFIRM decision. This step assures that the capper actually completed the down motion. If it did, the program goes back to the RETRACT CAPPER step. If the DOWN CONFIRM fails to see the input within 5 seconds, the program goes to the ERROR step.

The actual Quickstep program is very similar to the flow diagram. The Quickstep programming language divides a program into logical steps. The steps would correspond to the process blocks in the flow diagram. The decisions that were shown separately in the flow diagram can be done right in the steps of the program. Let's look at the actual program, but first let's see what a step looks like (see Figure 14-4).

Step number and name   [4] CHK_BOTTLE

Comments
```
;;; If the photo-sensor does not detect
;;; a bottle in place, branch to the ERROR
;;; step. Operator assistance is needed.
```

Changes to digital outputs
```
-------------------------------------------------
<NO CHANGE IN DIGITAL OUTPUTS>
-------------------------------------------------
```

One or more statements
```
monitor Bottle_Ready goto CAP_IT
goto Error
```

Figure 14-4. A typical Quickstep language step.

A step always has the same general format. First is the step number (4 in this case) followed by the step's name. The name of this step is CHK_BOTTLE. This identifies the step and can be used in the program to tell a program where to branch or move to. In this application all step names will be written in capital letters so that they are easily identifiable.

The next part of a step is optional. A user can add comments to clarify the program or step function. Anything after three semicolons is a comment. It is great programming practice to adequately comment a program. A completed Quickstep program with comments can actually become a valuable part of the system documentation.

The third part of a step is the section that makes any desired changes in the digital outputs. In this program all I/O names will begin with a capital letter.

The fourth part of a step is one or more statements. Statements can be used to monitor inputs, perform branches, and so on.

The bottle capper program will be divided into 6 steps, INIT, RETRACT_CAPPER, NEXT_BOTTLE, CHK_BOTTLE, CAP_IT, and ERROR. The actual program is shown below.

[1] INIT
;;; These are remarks that help to document the program. The INIT step ;;; turns all outputs off, retracts the capper and ensures that the alarm horn ;;; is off.

<TURN OFF ALL DIGITAL OUTPUTS>

----

profile Conveyor servo at position maxspeed=IndexSpeed accel=RampRate P=PropVal I=IntegralVal D=DerivVal

goto RETRACT_CAPPER

[2] RETRACT_CAPPER

;;; Raise the capper, then monitor for the Up_Confirm limit switch. IF we
;;; don't receive confirmation within five seconds, we will branch to ERROR
;;; and sound the alarm.

Capper_Up

----

monitor Up_Confirm goto NEXT_BOTTLE

delay 5 sec goto ERROR

[3] NEXT_BOTTLE

;;; Turn the servo 500 increments (steps) clockwise (cw) to advance the
;;; conveyor to position the next bottle. After the conveyor stops, branch to
;;; the CHK_BOTTLE step.

<NO CHANGE IN DIGITAL OUTPUTS>

Turn Conveyor cw 500 steps

monitor Conveyor:stopped goto CHK_BOTTLE

[4] CHK_BOTTLE

;;; If the photo-sensor does not see a bottle in place, branch to the ERROR
;;; step, need operator intervention.

<NO CHANGE IN DIGITAL OUTPUTS>

----

monitor Bottle_Ready goto CAP_IT

delay 100 ms goto ERROR

[5] CAP_IT

;;; This step caps the bottle, then branches to the RETRACT_CAPPER step
;;; when it detects the Down_Confirm limit switch.

Capper_Down

----

monitor Down_Confirm goto RETRACT_CAPPER

delay 5 sec goto ERROR

[6] ERROR

;;; This step sounds the alarm and halts program execution.

Alarm_On

Done

Now let's analyze each step of our program.

[1] INIT

;;; These are remarks that help to document the program. The INIT step ;;; turns all outputs off, retracts the capper and ensures that the alarm horn is off.

<TURN OFF ALL DIGITAL OUTPUTS>

----

profile Conveyor servo at position maxspeed=IndexSpeed accel=RampRate P=PropVal I=IntegralVal D=DerivVal
goto RETRACT_CAPPER

This is step 1 and its name is INIT. The user could have chosen any name for the step. Next the user wrote a few comments to explain what the step does.

The next section of the program turned off all outputs. This is to be sure that everything is off when we begin program execution.

The next section of the step is the statement section. The first statement was used to define the servo parameters. The statement is a *profile* statement. The profile statement is used to define the maximum speed, the acceleration value, and the proportional, integral and derivative values for the servo. Maxspeed was set to equal a user-defined variable (IndexSpeed). The other parameters were also assigned variables. Variables were used so that the user could change the value of the variables while in operation without rewriting the program.

[2] RETRACT_CAPPER

;;; Raise the capper, then monitor for the Up_Confirm limit switch. IF we ;;; don't receive confirmation within five seconds, we will branch to ERROR ;;; and sound the alarm.

Capper_Up

----

monitor Up_Confirm goto NEXT_BOTTLE
delay 5 sec goto ERROR

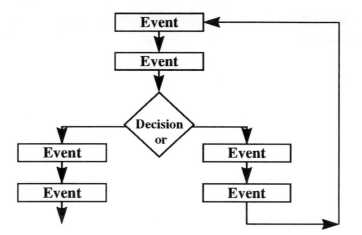

Figure 14-5. Flow diagram that shows how decisions can be used to control program flow. Courtesy Control Technology Corporation.

The second step is named RETRACT_CAPPER. Note that the programmer chose good step names. A good step name is one that is very descriptive about the purpose of the step. The programmer also added some comments to further explain what occurs in this step. In the output section the programmer typed Capper_Up. This is the name of an output. If you look at Figure 14-1 you will see that Capper_Up/_Down is the name of output 1. The programmer defined Capper_Up as output 1 being OFF. The programmer defined Capper_Down as output 1 being ON. I/O variables can be defined while programming by simply answering a few questions.

Figure 14-6 shows an example of a symbol definition screen. The symbol browser can be used to easily work with any of the symbolic names. Symbolic names are used to represent real-world I/O and variables. Symbolic names make programs very user-friendly and understandable. Figure 14-7 shows an example of the symbol browser. This is a part of the Quickstep software.

The list on the left of the figure shows the available types of symbols. Note that the programmer has chosen to work with the output type so the output symbols that exist are shown on the right of the figure. Notice the first two output symbols, Punch_On and Punch_Off. Punch_On was assigned an output state of On (1). Punch_Off was assigned a state of Off (0). This allows user-friendly symbol names to be used in the program.

For this application the programmer made sure the capper was up by putting Capper_Up in the output line. Capper_Up is equal to output 1 OFF.

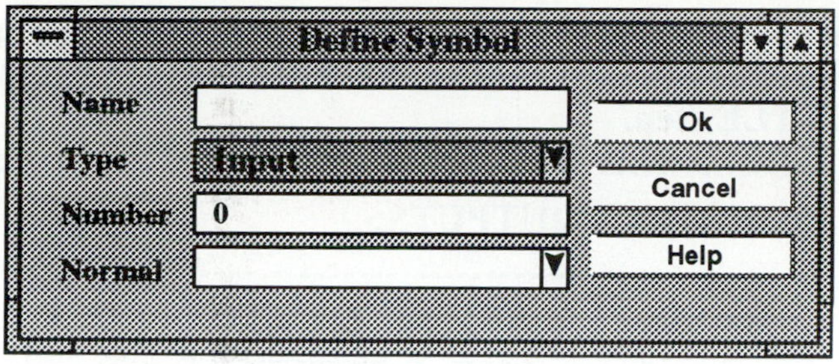

Figure 14-6. Symbol definition screen. Courtesy Control Technology Corporation.

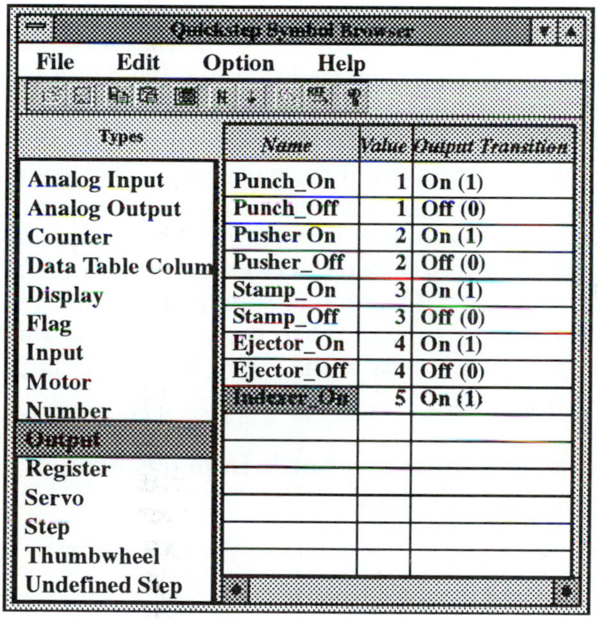

Figure 14-7. Symbol browser. Courtesy Control Technology Corporation.

In the statement portion of the step the programmer examined the state of a sensor. *Monitor* is a statement that causes the controller to look at an input condition. In this case the program looks at the state of input 2 (Up_Confirm). If it is on, the program will go to the step named NEXT_BOTTLE. If input 1 is not on, the program will continue to monitor the input for 5 seconds. If it does not detect the sensor to be on within 5 seconds the program will go to the step named ERROR.

[3] NEXT_BOTTLE
;;; **Turn the servo 500 increments (steps) clockwise (cw) to advance the**
;;; **conveyor to position the next bottle. After the conveyor stops, branch to**
;;; **the CHK_BOTTLE step.**

\<NO CHANGE IN DIGITAL OUTPUTS\>

Turn Conveyor cw 500 steps
monitor Conveyor:stopped goto CHK_BOTTLE

The third step is named NEXT_BOTTLE. This is the step that will move the servo motor, which moves the next bottle into position. The programmer again typed some comments to explain the purpose of the step.

The programmer stated that there were no changes to be made to the digital outputs in the output step.

The statement portion of the step was used to move the conveyor motor. The servo motor for the conveyor was given the name Conveyor. The statement Turn Conveyor turns the conveyor motor 500 steps (increments). Steps are equivalent to whatever encoder counts are used in the particular servo system. This is a closed-loop example. The velocity, acceleration and following error will be monitored with this simple command. Remember that the speed was set by the profile servo command in the first step (INIT). Stepper motors are programmed in essentially the same manner. The second statement (monitor Conveyor:stopped goto CHK_BOTTLE) monitors the move to see when it has finished. When the move is finished the program moves to the CHK_BOTTLE step.

[4] CHK_BOTTLE
;;; **If the photo-sensor does not see a bottle in place, branch to the ERROR**
;;; **step, need operator intervention.**

\<NO CHANGE IN DIGITAL OUTPUTS\>

----

monitor Bottle_Ready goto CAP_IT
delay 100 ms goto ERROR

The fourth step is named CHK_BOTTLE. The programmer typed a few remarks to explain the step. There were no changes made in the digital outputs. Note that this is quite important. We want to maintain the states of the outputs as they are now.

The programmer then uses statements to monitor input 1 (Bottle_Ready). Remember that Bottle_Ready/_Notready is input 1. The user defined Bottle_Ready as input 1 being ON. Bottle_Notready was defined as input 1 being OFF. If the input is ON, the program will go to the CAP_IT step. The next statement (delay) causes the program to monitor the input for 100 ms or until it comes on. If the input does not come on within 100 ms, the program will go to the ERROR step.

[5] CAP_IT
;;; This step caps the bottle, then branches to the RETRACT_CAPPER step
;;; when it detects the Down_Confirm limit switch.

Capper_Down
----
monitor Down_Confirm goto RETRACT_CAPPER
delay 5 sec goto ERROR

The fifth step is named CAP_IT. The programmer typed a few remarks to explain the purpose of the step. The programmer turned output 1 on in the output section. Remember that output one was named Capper_Down and that Capper_Down was defined as being output 1 ON.

The program then monitors input 3 (Down_Confirm) to see if it is ON. If it is ON, the program goes to the RETRACT_CAPPER step. If it is not on the delay statement causes the program to continue to monitor the input for five seconds or until it does turn on. If in five seconds the input has not become true, the program will go to the ERROR step.

[6] ERROR
;;; This step sounds the alarm and halts program execution.

Alarm_On

Done

The final step (6) is the ERROR step. This step simply turns output 2 ON. Output 2 was defined by the programmer to be Alarm_Horn_On. This also halts program execution.

INDUSTRIAL AUTOMATION CONTROLLER HARDWARE

Control Technology Corporation has several industrial automation controllers available. The simplest is a small controller that controls digital I/O. All of the controllers utilize the Quickstep programming language, so they are very easy to utilize. Figure 14-8 shows a starter kit that includes a CTC 2601 Automation

Controller. The 2601 Automation Controller is a high-performance control system whose small size conceals a range of resources and capabilities usually found only in much larger and more costly systems. A few of this controller's features include:

32 digital I/O.

Multitasking, with up to 28 parallel tasks running simultaneously.

988 storage registers, including 500 nonvolatile registers.

A data table capable of storing over 8000 numbers in a two-dimensional array.

Short-circuit and overcurrent-protected outputs.

Integrated RS-232 communications port.

Internal counters, which may be software-linked to any input.

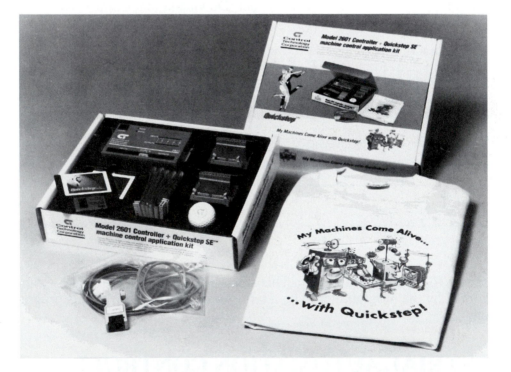

Figure 14-8. The Quickstep starter kit. It includes a 2601 32 I/O controller, cables, input/output simulator, Quickstep software, and manuals for under $600. Courtesy Control Technology Corporation.

The 2601 offers strong communications capabilities, including computer-based data collection, programming, and configuration, as well as the ability to actively transmit messages to displays and other remote subsystems. Because these messages can contain live data from the controller's registers, the user could create fully interactive operator controls using inexpensive LCD displays or remote terminals. The user could connect an operator interface to the serial port and have full interactive control of machine parameters. The user could also collect production data and communicate it though the communications port to a computer for logging or display. The next level is the 2600XM controller.

Software capabilities include the ability to simultaneously run up to 28 independent tasks, and the instruction set includes time delay, input monitoring, math and data manipulation commands. These commands make full use of such internal resources as the controller's nonvolatile and volatile registers, user-definable data table, input-linkable counters and 32 flags.

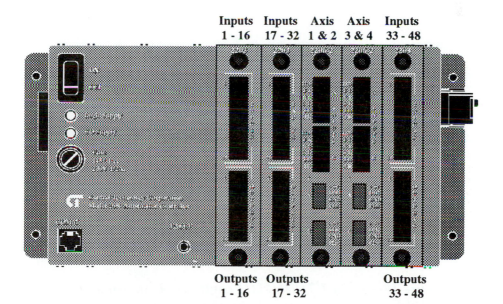

Figure 14-9. A 5-rack controller. In this configuration the controller has 3 digital I/O cards and two 2-axis servo cards. There is also a serial port on the lower left of the controller. Note that the user chooses the modules that are needed and plugs them into the slots. Courtesy Control Technology Corporation.

Figure 14-9 shows a modular-style controller. The user can choose modules to meet the application requirements. There are a wide variety of modules available. There are several types of digital and analog I/O modules available. The applications we have looked at utilized simple digital I/O and one servo motor. These controllers have virtually unlimited application analog capability as well. High-speed counter modules, thumbwheel input and display modules can be used. There are single or dual-axis stepper motor modules as well as single and dual-axis servo control modules. Communications modules are available to add additional RS-232 or Ethernet modules. Figure 14-10 shows a 2600XM controller. This controller has almost unlimited application capability. Ethernet or serial communications can be used to create networks of controllers. The Ethernet network allows the controllers to easily communicate with each other or to communicate with other computers on the network. This allows data gathering and control with SCADA software such as Wonderware.

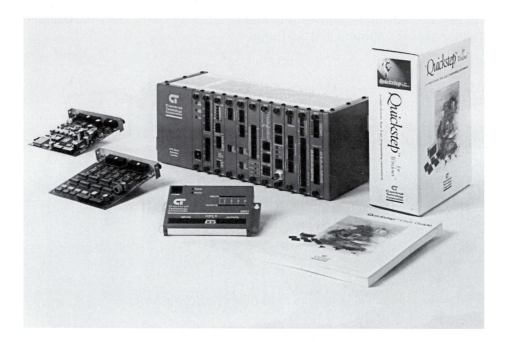

Figure 14-10. The CTC 2600XM controller is shown in the middle of the picture. The small 2601 32 I/O controller is shown in the front for a comparison. The 2600XM is available in several different rack sizes. This one has a 10-slot rack. The user can choose any combination of I/O to fill the rack. Several servo modules could be used to control a very complex multi-axis application. Two of the plug-in modules are shown to the lower left of the controller. Courtesy Control Technology Corporation.

The 2600XM controller is programmed using the same Quickstep programming language. There are also a variety of user displays available from simple to very versatile graphic touch screens.

The servo modules are very powerful and easy to use. The dual-axis servo modules can be used to independently control 2-axis or master-slave and/or electronic gearing type applications.

CTC also has a series of controllers called MultiPro Automation Controllers. Figure 14-11 shows a few of them. The MultiPro controller is aimed at very high speed machine control applications. The MultiPro controller uses multiple processors. Individual processors are dedicated to analog I/O, motion control, communications, and to the main controller. The overall control is exercised by the user's Quickstep program. The ease of use of the English-like Quickstep language makes it very easy to program even complex applications. There are models available for stepper or servo control.

All have serial communications built in and most have Ethernet communications built in. Figure 14-12 shows an example of how CTC controllers can be networked on an Ethernet network.

Figure 14-11. The Control Technology Corporation MultiPro family of Automation controllers. Courtesy of Control Technology Corporation.

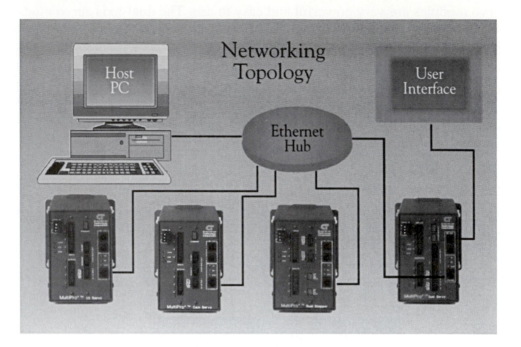

Figure 14-12. Networking topology for CTC controllers on an Ethernet network. Courtesy Control Technology Corporation.

Multitasking

Thus far we have only looked at linear processes with Quickstep programming.

Multitasking is also possible. This means that the controller can be programmed to control multiple processes that occur simultaneously. Figure 14-13 shows an example of multitasking. This is just a block diagram of the steps. Each step would have to be defined of course. Each of the rectangles represent steps in the program. In this example the START_CYCLE step has a do statement that causes the program to move to step ONload and step OFFload. When the do statement is used it means that the program should execute all steps that are listed in the parentheses after the do simultaneously.

The second example of concurrent multitasking is shown in Figure 14-14. The top rectangle labeled [42] START_CYCLE has a statement that says do (MOVE_PART PROCESS PART) goto next. Both sequences will then process at the same time. This example shows that data from one step can be shared with other steps. Step 13 sets Flag_1 to a 1. Step 22 contains a statement to monitor Flag_1. This allows concurrent tasks to share information that may be crucial for timing and/or correct sequencing. Flags may be set to a 1 or a 0. A set statement can be used to set flags to the desired state. Monitor statements can then be used to check the state of a flag and make decisions based on the flag's state.

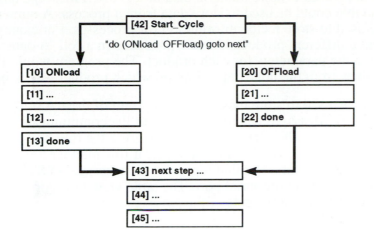

Figure 14-13. Example of multitasking. Courtesy Control Technology Corporation.

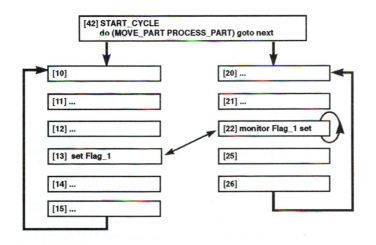

Figure 14-14 . Use of a flag to communicate between steps. Courtesy Control Technology Corporation.

There are other memory resources available for the user. Numeric registers can be used to hold numbers that range in size from 0 to 65,535. Registers 1 to 500 are volatile. Their contents are lost when power is removed from the controller. Registers 501 to 1000 are nonvolatile. These registers maintain their data even if power is removed for up to 10 years. Registers 1001 to 24999 are special-purpose registers. These registers are used for advanced programming techniques and give the user access to controller resources such as inputs, outputs, servo parameters, and communications capabilities. The *store* command is used to store data to registers. *If* instructions can be used to test registers.

Data tables can be used to store numeric data or an ASCII message table. A numeric data table could be used to store data from a process. A numeric table could also be used to store recipe information for processes. For example, there might be several different products that are produced in a cell. A data table could be used to store the parameters for each product. The program would then read the needed values from the table depending on which product was to be produced.

Tables can also be used to store messages in the form of their ASCII equivalents (see Figure 14-15). Messages can then be sent out the communications port to a computer or display or other device by simply storing the row number to a special-purpose register (reg_12001). For example, if row number 5 was stored to reg_12001, the ASCII characters that represent the message SYSTEM SHUT DOWN would be sent out the serial port. The table view in Figure 14-15 only shows the first 8 ASCII equivalents of the message. The user could scroll the screen to see the rest.

This chapter has been a very brief introduction to this topic. Industrial automation controllers have almost unlimited potential. It should be remembered that PLCs, industrial computers, and industrial automation controllers are all just microprocessor-based computers. Industrial automation controllers provide an easily implemented solution for integrating complex servo, process control and communications into a single environment. Their other strength will continue to be their ease of use and programming.

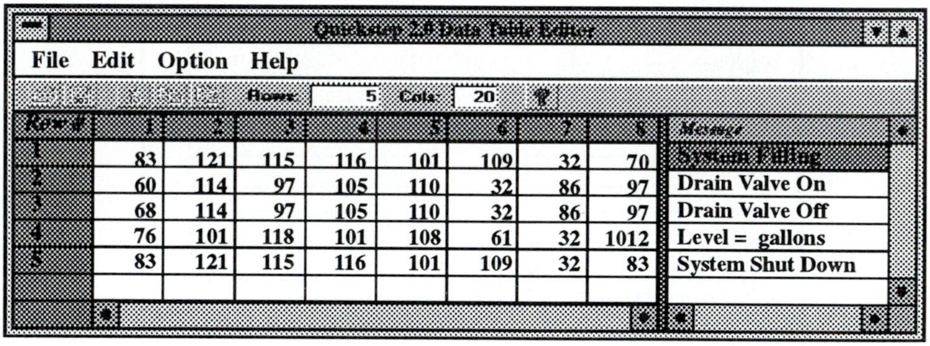

Row #	1	2	3	4	5	6	7	8	Message
1	83	121	115	116	101	109	32	70	System Filling
2	60	114	97	105	110	32	86	97	Drain Valve On
3	68	114	97	105	110	32	86	97	Drain Valve Off
4	76	101	118	101	108	61	32	1012	Level = gallons
5	83	121	115	116	101	109	32	83	System Shut Down

Figure 14-15. Data table used as an ASCII message table. Courtesy Control Technology Corporation.

Questions

1. What is an industrial automation controller? Compare and contrast it with PLCs and industrial computers.

2. What are some of the types of applications that industrial automation controllers are particularly well suited to?

3. List and explain the four parts that a step can contain.

4. Explain the following step.

   [12] FILL_TANK
   ;;; If the photo-sensor does not see a bottle in place,
   ;;; branch to the ERROR step,
   ;;; operator intervention required

   Valve_On
   ----
   monitor Fill_Level goto MIX
   delay 20 sec goto ALARM

5. Draw a flow diagram for the step shown in question 4.

6. Draw a flow diagram that would represent the following application: Fill a tank, mix the contents, heat the contents, and drain the tank. The I/O is shown in Figure 14-16.

Tank Application I/O		
Tagname (register name)	Controller Resource	I/O Type
Fill_Valve_Open/_Close	Output 1	Digital Output
Heat_Coil	Output 2	Digital Output
Mix_Motor	Output 3	Digital Output
Drain_Valve_Open/_Close	Output 4	Digital Output
Tank_Full/_Not_Full	Input 1	Digital Input
Temp_OK/Temp_Low	Input 2	Digital Input

Figure 14-16. I/O for the tank application.

7. Write the program to accomplish the tank application from Question 6 using the Quickstep language. Make sure you use the correct format for steps and adequately comment your program.

8. Explain how multitasking can be done in the Quickstep language.

Chapter 15

Single-Board Controllers

Single-board controllers are becoming more widely used for industrial control. This chapter will examine the programming and use of embedded controllers.

OBJECTIVES

Upon completion of this chapter, the student will be able to:

Describe the use of a single-board computer.

Compare and contrast the advantages and disadvantages of single-board controllers vs. programmable logic controllers.

Define such terms as **embedded controller, expansion board, master/slave,** *and so on.*

Describe how single-board computers can be networked.

Define multitasking.

OVERVIEW

Relay ladder logic (RLL) was intended to replace the relays, timers, and counters common to control systems of the 1960s and 1970s with a more flexible software implementation. RLL was very successful in achieving and surpassing those modest goals. RLL is still a very powerful tool when applied to applications it is appropriate for.

The industrial world has become more complex. It is no longer sufficient for a machine to endlessly perform the same series of actions. Systems must now be optimizing, in order to maximize production. Many systems must keep basic statistics, which are a requirement for quality assurance. Many systems need to interact with one or more third-party devices whose interfaces are alien to the controlling PLC. More and more systems need to interface with enterprisewide networks, which provide management with up-to-the-minute information.

RLL was not designed for these complications. Some RLLs have been retrofitted with features to handle some of these situations. At best many of these additions have been awkward and slow. They have also been difficult or nearly impossible to use. RLL was simply not designed to solve all of today's industrial problems. These requirements have made other types of control devices gain rapidly in use. Single-board controllers are one of these technologies.

SINGLE-BOARD CONTROLLERS

Single-board controllers are gaining acceptance in industrial applications. They are also called embedded controllers. They are widely used for machine control. Original equipment manufacturers (OEMs) use this technology to control machines that they build. As more and more people are exposed to this technology it will become more and more accepted and prevalent in industrial control.

Many computers do not look like desktop PCs simply because you cannot see them at all. These invisible computers are hidden or embedded within a larger system or product. Embedded controllers provide better control for a larger system. The components of many embedded computers fit on a single printed circuit board, making an embedded computer compact and economical. The stringent real-world, real-time requirements of control applications mean that the hardware and software of embedded computers are distinctly different from ordinary desktop PCs.

There are many single-board controllers available. Many choices of programming languages are also available. Programming languages include C, BASIC, assembly, and others. C, assembly and BASIC are probably the most commonly used.

RLL was designed to perform very specific tasks and does this very well. C, on the other hand, was designed to be a general-purpose programming language. Not only does C dominate the development of applications for personal and corporate computing, it also dominates the development of single-board controllers. Thus, C is used to develop programs for everything from word processors and databases to microwaves and TVs. C is everywhere.

C's general-purpose nature is the reason for its popularity. Because it makes no assumptions about its environment or what tasks it will perform, programs written in C use resources more efficiently than equivalent RLL systems.

For example, most PLCs provide the user with a fixed number of auxiliary relays, a fixed number of registers, a fixed number of timers, and so on. Eventually every RLL programmer encounters a situation where this fixed allocation of resources is not acceptable for the application. A particular program may require only a few auxiliary relays and timers but more than the available number of registers. The most aggravating aspect of this situation is that the PLC often has the needed memory, but unfortunately has allocated it to unneeded auxiliary relays and timers.

In C this is not a problem. C only allocates resources as requested. As long as the programmer doesn't attempt to allocate more resources than are available, C lets him/her decide how to best use them.

Another important advantage of C is a definite focus of execution. Remember, RLL was written to simulate hardware. Since hardware inherently operates in parallel, RLL had to simulate this parallel operation. This is accomplished by executing every rung of the RLL program on each scan. While this achieves the desired effect, it has one major drawback: Scan times increase as the size and complexity of the program increases. Those attempting to use RLL programs to perform complex tasks can find scan times increasing well beyond 10 milliseconds, which can be too slow for some applications.

C, on the other hand, has the notion of a program counter, which starts at the main (the first function executed in a C program). The program counter works its way through the code, directed by the flow control statements written by the programmer. At any given time, a processor running a C program is executing only one sequence of instructions. While doing this, the rest of the program sits idly by, waiting for its turn to execute. This single focus of execution gives the programmer incredible power to decide what aspects of the application need the most intense attention and which need to be checked periodically. Without such control, handling of high-speed events (such as the deployment of airbags) would be impossible.

C is powerful. But, like all powerful tools, it also has dangers. And, like most tools, its strengths are often its dangers.

> *C's flexibility allows so many solutions to a problem that beginners often get lost in its possibilities. RLL is so restrictive that the simplest problems often seem to solve themselves.*

C's ability to focus execution on a small portion of code also means that it is easy to write programs where other things are not done often enough or are skipped entirely. RLL executes everything on every scan with no chance of code not executing.

C is less intuitive to beginners and a little harder to learn than RLL. All of RLL can be learned relatively quickly. With C, a useful set of basics can be learned quickly, but the more complicated aspects of the language can be quite confusing even to programmers with moderate experience.

Even with these problems, C is well worth the investment. C can be used to solve most problems and is supported on all major programming platforms. Ultimately, a good C programmer will always outperform an equally competent counterpart who uses RLL.

Software programs that an embedded controller executes are generally developed on a PC using a software development system. For many embedded controllers, the development system consists of a software editor, a compiler, and a debugger, all of which are separate items that are not always guaranteed to cooperate. An alternative to utilizing separate editors, compilers, and debuggers is to use a software development system that integrates all the components into one software development tool. This alleviates potential problems, guaranteeing all three development tools will work properly. An integrated software development system, such as Z-World's Dynamic C, provides a direct and fast means of developing and debugging programs for an embedded controller. And because the target controller can be used during program development, the software and the hardware can be considered as one unit.

Single-board controllers are available with a wide variety of capabilities from very simple digital I/O to very complex motion control capability. Single-board controllers are very flexible. Their capabilities are really only limited by the ability of the programmer. Programming has become easier as companies have added special functions to accomplish many standard tasks. It should be remembered that a PLC or a computer is just a microprocessor with memory and associated peripheral hardware. A single-board controller is no different. It is a microprocessor with memory and associated peripheral hardware for I/O. The PLC is programmed with ladder logic and the single-board controller with some other language.

Single-board controllers will usually offer a price advantage over PLCs. This will be especially true when multiple machines are to be built. PLCs still offer the easiest programming language for those familiar with electrical diagrams. In this chapter we will consider one brand of controller. It is a controller made by Z-World, which manufactures a wide range of C-programmable controllers. C-programmable controllers enable relatively easy design of control systems requiring multitasking or floating-point math calculation. This chapter will examine the use of single-board controllers by examining the use of the Z-World PK2200 series controller. They offer this controller in a starter kit complete with software, a single-board controller, I/O simulator, and communications cable (see Figure 15-1). The controller and simulator are shown in Figure 15-2. The controller is programmed in a personal computer by using a C language program that comes with the controller. The program is then downloaded to the controller and is ready to run.

I/O Capabilities of the PK2200

Single-board controllers have extensive I/O capabilities. They offer analog, digital, communications, operator I/O, and many other capabilities. The PK2200 controller has the following I/O capabilities.

Figure 15-1. A PK2270 starter kit. Courtesy Z-World.

Digital Inputs

Sixteen digital inputs, suitable for detecting contact closures and CMOS levels, are protected in the range -48V to +48V. The logic threshold is at 2.5V nominal. Built-in pull-up or pull-down resistors are selectable with jumpers. These can be used to configure the inputs as either sinking or sourcing. Additional boards can be added to increase the number of inputs.

Here is an example of the use of an input in C code.

waitfor (zDigIn (5) == 1);

This statement will cause the program to wait until input 5 becomes a 1 (ON). The waitfor comand causes the wait. zDigIn is a function that gets the value of the input it references, in this case 5. The == is used in C code to check for equality. The semicolon is normally used at the end of every line of code in C. So, if digital input 5 is equal to 1 the program will continue. If not, the program will wait until digital input 5 becomes true (1).

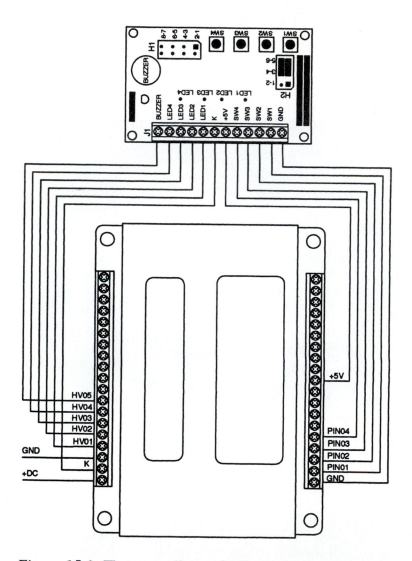

Figure 15-2. The controller and I/O simulator. Note the four user switches marked SW1, SW2, SW3, and SW4, and the outputs labeled LED1, LED2, LED3, and LED4. Note also that the actual inputs on the single-board controller are labeled PIN01 to PIN04 and the outputs are labeled HV01 to HV05. Also note that there are more I/O available but the I/O simulator only makes use of 4 inputs and 5 outputs. Courtesy Z-World.

Digital Outputs

Fourteen digital output lines are capable of controlling inductive loads such as relays, stepping motors and solenoids. The drivers have built-in protective diodes that return inductive spikes to the power supply. Each line can normally sink up to 500 mA at 48V. Additional boards can be added to increase the number of inputs. Various types of digital outputs are available including relay and high voltage.

Here is an example of the use of an output function in C code.

zDigOut (3,1);

This statement uses the *zDigOut* function. This statement turns on output 3 (1). If the statement was zDigOut (3,0);, the output would be turned off.

Watchdog Timer

When enabled by a jumper, the watchdog timer resets the controller after a short period (about 1.6 seconds) if running software has failed to reset the watchdog. This feature allows your system to recover should a program "hang" or stop operating correctly.

Handling Power Failures

Even if the code is perfect, a controller's program may crash because of conditions that are beyond the control of the software engineer: complete power failures, brownouts and spikes on the incoming power lines, or even human error.

Software cannot reliably identify or handle all hardware related failures because, after all, the software runs on the failing hardware. Detecting hardware failures requires special-purpose hardware. For example, a watchdog timer is an independent hardware device that monitors the health of the software. The watchdog timer will reset the controller unless it receives a signal from the software within a specified time, usually about 1 second.

On most boards, when input voltage falls below an acceptable threshold, a non-maskable power-failure interrupt takes place. The program can have a function to perform a short shutdown procedure prior to reset.

Memory

RAM (random access memory), or SRAM (static random access memory), is a common form of memory. Although this type of data memory is inexpensive, it is volatile. Data residing in RAM disappears when power is removed. This behavior is not acceptable in control applications. Nonvolatile memory in the form of EEPROM (electrically eraseable programmable read-only memory), battery-backed RAM, or flash EPROM, provides varying degrees of security for mission-critical data. The type of low-power, read-write SRAM (static random access memory) used in embedded controllers will retain its data in a sleep mode as long as a modest power level is applied. Consequently, a single lithium battery can possibly keep data alive many years, if necessary. Once written to EEPROM, data will not be lost no matter how long power is removed. However, given the EEPROM's combination of slow write speeds, high cost, and small capacity, the best use for an EEPROM is storing small amounts of mission-critical data at infrequent intervals.

Nonvolatile Memory

Flash EPROM is a variation of EEPROM that is becoming more and more popular. Clever designers have reduced the amount of on-chip circuitry needed for writing to flash EPROM, lowering cost substantially while retaining the inherent advantages of EEPROM. Flash EPROM is much less expensive than an equivalent EEPROM and not much more expensive than an equivalent EPROM. Flash EPROM costs slightly more than EPROM, but flash EPROM can be reprogrammed without removing it from the controller, eliminating the need to remove the chip from the controller. This eliminates a separate programming step, making software development much easier using flash EPROM. And field upgrades become simple and efficient.

Power Failures

Personal computers react to a power failure by simply shutting down and losing any data that may be resident in the RAM. This is not acceptable for an embedded controller supervising a critical process. A special-purpose circuit in an embedded controller constantly monitors the input voltage as well as the regulated supply for the controller's internal components. A circuit called a supervisor will give the controller advanced warning of an impending power failure, allowing the controller to properly shut down. If the controller has a backup battery, the supervisor can switch critical components, such as RAM, over to battery power. If the controller has some form of nonvolatile memory, the data can then be saved.

Mission-Critical Data

A common problem in industrial applications is retaining mission-critical data during power outages or other emergencies. To understand how an embedded controller handles outages, one needs to know the nature of interrupts. When a controller's microprocessor recognizes an interrupt, the microprocessor almost immediately shelves what it is doing and jumps to the appropriate interrupt service routine. Software engineers use this built-in, very fast hardware and software facility in three ways:

To respond quickly to unscheduled or anomalous events.

To acknowledge some input, but defer its processing until later.

To respond to interrupts generated by some regular timing mechanism to achieve the regulated, orderly operation of some sequence.

Real-Time Clock

In many applications, software engineers use simple hardware counters to keep track of time intervals. But if the application needs to log data or schedule operations by time and date, then a special-purpose real-time clock may be beneficial because it relieves the controller's microprocessor of much of the timekeeping overhead.

Communications

The Electronic Industry Association's RS-232 is the oldest and still most widely used standard for exchanging data between two devices. A vast array of equipment has RS-232 ports available. RS-232 serial communication transmits data one bit at a time over a single data line. Single data bits are grouped together into a byte and transmitted at a set interval, or baud rate. Serial transmission can be over a simple connection consisting of 3 wires of transmit, receive, and ground signals.

A benefit of RS-232 is the ability to coordinate communication between a fast controller and a slow serial device using the RS-232 RTS and CTS hardware signals for handshaking. This ensures both devices are in sync so data is not lost. Also very common in control applications is RS-485. You can think of RS-485 as an industrial-strength version of RS-232. Among the numerous improvements are more noise-resistant balanced lines (2 conductors), longer transmission distances (up to 4,000 feet at 9600 baud versus 30 to 50 feet for RS-232), and multidrop connections between several devices instead of just two. Unlimited distance is possible with the use of modems.

Operator Interfaces

Embedded applications often do not need and cannot use a typical PC monitor, keyboard, and mouse. Instead, a simple keypad and LCD (see Figure 15-3) will often be more appropriate and generally less expensive. A keypad is an array of switches. A controller senses keypad presses just as it would any other switch closure. LCD interfaces require a number of parallel output lines and software drivers to send codes to the LCD.

A 2x20 LCD, a 12-button keypad, and a beeper comprise the operator interface for this controller. The programing software has functions available to make it easy to utilize the display and entry keypad in applications. The operator interface is driven by Z-World's FiveKey system (in an EasyStart C library). This software allows the operator display to be easily programmed for operator input and output. The operator can change system parameters, scanning rapidly through multiple menus and submenus, with only five keys. The program can monitor and make sure that the operator only enters values that are acceptable for the application. Besides allowing an operator to change system parameters, it will periodically display messages or parameters of your choice. A sixth key can be used as a help key. The user can easily customize the keypad legends with common desktop publishing or word processing software. The other six keys are available for the user's application.

Figure 15-3. A PK2270 controller. Courtesy Z-World.

SIMPLE TANK APPLICATION

Consider the tank system shown in Figure 15-4. There are two level switches, one for when the tank level gets too full and one for when the tank level becomes too low. There is one output in this simple system The output will control the pump. For this example we will connect the low-level switch to input 1 (PIN01) and the high-level switch to input 2 (PIN02). These are the first two inputs on the controller. The pump is connected to the first output (HV01) on the controller.

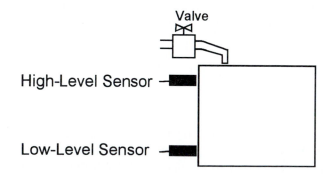

Figure 15-4. Simple tank level application.

Imagine if this were a system that a worker operated. Our instructions to an operator would be:

> ***If the tank is almost empty open the valve.***
> ***If the tank is almost full close the valve.***

This is obviously a simple system. An operator could easily perform this operation. There are some problems, however. What does "almost empty" or "almost full" mean? Will it mean the same to every operator? Obviously not. There will also be times when the tank overflows and other times when the tank runs empty. This may be caused by boredom and tedium on the operator's part. It would also be an expensive proposition to have an operator watch a tank level. Figure 15-5 shows the I/O that are used in this application. There are two inputs and one output.

I/O Type	Use
Input 1 (PIN01)	Low Level Switch
Input 2 (PIN02)	High Level Switch
Output 1 (HV01)	Pump

Figure 15-5. The I/O that are used in the tank application.

Figure 15-6 shows what the program would look like if we wrote it in English IF statements. The I/O states are shown in parentheses after each statement.

if the tank is nearly empty (PIN01 is HIGH)
turn on the pump (enable HV01)
If the tank is almost full (PIN02 is HIGH)
turn off the pump (HV01)

Figure 15-6. The program written with IF statements.

Figure 15-7 shows what the program would be in C language. Let's examine a line of code. The *if* statement evaluates zDigIn (PIN01) to see if it is equal to (==) 1. In the C language == is used to check for equality. PIN01 is input 1. Input 1 is our low-level switch. If it is a high (1) the if statement is true. If the first part of the statement is true, the rest of the statement will be executed. In this case the second portion of the statement is zDigOut (HV01,1). This statement turns output HV01 on.

The second statement in Figure 15-7 examines input 2 (high-level switch) to see if the tank level is too high. If it is on the second portion of the statement turns output 1 (pump) off.

if (zDigIn (PIN01) == 1 zDigOut (HV01,1);
if (zDigIn (PIN01) == 1 zDigOut (HV01,0);

Figure 15-7. Examples of input and output functions.

This was a simplification of the program to gain an understanding of what a program would look like. The previous code would have had to be in a loop so that the controller continuously monitored the tank level. A more complete program is shown in Figure 15-8.

MAIN

main_init();

MasterInit();

BEGINTASKS

   if (zDigIn (PIN01) ==1 zDigOut (HV01,1);

   if (zDigIn (PIN01) ==1 zDigOut (HV01,0);

ENDTASKS

Figure 15-8. Example of an input/output C program.

Everything between the BEGINTASKS and ENDTASKS executes endlessly. This would be the loop that would assure that the tank level is monitored continuously. The MAIN statement is used to begin all C programs. The main_init(); and MasterInit(); statements are functions that would initialize the system.

The EasyStart-C™ Framework

Z-World has created a framework that the programmer can use to create a program for an application. The framework is shown in Figure 15-9. This can be used as a template by the programmer. The programmer mainly just develops the tasks that need to be executed as well as any variables that need to be used.

#use [library name goes here] // The programmer includes the name of libraries here that will be needed in the program. Libraries are collections of functions that the programmer can use to make writing the program easier.

Main

   Define variables and constants here.

main_init();

   Put your initializations here.

zInitmaster; needed only if I/O operations will be performed.

BEGINTASKS // Tasks in this area execute continuously.

   Write one or more tasks here.

ENDTASKS

Figure 15-9. The EasyStart-C™ framework.

INDEXING, SEALING AND PERFORATING APPLICATION

This application involves a machine that indexes and perforates material. This machine produces plastic bags. The bags are wound on a continuous roll (see Figure 15-10). They are sealed on one end and perforated for easy removal by the user. A roll of garbage bags is an example of this type of product.

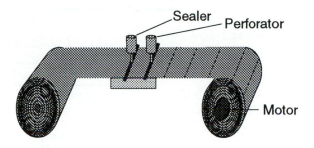

Figure 15-10. Drawing of the indexing and perforating application.

The Process

The motor is used to advance the material the correct length for cutting (perforating) and sealing. The machine consists of pneumatic cylinders to extend and retract the perforating knife and sealer. The sealer is a hot wire that is used to seal one end of each bag. The cutter knife is then advanced to perforate the bag. The controller ensures the safety of the machine and operator by monitoring inputs for knife position, operator location, material shortage, and preventive maintenance schedules. The machine also records and displays dates and times of operation. Figure 15-11 shows a diagram of the single-board controller and I/O. The inputs are shown on the left and the outputs are shown on the right of the controller. Figure 15-12 is the program to control this application.

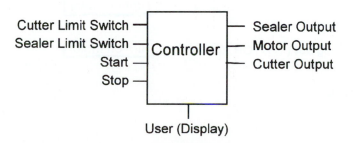

Figure 15-11. A block diagram of the I/O and controller for the indexing, sealing and cutting application.

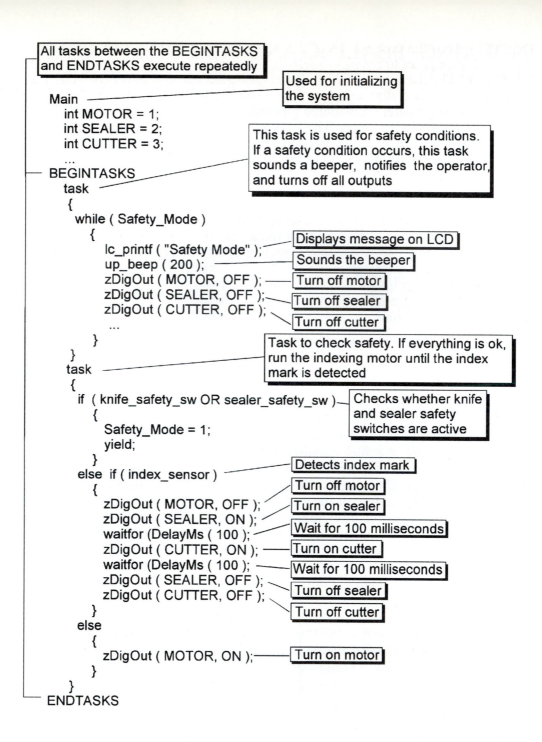

All tasks between the BEGINTASKS and ENDTASKS execute repeatedly

Used for initializing the system

```
Main
    int MOTOR = 1;
    int SEALER = 2;
    int CUTTER = 3;
    ...
BEGINTASKS
    task
    {
      while ( Safety_Mode )
        {
            lc_printf ( "Safety Mode" );
            up_beep ( 200 );
            zDigOut ( MOTOR, OFF );
            zDigOut ( SEALER, OFF );
            zDigOut ( CUTTER, OFF );
            ...
        }
    }
    task
    {
      if ( knife_safety_sw OR sealer_safety_sw )
        {
            Safety_Mode = 1;
            yield;
        }
      else  if ( index_sensor )
        {
            zDigOut ( MOTOR, OFF );
            zDigOut ( SEALER, ON );
            waitfor (DelayMs ( 100 );
            zDigOut ( CUTTER, ON );
            waitfor (DelayMs ( 100 );
            zDigOut ( SEALER, OFF );
            zDigOut ( CUTTER, OFF );
        }
      else
        {
            zDigOut ( MOTOR, ON );
        }
    }
ENDTASKS
```

This task is used for safety conditions. If a safety condition occurs, this task sounds a beeper, notifies the operator, and turns off all outputs

Displays message on LCD

Sounds the beeper

Turn off motor

Turn off sealer

Turn off cutter

Task to check safety. If everything is ok, run the indexing motor until the index mark is detected

Checks whether knife and sealer safety switches are active

Detects index mark

Turn off motor

Turn on sealer

Wait for 100 milliseconds

Turn on cutter

Wait for 100 milliseconds

Turn off sealer

Turn off cutter

Turn on motor

Figure 15-12. Program for the indexing, sealing and perforating application.

This first part of the program creates three integer variables and assigns them a number. Motor is set to equal 1, Sealer is set to equal 2 and Cutter is set to equal 3. MOTOR, SEALER, and CUTTER are all digital outputs. MOTOR is digital output 1, SEALER is digital output 2, and CUTTER is digital output 3. CUTTER is the perforating blade.

```
MAIN
int MOTOR = 1;
int SEALER = 2;
int CUTTER = 3;
```

This first task is used for safety. If a dangerous condition occurs the task will display Safety Mode and the display of the single-board controller, sound the beeper and then turn the motor, sealer and cutter off. Note that the names MOTOR, SEALER and CUTTER were used in the output statements. The actual number of the output could have been used also but the use of a name makes the program much more understandable. The lc_printf is a function that is used to display information to the controller display. The up_beep function is used to sound the beeper.

```
BEGINTASKS
  task
    {
    while (Safety_Mode)
      {
      lc_printf ("Safety Mode");
       up_beep (200);
      zDigOut (MOTOR, OFF);
      zDigOut (SEALER, OFF);
      zDigOut (CUTTER, OFF);
      }
    }
```

The first part of the next task checks to be sure there are no safety problems (knife safety switch or sealer safety switch). The if statement looks at the two safety switches and if either of them are ON, Safety_Mode is set to 1 and the yield causes the program to leave this task and return to the first task which then notifies the operator that there is an unsafe condition. If neither switch is ON the program continues to the next lines of code.

```
  task
    {
    if (knife_safety_sw OR sealer_safety_sw)
      {
      Safety_Mode = 1; yield;
      }
    }
```

The next portion of the task actually runs the process. This portion begins with an else if statement. This is used to detect when the plastic material has moved to the correct position. A sensor is used to look for the index mark on the material. When the index sensor is ON, the material is in position so the motor is turned off and the sealer is turned on to seal the bag. Then the waitfor function causes the program to delay for 100 milliseconds. Next the cutter is turned on and the bag is cut. The *waitfor* causes another delay of 100 milliseconds and then the sealer is turned off and the cutter is turned off.

```
else if (index_sensor)
    {
    zDigOut (MOTOR, OFF);
    zDigOut (SEALER, ON);
    waitfor (DelayMs(100));
    zDigOut (CUTTER, ON);
    waitfor (DelayMs(100));
    zDigOut (SEALER, OFF);
    zDigOut (CUTTER, OFF);
    }
```

If the index sensor was not on, the next portion of the code executes. The *else* statement executes if the previous if statements have not been true. This else is used to turn the motor on. This code turns the motor on and then program execution goes to the top of the program again to check for safety conditions and to watch for the index mark on the material. The motor then turns and advances the material until the index sensor senses the index mark on the plastic.

```
else
    {
    zDigOut (MOTOR, ON);
    }
ENDTASKS
```

Expansion Boards

There are a wide variety of expansion boards available for embedded controllers. These expansion boards can be used to add additional digital or analog I/O to a single-board controller. Figure 15-13 shows an example of a expansion card.

Figure 15-14 shows an example of how expansion boards can be connected with a ribbon connector to the PLCBus port on the controller. The figure also shows an example of a single board controller's I/O capability being expanded by attaching an analog card, a relay expansion board, a digital I/O card and a digital-to-analog expansion card.

Z-World's PLCBus expansion boards connect directly to your controller through a 26-wire ribbon connector that functions as a "flexible backplane" for your system. The PLCBus lets you daisy-chain multiple cards as needed, providing power and flexibility for your system. Cards are attached by connecting a 26-ribbon cable between the P1 connector on the single-board controller and P2 on the expansion card. Additional cards are attached by connecting the ribbon connector from P1 on the card to P2 on the next expansion card. Four expansion boards are available: the XP8500, the XP8600, the XP8300 and the XP8200. The XP8500 card is a 4 channel input card, the XP8300 card is a 6-relay output card, the XP8200 is a digital I/O card, and the XP8600 is a digital output card (see Figure 15-13). The number of expansion cards is limited to no more than six for this single-board controller.

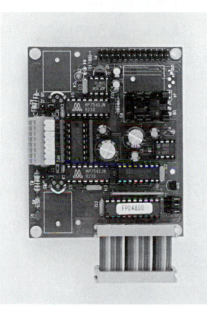

Figure 15-13. An example of an XP8600 expansion card. Courtesy Z-World Inc.

Analog I/O

Z-World provides analog input though the XP8500 card. The XP8500 card is an 11-channel, 12-bit analog-to-digital (A/D) converter card. This means that an analog signal can be converted to an integer value between 0 and 4095. There is signal conditioning for 4 of the analog inputs. The card supports unipolar and bipolar input ranges from 0.1 to 10 volts, with thousands of range and offset combinations. Seven additional 0 to 2.5V nonsignal conditioned 12-bit inputs are also available.

Analog outputs are provided through an XP8600 D/A card. The XP8600 has two 12-bit analog outputs. The XP8600 adds two 12-bit digital-analog conversion channels to the controller. Each channel can be set up either as a voltage source (0 to 10V) or a current source (0 to 20 mA).

Below is an example of an analog statement.

zAnOut (DAC_0 + 1, 725);

The zAnOut function is used to send the number 725 to analog output 1. In the argument (DAC_0+1,725), *DAC_0* is the name of the analog card that is attached to the single-board controller and the *1* is the analog output that the number *725* is being sent to.

Below is an example of the use of an analog input statement.

var1 = zAnIn (ADC_1_RAW + 2);

This function would read the "raw" 12-bit value of input 2 from analog card number 1 (ADC_1_RAW) and assign it to var1 (a variable). Var1 could then be used in the program.

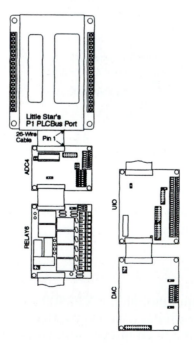

Figure 15-14. A single-board controller expanded with an analog card, a relay expansion board, a digital I/O card and a digital-to-analog expansion card. Courtesy Z-World Inc..

Digital I/O Expansion

Digital I/O can be expanded also. An output card that utilizes relays for the output device is available. The card is called XP8300. The XP8300 board adds six 1-amp relays (SPDT) to a system. Each relay is protected by a metal oxide varistor (MOV) and a 10-amp fuse. An LED indicates the status of each relay. The connectors provide N.O. (normally open) and common terminals for all six relays and N.C. (normally closed) terminals for two of the relays.

There is also a digital I/O expansion card available, the XP8200 card. The card has 16 "universal" inputs and 6 high-current driver outputs capable of driving inductive loads up to 500 mA. The universal inputs accept voltages ranging from -48V to +48V. Software for the XP8200 interprets the input voltage by comparing it to a programmable threshold and reporting either a digital 1 or 0.

Networking

There are cases where it is advantageous or necessary to run more than one controller in an application. If there are many inputs and outputs in an application, it may be disadvantageous to run all of the needed wiring across the whole machine. It may cause safety concerns or it may be economically infeasible. More than one controller may also be needed if large numbers of I/O are needed. In these cases it makes sense to use a network of controllers. Each controller can be placed close to the I/O it controls.

Z-World single-board controllers can be linked over a standard RS-485 network. One of the controllers is used as a master and the rest of the controllers become slaves (see Figure 15-15). The application program you develop runs on the master controller and controls all of the system I/O over the network. The slave controllers run a standard program that you do not have to develop. Slaves are addressed in one of two ways. An ASCII terminal can be connected to each controller and an address can be assigned or one slave at a time can be put on the network and addressed using the personal computer via the master controller and standard program running in the slave.

The program in the master is written in exactly the same manner as before. The only thing that changes is that the I/O addresses become a three-part address. The address consists of a node, a device, and a channel. The Z-World controller uses a master/slave polling access scheme. The master requests data from the slave nodes as needed and they respond when called. The master node also polls each slave node once each second to make sure they are still there.

The node portion of the address is the address of the desired controller. The device portion of the address describes the type of card that is being talked to, for example, zDigOut (Node + Device + Channel, output). The node address for the master is 0. Slave controllers are given addresses of between 1 and 15. The channel would be the actual I/O point desired. Output in the statement determines whether the output will be set to on or off. This could be an integer constant (0 or any other integer) or a variable.

Wiring the Network

The wiring is very simple. All of the nodes (master and slaves) are connected by a twisted-pair cable. It only takes two wires to communicate. Remember that the twist in twisted-pair wiring offers some noise immunity. Each node must still be provided with 12 to 24 volts to operate. A four-wire cable can be used to connect nodes. In that case two wires provide the power for operation and two are used for RS-485 communication.

Figure 15-16 shows an example of a networked and expanded system. Note that three single-board controllers have been networked. One of them is the master and the other two are slaves. Each of the controllers has been expanded with additional I/O.

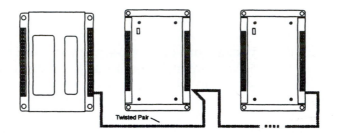

Twisted Pair

Figure 15-15. An example of networking single-board control-
lers. In this case three have been networked. They are connected
with simple twisted-pair wiring. Courtesy Z-World.

Multitasking

Simple computers usually execute one program at a time. Embedded computers often execute more than one task at a time through the use of multitasking software. Multitasking simply means that a computer can perform multiple tasks simultaneously. In reality, a single processor can only execute one instruction at a time. Multiple tasks interleave their execution, only appearing to execute together. The tasks are actually sharing the processor's time. Two common types of multitasking are preemptive and cooperative.

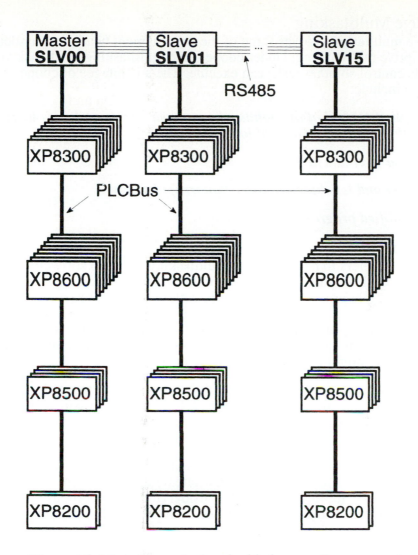

*Figure 15-16. A theoretical embedded controller system that
has been networked and expanded. Courtesy Z-World.*

Preemptive Multitasking

Preemptive multitasking means that some top-priority event, usually an alarm, a
timer interrupt, or a supervisory task (often referred to as a kernel), takes control
of the processor from the task currently running and turns it over to another task.
In this scenario, program tasks compete for the processor's time. Tasks have no
control over when they may be preempted, and usually have no information about
other tasks. Cooperation, coordination, and communication among tasks that get
asynchronously preempted is a major application problem. Preemptive multitasking requires special attention during software development.

Cooperative Multitasking

Cooperative multitasking solves some of the development headaches associated with preemptive multitasking. Under cooperative multitasking, each task voluntarily yields control so other tasks can execute. The advantages of cooperative multitasking include:

> *Explicit control of the points where a task begins and ends logical subsections of its overall job.*
>
> *Easier to create tasks.*
>
> *Lower and less indeterminate "interrupt latency."*
>
> *Simplified programming.*

Networking/Multitasking

Multiprocessing refers to hardware and multitasking refers to software, yet the two often go together and have much in common. Multiprocessing means, literally, having multiple processors. Some applications may require more input and output ports than a single controller can provide, but the application may not support two separate programs running on two separate controllers. Or, in other cases, running numerous input and output lines throughout a machine to and from a single controller might be physically unsafe or impractical.

Although software can execute across multiple processors in many different fashions, a simple and straightforward approach to multiprocessing is to create a multiprocessing system by establishing a network of controllers communicating over industry-standard RS-485 serial lines.

Z-World's programming language also offers extensions to the C language. One of the most powerful is the costatement. Costatements are a extension to C that facilitate cooperative multitasking. Let's consider an example. Industrial applications often involve a simple, linear sequence of events such as the one shown in Figure 15-17.

> *Start:*
>
> *1. Wait for cycle start push-button to be pressed.*
>
> *2. Turn on pump 1.*
>
> *3. Wait for 30 seconds.*
>
> *4. Turn on pump 2.*
>
> *5. Wait 45 seconds.*
>
> *6. Turn off pump 1 and pump 2.*
>
> *Go back to start.*

Figure 15-17. A linear sequence.

One way to do this task would be to create the wait states by making the controller wait for 30 seconds in step 3 and then wait for 45 seconds in step 5. If it is done this way the controller cannot perform any other tasks while it is waiting. All of this wait time is essentially wasted.

Another way to think of this would be a single task like baking a cake. The cake baking task has several steps (see Figure 15-18). Each step takes time. The baker is initially very busy as he finds the ingredients, measures them, mixes them, and so on. But when the baker puts the cake in the oven he has some time with nothing to do. If there is nothing else to bake he waits and occasionally looks at the temperature and time to be sure he takes the cake out in time. In this example the baker has a lot of wasted time as he waits for the cake to bake.

The second type of situation involves several tasks, all of which must operate concurrently. Imagine a cook in a restaurant. A waiter brings the order from table 12 to the kitchen (see Figure 15-19). There are four orders from the table. All are received at the same time in this case and all should be finished at about the same time so that everyone receives hot food. The cook must co-process these orders. In the previous example the baker could relax during wait times (the cake in the oven). The cook, on the other hand, must use wait times in one task to work on the others.

It is often the case in industrial applications that concurrent tasks are required. If there were other tasks that need to be run we would need to write a program that could process several tasks concurrently (see Figure 15-19). When there is idle time in one task the processor can be servicing other tasks. Each task can relinquish its control when it is waiting. This allows other tasks to be processed. Every task's work is done in the idle time of others.

A more industrial example of concurrent processing would be a worker in a machining cell. The cell has several machines. The worker loads a part into a machine and the machine starts its cycle. The worker has free time so she moves to another machine and loads or unloads a part. The worker then has time to inspect a part from another machine. The worker can thus stay very productive by moving between the machines when needed. The worker is like the processor. The machines are like the tasks. The worker is free to work among the machines as needed and so is much more productive than having to start one machine and stay until the sequence is complete before moving to the next machine.

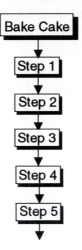

Figure 15-18. An example of a single-task sequence of operations. The task in this case is to bake a cake. Note that there are several steps in the process.

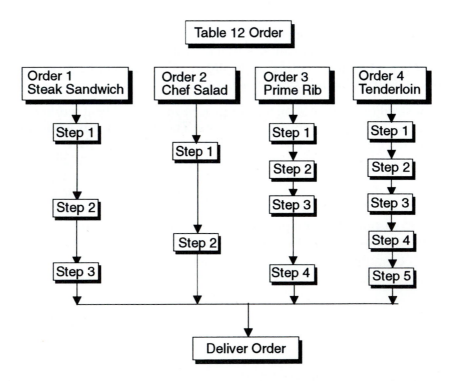

Figure 15-19. Several tasks that need to be co-executed.

A ladder logic program is usually hard to modularize. It is essentially one long program. Every rung of logic is evaluated every time unless MCRs or ZCLs are used. Even portions of the ladder diagram that are not active during a portion of a sequence are examined every scan. This is usually not a problem in most applications, but it does make the logic quite confusing.

It would normally be quite difficult to write C code for concurrent tasks. Z-World has created C language extensions to make writing programs with concurrent tasks easy. They are called costates (see Figure 15-20). They also help to modularize a program. This helps break a program into manageable and understandable blocks.

```
costate {task1}
  {
    ( C code for task 1 )
  }
costate {task2}
  {
    ( C code for task 2 )
  }
costate {task3}
  {
    ( C code for task 4 )
  }
}
```

Figure 15-20. Simplified example of how a costatement can be used.

Costatements are very powerful because they can wait for events, conditions, or the passage of time. They can yield temporarily to other costatements. They can also abort their own operation. They can then resume operation from the point at which they left the costate. Costatements are also very powerful for the programmer because they take the complexity out of programming concurrent activities and they help modularize programs.

One can see that single-board controllers offer many advantages for industrial control. Today's single-board controllers are very fast, powerful, and offer almost unlimited I/O capability. Networking, communications, and operator I/O have been made simple for the programmer through the use of functions.

Questions

1. Explain what a single-board controller is.

2. What are the most common languages used to program single-board controllers?

3. List at least three advantages of single-board controllers.

4. Explain how single-board controllers can be expanded.

5. Explain how single-board controllers can be networked.

6. What are costatements and what are they used for?

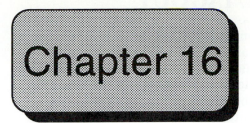

Chapter 16

Lockout/Tagout

Procedures to ensure the safety of personnel involved in maintenance of equipment are essential. This chapter will cover lockout/tagout procedures. Troubleshooting and maintenance of systems becomes more crucial as systems become more automatic, more complex, and more expensive. Maintenance personnel must understand how to use lockout/tagout procedures to ensure their safety while maintaining machines.

OBJECTIVES

Upon completion of this chapter, the reader will be able to:

List and explain the sources of energy that are typically found in a industrial environment.

Define terms such as **affected employee, authorized employee, energy source, stored energy, notification,** *and so on.*

Define the term **lockout.**

Define the term **tagout.**

Describe the typical steps that must be included in a typical lockout/tagout procedure.

Explain why it is so crucial to follow lockout/tagout procedures.

OVERVIEW OF LOCKOUT/TAGOUT

On October 30, 1989, the Lockout/Tagout Standard, 29 CFR 1910.147, went into effect. It was released by the Department of Labor. The standard is titled "The control of hazardous energy sources (lockout/tagout)." The standard was intended to reduce the number of deaths and injuries related to servicing and maintaining machines and equipment. Deaths and tens of thousands of lost work days are attributable to maintenance and servicing activities each year.

The lockout/tagout standard covers the servicing and maintenance of machines and equipment in which the unexpected startup or energization of the machines or equipment or the release of stored energy could cause injury to employees. The standard is intended to cover energy sources such as electrical, mechanical, hydraulic, pneumatic, chemical, thermal and other energy sources.

Electrical energy is present in almost every piece of industrial equipment. Mechanical energy is present in moving part such as spindles, flywheels, and so on. Mechanical energy is also present in the form of stored energy such as heavy objects that may fall or springs that may release their energy. Hydraulic energy is present in equipment like lifts, presses, and so on. Pneumatic energy is just another form of hydraulic energy. Chemical energy comes from reactions caused by mixing chemicals. Chemical energy can cause severe burns, rashes, and so on. Thermal energy is heat or cold. The standard establishes minimum standards for the control of such hazardous energy. Normal production operations, cords and plugs under exclusive control, and hot tap operations are not covered by the standard.

Hot tap operations are those involving transmission and distribution systems for substances such as gas, steam, water or petroleum products. A hot tap is a procedure used in the repair maintenance and services activities which involves welding on a piece of equipment such as a pipeline, vessel, or tank, under pressure, in order to install connections or appurtenances. Hot tap procedures are commonly used to replace or add sections of pipeline without the interruption of service for air, gas, water, steam, and petrochemical distribution systems. The standard does not apply to hot taps when they are performed on pressurized pipelines, provided that the employer demonstrates that continuity of service is essential; shutdown of the system is impractical; and that documented procedures are followed, and special equipment is used which will provide proven, effective protection for employees.

Employers are required to establish a program consisting of an energy control (lockout/tagout) procedure and employee training to ensure that before any employee performs any servicing or maintenance on a machine or equipment where the unexpected energizing, start up or release of stored energy could occur and cause injury, the machine or equipment shall be isolated, and rendered inoperative. The employer is also required to conduct periodic inspection of the energy control procedure at least annually to ensure that the procedures that were developed and the requirements of this standard are being followed.

The standard defines an energy source as any source of electrical, mechanical, hydraulic, pneumatic, chemical, thermal, or other energy. Machinery or equipment is considered to be energized if it is connected to an energy source or contains residual or stored energy. Stored energy can be found in pneumatic and hydraulic systems, springs, capacitors, and even gravity.

Servicing and/or maintenance includes activities such as constructing, installing, setting up, adjusting, inspecting, modifying, and maintaining and/or servicing machines or equipment. These activities include lubrication, cleaning or unjamming of machines or equipment and making adjustments or tool changes, where the employee may be exposed to the unexpected energization or startup of the equipment or release of hazardous energy.

Normal production operations are excluded from lockout/tagout restrictions. Normal production operation is the utilization of a machine or equipment to perform its intended production function. Minor tool changes, machine adjustments, and other repetitive, routine activities are excluded from lockout/tagout procedures. These activities are still subject to appropriate guarding and safety procedures. Any work performed to prepare a machine or equipment to perform its normal production operation is called setup.

If an employee is working on cord and plug connected electrical equipment for which exposure to unexpected energization or start up of the equipment is controlled by the unplugging of the equipment from the energy source, and the plug is under the exclusive control of the employee performing the servicing or maintenance, this activity is also excluded from requirements of the standard.

Only authorized employees may lock out machines or equipment. An authorized employee is one who has been trained and has the authority to lock or tag out machines or equipment in order to perform servicing or maintenance on that machine or equipment.

An energy isolating device is a mechanical device that physically prevents the transmission or release of energy. Energy isolating devices include the following: manually operated electrical circuit breakers, disconnect switches, manually operated switches by which the conductors of a circuit can be disconnected from all ungrounded supply conductors and, in addition, no pole can be operated independently; line valves; a block; and any similar device used to block or isolate energy. Push-buttons, selector switches and other control circuit type devices are not energy isolating devices.

An energy isolating device is capable of being locked out if it has a hasp or other means of attachment to which, or through which, a lock can be affixed, or it has a locking mechanism built into it. Other energy isolating devices are capable of being locked out, if lockout can be achieved without the need to dismantle, rebuild, or replace the energy isolating device or permanently alter its energy control capability.

Since January 2, 1990 there has been a requirement that whenever new machines or equipment are installed, energy isolating devices for such machines or equipment must be designed to accept a lockout device.

LOCKOUT

Lockout is the placement of a lockout device on an energy isolating device, in accordance with an established procedure to ensure that the energy isolating device can ensure that the equipment being controlled cannot be operated until the lockout device is removed.

A lockout device utilizes a positive means such as a lock (see Figure 16-1) to hold an energy isolating device in the safe position and prevent the energizing of a machine or equipment. A lock may be either key or combination type. If an energy isolating device is incapable of being locked out, the employer's energy control program shall utilize a tagout system.

Notification of Employees

Affected employees must be notified by the employer or authorized employee of the application and removal of lockout devices or tagout devices. Notification shall be given before the controls are applied, and after they are removed from the machine or equipment. Affected employees are defined as employees whose job requires them to operate or use a machine or equipment on which servicing or maintenance is being performed under lockout or tagout, or whose job requires them to work in an area in which such servicing or maintenance is being performed.

Figure 16-1. A typical machine disconnect.

A lock may be used without a tag if the following conditions are met: all employees who will be exposed to the hazards are familiar with the procedure, only one piece of equipment is deenergized, and the lockout period does not extend beyond one work shift.

TAGOUT

Tagout is the placement of a tagout device on an energy isolating device, in accordance with an established procedure, to indicate that the energy isolating device and the equipment being controlled may not be operated until the tagout device is removed. Tagout shall be performed only by the authorized employees who are performing the servicing or maintenance.

A tagout device is a prominent warning device, such as a tag and a means of attachment, which can be securely fastened to an energy isolating device in accordance with an established procedure, to indicate that the energy isolating device and the equipment being controlled may not be operated until the tagout device is removed from each energy isolating device by the employee who applied the device.

When the authorized employee who applied the lockout or tagout device is not available to remove it, that device may be removed under the direction of the employer, provided that specific procedures and training for such removal have been developed, documented and incorporated into the energy control program.

Tagout devices, where used, must be affixed in such a manner as to clearly indicate that the operation or movement of energy isolating devices from the safe or off position is prohibited. Where tagout devices are used with energy isolating devices designed with the capability of being locked, the tag attachment must be fastened at the same point at which the lock would have been attached. Where a tag cannot be affixed directly to the energy isolating device, the tag must be located as close as safely possible to the device, in a position that will be immediately obvious to anyone attempting to operate the device.

TRAINING

Training must be provided by the employer to ensure that the purpose and function of the energy control program are understood by employees and that employees have the knowledge and skills required for the safe application, usage, and removal of energy controls. Employees should be trained to:

> *recognize hazardous energy sources*
>
> *recognize the type and magnitude of the energy available in the workplace*
>
> *know methods and means necessary for energy isolation and control*
>
> *understand the purpose and use of the lockout/tagout procedures*

All other employees whose work operations are or may be in an area where lockout tagout procedures may be used, shall be instructed about the procedure and about the prohibition against attempting to restart or reenergize machines or equipment that are locked out or tagged out.

When tagout procedures are used, employees must be taught about the following limitations of tags:

> *Tags are really just warning devices and do not provide physical restraint on devices.*

> *When a tag is attached it is not to be removed without authorization of the authorized person responsible for it, and it is never to be by-passed, ignored, or otherwise defeated.*

> *Tags must be legible and understandable by all authorized employees, affected employees, and all other employees whose work operations are or may be in the area.*

> *Tags may create a false sense of security. Their meaning needs to be understood by all.*

Retraining

Retraining shall be provided for all authorized and affected employees whenever there is a change in their job assignments, a change in machines, equipment or processes that present a new hazard, or when there is a change in the energy control procedures.

Additional retraining shall also be conducted whenever a periodic inspection reveals, or whenever the employer has reason to believe that there are deviations from or inadequacies in the employee's knowledge or use of the energy control procedures.

The retraining shall reestablish employee proficiency and introduce new or revised control methods and procedures, as necessary.

The employer shall certify that employee training has been accomplished and is being kept up to date. The certification shall contain each employee's name and dates of training.

REQUIREMENTS FOR LOCKOUT/TAGOUT DEVICES

Lockout and tagout devices must be singularly identified, must be the only device(s) used for controlling energy and must not be used for other purposes.

They must be durable, which means they must be capable of withstanding the environment to which they are exposed for the maximum period of time that exposure is expected.

Tagout devices must be constructed and printed so that exposure to weather conditions or wet and damp locations will not cause the tag to deteriorate or the message on the tag to become illegible. Figure 16-2 shows an example of a tag.

Tags must not deteriorate when used in corrosive environments such as areas where acid and alkali chemicals are handled and stored.

Lockout and tagout devices must be standardized within the facility in at least one of the following criteria: color, shape, or size. Print and format must also be standardized for tagout devices.

Lockout devices must be substantial enough to prevent removal without the use of excessive force or unusual techniques, such as with the use of bolt cutters or other metal cutting tools.

Tagout devices, including their means of attachment, shall be substantial enough to prevent inadvertent or accidental removal. Tagout devices must be attached with a nonreusable attachment. They must be attachable by hand, self-locking, and nonreleasable with a minimum unlocking strength of at least 50 pounds. They should have the general design and basic characteristics of being at least equivalent to a one-piece, all-environment-tolerant nylon cable tie.

Lockout devices and tagout devices must identify the employee who applied the devices.

Figure 16-2. A typical tag.

Tagout devices must warn against hazardous conditions if the machine or equipment is energized and must include a clear warning such as: "Do Not Start. Do Not Open. Do Not Close. Do Not Energize. Do Not Operate."

APPLICATION OF CONTROL

The established procedures for the application of lockout or tagout devices shall cover the following elements and actions and shall be done in the following sequence:

1. Notify all affected employees that a lockout or tagout system is going to be used. They must also understand the reason for the lockout. Before an authorized or affected employee turns off a machine or equipment, the authorized employee must understand the types and magnitudes of the energy, the hazards of the energy to be controlled, and the method or means to control the energy for the machine or equipment being serviced or maintained.

2. The machine or equipment shall be turned off or shut down using the procedures established for the machine or equipment. An orderly shutdown must be utilized to avoid any additional or increased hazards to employees as a result of the equipment stoppage.

3. All energy isolating devices that are needed to control the energy to the machine or equipment shall be physically located and operated in such a manner as to isolate the machine or equipment from the energy sources.

4. Lockout or tagout devices shall be affixed to each energy isolating device by authorized employees. Lockout devices, where used, shall be affixed in a manner that will hold the energy isolating devices in a safe or off position.

5. Following the application of lockout or tagout devices to energy isolating devices, stored energy must be dissipated or restrained by methods such as repositioning, blocking, bleeding down, and so on. If there is a possibility of reaccumulation of stored energy to a hazardous level, verification of isolation shall be continued until the servicing or maintenance is completed, or until the possibility of such accumulation of energy no longer exists.

6. Prior to starting work on machines or equipment that have been locked out or tagged out, the authorized employee shall verify that the machine or equipment has actually been isolated and deenergized. This is done by operating the push-button or other normal operating controls to make certain that the equipment will not operate.

Caution: Make sure you return the operating controls to the neutral or off position after the test.

The machine is now locked out or tagged out.

Before lockout or tagout devices are removed and energy is restored to the machine or equipment, procedures shall be followed and actions taken by the authorized employees to ensure the following:

The work area shall be inspected to ensure that nonessential items have been removed and to ensure that machine or equipment components are operationally intact.

The work area shall be checked to ensure that all employees have been safely positioned or removed.

Before lockout or tagout devices are removed and before machines or equipment are energized, affected employees shall be notified that the lockout or tagout devices have been removed.

Each lockout or tagout device must be removed from each energy isolating device by the employee who applied the device. The only exception to this is when the authorized employee who applied the lockout or tagout device is not available to remove it. That device may be removed under the direction of the employer, provided that specific procedures and training for such removal have been developed, documented and incorporated into the employer's energy control program. The employer must demonstrate that the specific procedure includes at least the following elements:

The employer must verify that the authorized employee who applied the device is not at the facility.

All reasonable efforts must be made to contact the authorized employee to inform him/her that his/her lockout or tagout device has been removed.

The authorized employee must be made aware that his/her lockout tagout device was removed before he/she resumes work at that facility.

Testing of Machines, Equipment or Components

There may be situations in which lockout or tagout devices must be temporarily removed from the energy isolating device and the machine or equipment energized to test or position the machine, equipment or component. In this case the following sequence of actions must be followed:

1. Clear the machine or equipment of tools and materials.

2. Remove employees from the machine or equipment area.

3. Remove the lockout or tagout devices as specified in the standard.

4. Energize and proceed with testing or positioning.

5. Deenergize all systems and reapply energy control measures in accordance with the standard to continue the servicing and/or maintenance.

Outside Personnel Working in the Plant

When outside servicing personnel (contractors, etc.) are to be engaged in activities covered by the lockout/tagout standard, the on-site employer and the outside employer must inform each other of their respective lockout or tagout procedures. The on-site employer must ensure that his/her employees understand and comply with the restrictions and prohibitions of the outside employer's energy control program.

Group Lockout or Tagout

When servicing and/or maintenance is performed by a group of people, they must use a procedure that protects them to the same degree that a personal lockout or tagout procedure would. The lockout/tagout standard specifies requirements for group procedures. Primary responsibility is vested in an authorized employee for a set number of employees. These employees work under the protection of a group lockout or tagout device (see Figure 16-3). The group lockout device assures that no one individual can start up or energize the machine or equipment. All lockout or tagout devices must be removed to reenergize the machine or equipment. The authorized employee who is responsible for the group must ascertain the exposure status of individual group members with regard to the lockout or tagout of the machine or equipment. When more than one crew, craft, department, etc. is involved, overall job-associated lockout or tagout control responsibility is assigned to an authorized employee. This employee is designated to coordinate affected work forces and ensure continuity of protection. Each authorized employee must affix a personal lockout or tagout device to the group lockout device, group lockbox, hasp (see Figure 16-2), or comparable mechanism when he or she begins work, and shall remove those devices when he or she stops working on the machine or equipment being serviced or maintained.

Figure 16-3. An open and a closed hasp. The hasp allows multiple personnel to lock out machines or equipment.

Shift or Personnel Changes

Specific procedures must be utilized during shift or personnel changes to ensure the continuity of lockout or tagout protection. This ensures the orderly transfer of lockout or tagout device protection between off-going and oncoming employees, to minimize exposure to hazards from the unexpected energization or startup of the machine or equipment, or the release of stored energy.

SAMPLE LOCKOUT PROCEDURE

The following is a sample lockout procedure. Tagout procedures may be used when the energy isolating devices are not lockable, provided the employer complies with the provisions of the standard which require additional training and more rigorous periodic inspections. When tagout is used and the energy isolating devices are lockable, the employer must provide full employee protection and additional training and more rigorous periodic inspections are required. When more complex systems are involved, more comprehensive procedures may need to be developed, documented, and utilized.

Lockout Procedure for Machine #37

Note: This would normally name the machine when multiple procedures exist. If only one exists it would normally be the company name.

Purpose

This procedure establishes the minimum requirements for the lockout of energy isolating devices whenever maintenance or servicing is done on machine #37. This procedure must be used to ensure that the machine is stopped, isolated from all potentially hazardous energy sources and locked out before employees perform any servicing or maintenance where the unexpected energization or startup of the machine or equipment or release of stored energy could cause injury.

Employee Compliance

All employees are required to comply with the restrictions and limitations imposed upon them during the use of this lockout procedure. Authorized employees are required to perform the lockout in accordance with this procedure. All employees, upon observing a machine or piece of equipment that is locked out to perform servicing or maintenance shall not attempt to start, energize, or use that machine or equipment.

A company may want to list actions that may be taken in the event of an employee violating the procedure.

Lockout Sequence

1. Notify all affected employees that servicing or maintenance is required on the machine and that the machine must be shut down and locked out to perform the servicing or maintenance.

> *The procedure should list the names and/or job titles of affected employees and how to notify them.*

2. The authorized employee must refer to the company procedure to identify the type and magnitude of the energy that the machine utilizes, must understand the hazards of the energy, and must know the methods to control the energy.

> *The types and magnitudes of energy, their hazards and the methods to control the energy should be detailed here.*

3. If the machine or equipment is operating, shut it down by the normal stopping procedure (press the stop button, open switch, close valve, etc.).

> *The types and locations of machine or equipment operating controls should be detailed here.*

4. Deactivate the energy isolating devices so that the machine or equipment is isolated from the energy sources.

> *The types and locations of energy isolating devices should be detailed here.*

5. Lock out the energy isolating devices with assigned individual locks.

6. Stored or residual energy (such as that in capacitors, springs, elevated machine members, rotating flywheels, hydraulic systems, and air, gas, steam, or water pressure, etc.) must be dissipated or restrained by methods such as grounding, repositioning, blocking, bleeding down, etc.

> *The types of stored energy, as well as methods to dissipate or restrain the stored energy should be detailed here.*

7. Ensure that the equipment is disconnected from the energy sources by first checking that no personnel are exposed, then verify the isolation of the equipment by operating the push-button or other normal operating controls or by testing to make certain the equipment will not operate. Caution: Return operating controls to neutral or off position after you verify the isolation of the equipment.

> *The method of verifying the isolation of the equipment should be detailed here.*

8. The machine or equipment is now locked out.

Returning the Machine or Equipment to Service

When the servicing or maintenance is completed and the machine or equipment is ready to return to normal operating condition, the following steps shall be taken.

1. Check the machine or equipment and the immediate area around the machine to ensure that nonessential items have been removed and that the machine or equipment components are operationally intact. Check the work area to ensure that all employees have been safely positioned or removed from the area.

2. After all tools have been removed from the machine or equipment, guards have been reinstalled and employees are in the clear, remove all lockout or tagout devices. Verify that the controls are in neutral and reenergize the machine or equipment. Note: The removal of some forms of blocking may require reenergization of the machine before safe removal. Notify affected employees that the servicing or maintenance is completed and the machine or equipment is ready for use.

SAMPLE LOCKOUT/TAGOUT CHECKLIST

Notification

I have notified all affected employees that a lockout is required and the reason for the lockout.

Date _____ Time _____ Signature _____

Shutdown

I understand the reason the equipment is to be shut down following normal procedures.

Date _____ Time _____ Signature _____

Disconnection of Energy Sources

I operated the switches, valves and other energy isolating devices so that each energy source has been disconnected or isolated from the machinery or equipment. I have dissipated or restrained all stored energy such as springs, elevated machine members, capacitors, rotating flywheels, pneumatic and hydraulic systems, etc.

Date _____ Time _____ Signature _____

Lockout

I have locked out the energy isolating devices using my assigned individual locks.

Date _____ Time _____ Signature _____

Safety Check

After ensuring that no personnel are exposed to hazards I have operated the start button and other normal operation controls to ensure that all energy sources have been disconnected and that the equipment will not operate.

Date _____ Time _____ Signature _____

The machine is now locked out.

Questions

1. List and explain the sources of energy that are typically found in a industrial environment.

2. What is an affected employee?

3. What is an authorized employee?

4. Define the term lockout.

5. Define the term tagout.

6. Describe the typical steps in a lockout/tagout procedure.

7. When is it OK not to use a tag?

8. Write a lockout/tagout procedure for a cell that contains electrical and pneumatic energy.

Chapter 17

Installation and Troubleshooting

Proper installation is crucial in automated systems. The safety of people and machines is at stake. Troubleshooting and maintenance of systems becomes more crucial as systems become more automatic, more complex, and more expensive. An enterprise cannot afford to have a system down for any length of time. The technician must be able to find and correct problems quickly.

OBJECTIVES

Upon completion of this chapter, the student will be able to:

> *Describe safety considerations that are crucial in troubleshooting and maintaining systems.*
>
> *Explain such terms as* **noise, snubbing, suppression,** *and* **single-point ground.**
>
> *Explain correct installation techniques and considerations.*
>
> *Explain proper grounding techniques.*
>
> *Explain noise reduction techniques.*
>
> *Explain a typical troubleshooting process.*

INSTALLATION AND TROUBLESHOOTING

Installation, troubleshooting and maintenance of an automated system is one of the most crucial phases of any project. It is the point at which the hopes and fears of the engineers and technicians are realized. It can be a frustrating, exciting, and rewarding time. Installation, troubleshooting, maintenance and operation of any automated system is highly dependent on the quality of the associated documentation. Documentation is unfortunately often an afterthought. Documentation should be developing as the system is developed. The system must be accurately and completely documented. Downtime can be significantly reduced with good documentation. Fortunately programming software for PLCs and other control devices has extensive documentation capabilities. Documentation should include the following:

> *a description of the overall system;*
>
> *a block diagram of the entire system;*
>
> *a program listing including cross referencing and clearly labeled I/O;*
>
> *a printout of the PLC or control device's memory showing I/O and variable usage;*
>
> *a complete wiring diagram;*
>
> *a description of peripheral devices and their manuals;*
>
> *an operator's manual including startup and shutdown procedures;*
>
> *notes concerning past maintenance.*

INSTALLATION

Proper installation of a PLC is crucial. The PLC must be wired so that the system is safe for the workers. The system must also be wired so that the devices are protected from overcurrent situations. Proper fusing within the system is important. The PLC must also be protected from the application environment. There are usually dust, coolant, chips, and other contaminants in the air. The proper choice of enclosures can protect the PLC. Figure 17-1 shows a block diagram of a typical installation.

Enclosures

PLCs are typically mounted in protective cabinets. The National Electrical Manufacturers Association (NEMA) has developed standards for enclosures (see Figure 17-2). Enclosures are used to protect the control devices from the environment of the application. A cabinet type is chosen based on how severe the environment is in the application. Cabinets typically protect the PLC from airborne contamination. Metal cabinets can also help protect the PLC from electrical noise.

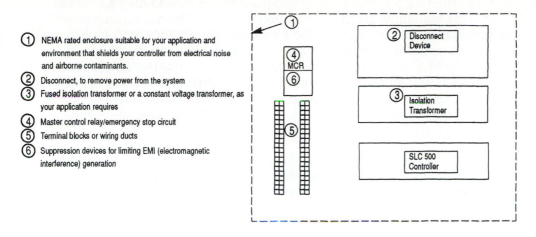

① NEMA rated enclosure suitable for your application and
environment that shields your controller from electrical noise
and airborne contaminants.

② Disconnect, to remove power from the system

③ Fused isolation transformer or a constant voltage transformer, as
your application requires

④ Master control relay/emergency stop circuit

⑤ Terminal blocks or wiring ducts

⑥ Suppression devices for limiting EMI (electromagnetic
interference) generation

Figure 17-1. Block diagram of a typical PLC control cabinet.
Courtesy Rockwell Automation/Allen-Bradley Company Inc..

Heat is generated by devices. One must make sure that the PLC and other devices to be mounted in the cabinet can perform at the temperatures required. Remember that it will be even hotter in the cabinet than in the application because of the heat generated within the cabinet. In some applications substantial heat may be generated by other devices in the system or cabinet. In this case, blower fans should be placed inside the enclosure to increase air circulation and reduce hot spots within the enclosure. These fans should filter the incoming air so that contaminants are not introduced to the cabinet and components.

Protection Against	Enclosure Type												
	1	2	3	3r	3s	4	4x	5	6	6p	11	12	13
Accidental contact with enclosed equipment	*	*	*	*	*	*	*	*	*	*	*	*	*
Falling dirt	*	*				*	*	*	*	*	*	*	*
Falling liquids, light splashing		*				*	*		*	*	*	*	*
Dust, lint, fibers, (non-combustible, non-ignitable)						*	*	*	*	*		*	*
Windblown dust			*		*	*	*		*	*			
Hose down and splashing water						*	*		*	*			
Oil and coolant seepage												*	*
Oil or coolant spraying or splashing													*
Corrosive agents							*			*	*		
Occasional temporary submersion									*	*			
Occasional prolonged submersion										*			

Figure 17-2. Comparison of the features of NEMA cabinets
versus enclosure type.

The main consideration is that there is adequate space around the PLC in the enclosure. This will allow air to flow around the PLC. The manufacturer will provide installation requirements in the hardware manual for the PLC. Figure 17-3 shows an example for a AB SLC 500. Note that the dimensions shown from each side of the PLC to the cabinet are minimum distances. The actual distance between the PLC and the cabinet should be greater than those shown. Check the specifications for the PLC that will be used to see what temperatures it can operate in. In most cases fans will be unnecessary. Proper clearances around devices will normally be sufficient for heat dissipation.

When drilling holes in the cabinet for mounting components be very careful not to allow chips to fall into the PLC or other components. Metal chips or wire clippings could cause equipment to short circuit or could cause intermittent or permanent problems.

Wiring

Proper wiring of a system involves choosing the appropriate devices and fuses (see Figure 17-4). Normally, three-phase power (typically 480 V) will be used in manufacturing. This will be the electrical supply for the control cabinet.

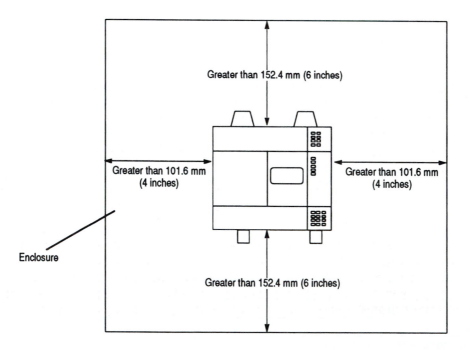

Figure 17-3. Proper mounting for an AB SLC 500. Courtesy Rockwell Automation/Allen-Bradley Company Inc.

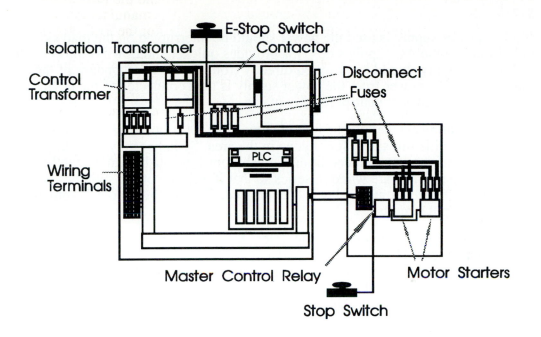

Figure 17-4. Block diagram of a typical control cabinet wiring scheme.

The three-phase power is connected to the cabinet via a mechanical disconnect. This disconnect is mechanically turned on or off by the use of a lever on the outside of the cabinet. This disconnect should be equipped with a lockout. This means that the technician should be able to put a lock on the lever to prevent anyone from accidentally applying power while it is being worked on. This three-phase power must be fused. Normally, a fusible disconnect is used. This means that the mechanical disconnect has fusing built in. The fusing is to make sure that too much current cannot be drawn. Figure 17-4 shows the fuses installed after the disconnect.

The three-phase power is then connected to a contactor. The contactor is used to turn all power off to the control logic in case of an emergency. The contactor is attached to hard-wired emergency circuits in the system. If someone hits an emergency stop switch, the contactor drops out and will not supply power. The hardwired safety should always be used in systems. This is called a master control relay.

Master Control Relay

A hard-wired master control relay (MCR) is used for emergency controller shutdown (see Figure 17-5). The MCR must be able to inhibit all machine motion by removing power to the machine I/O devices when the MCR relay is deenergized.

If a DC power supply is used the power should be interrupted on the load side rather than on the AC supply side. This provides quicker shutdown of power. The DC power supply should be powered directly from the fused secondary of the transformer. DC power from the power supply is then connected to the devices through a set of master control relay contacts.

Emergency stop switches can be placed at multiple locations. These should be chosen to provide full safety for anyone in the area. Switches may include limit-type switches for overtravel conditions, mushroom-type push-button switches, and other types. They are wired in series so that if any one of the switches is activated the master control relay is deenergized. This removes power from all input and output circuits.

Never alter or bypass these circuits to defeat their function. Severe injury and/or damage can occur. These switches are designed to "fail safe." If they fail they should open the master control circuit and disconnect power. It is possible that a switch could short out and not offer any protection anymore. Switches should be tested periodically to be sure they will still stop all machine motion if used.

The main power disconnect switch should be placed so that it is in a convenient and easily accessible place for operators and maintenance personnel. The disconnect should be placed so that power can be turned off before the cabinet is opened.

The master control relay is not a substitute for a disconnect to the PLC. Its purpose is to quickly deenergize I/O devices. Figure 17-6 shows a cutoff system with a hard-wired emergency switch connected to the master control relay. The figure also shows the wiring diagram. Note that the system provides for a mechanical disconnect for output module power and a hard-wired stop switch for disconnecting the system power. The programmer should also provide for a orderly shutdown in the PLC control program.

The three-phase power must then be converted to single-phase for the control logic. Power lines from the fusing are connected to transformers. In the case of Figure 17-7 there are two transformers: an isolation transformer and a control transformer. The isolation transformer is used to clean up the power supply for the PLC. Isolation transformers are normally used when there is high-frequency conducted noise in or around power distribution equipment. Isolation transformers are also used to step-down the line voltage.

The control transformer is used to supply other control devices in the cabinet. The lines from the power supplies are then fused to protect the devices they will supply. These individual device circuits should be provided with their own fuses to match the current draw. DC power is usually required also. A small power supply is typically used to convert the AC to DC.

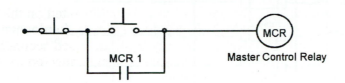

Figure 17-5. Example of a master control relay circuit.

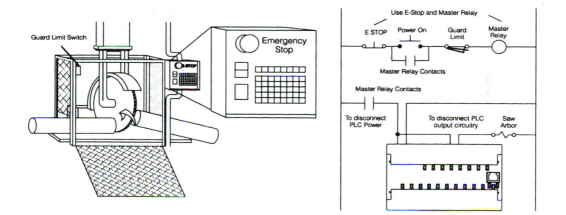

Figure 17-6. The use of a hardwired E-Stop and master control relay. Courtesy PLC Direct by Koyo.

Motor starters are typically mounted in separate enclosures. This protects the control logic from the noise these devices generate.

Within the cabinet certain wiring conventions are typically used. Red wiring is normally used for control wiring, black wiring is used for three-phase power, blue wire is used for DC, and yellow wire is used to show that the voltage source is separately derived power (outside the cabinet).

Signal wiring should be run separately from 120 V wiring. Signal wiring is typically low voltage or low current and could be affected by being too close to high-voltage wiring. When possible, run the signal wires in separate conduit. Some conduit is internally divided by barriers to isolate signal wiring from higher voltage wiring.

Voltage is supplied to the PLC through wiring terminals. The user can often configure the PLC to accept different voltages (see Figure 17-8). The PLC in this figure can be configured to accept either 220 or 110 V. In this case the shorting bar must be installed if 110 V will be used.

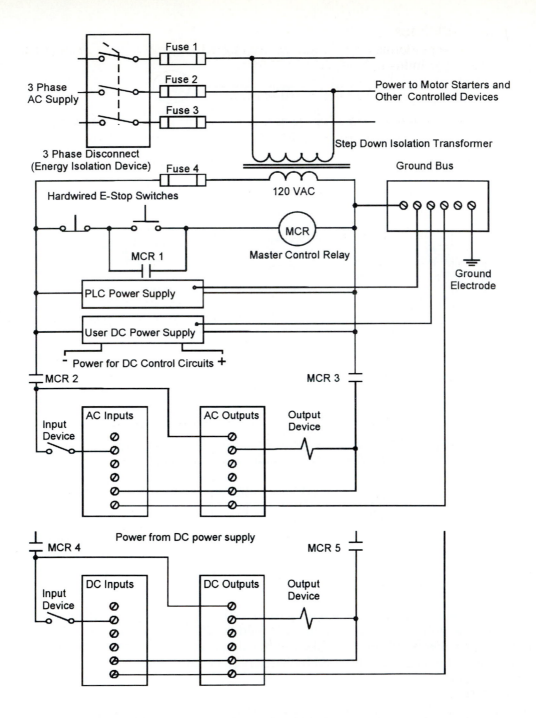

Figure 17-7. Control wiring diagram.

Wiring Guidelines

Here are some guidelines that should be considered when wiring a system, courtesy of Siemens Industrial Automation, Inc.

> *Always use the shortest possible cable.*
>
> *Use a single length of cable between devices. Do not connect pieces of cable to make a longer cable. Avoid sharp bends in wiring.*
>
> *Avoid placing system and field wiring close to high-energy wiring.*
>
> *Physically separate field input wiring, output wiring, and other types of wiring.*
>
> *Separate DC and AC wiring when possible.*
>
> *A good ground must exist for all components in the system (0.1 ohm or less).*
>
> *If long return lines to the power supply are needed, do not use the same wire for input and output modules. Separate return lines will minimize the voltage drop on the return lines of the input connections.*
>
> *Use cable trays for wiring.*

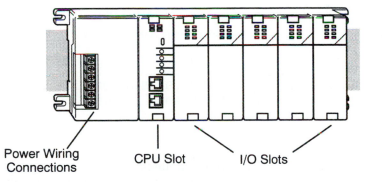

Power Wiring Connections   CPU Slot   I/O Slots

Figure 17-8. Diagram of PLC power supply wiring, CPU, and I/O. Courtesy PLC Direct by Koyo.

Grounding

Proper grounding is essential for safety and proper operation of a system. Grounding also helps limit the effects of noise due to electromagnetic interference (EMI). Check the appropriate electrical codes and ordinances to assure compliance with minimum wire sizes, color coding, and general safety practices.

Connect the PLC and components to the subpanel ground bus. See Figure 17-9. Ground connections should run from the PLC chassis and the power supply for each PLC expansion unit to the ground bus. The connection should exhibit very low resistance. Connect the subpanel ground bus to a single-point ground, such as a copper bus bar to a good earth ground reference. There must be low impedance between each device and the single-point termination. A rule of thumb would be less than 0.1 ohm DC resistance between the device and the single-point ground. This can be accomplished by removing the anodized finish and using copper lugs and star washers.

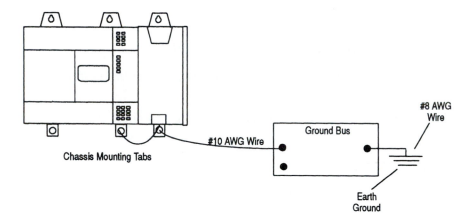

Figure 17-9. Grounding of an SLC 500 controller with a 2-slot expansion chassis. Courtesy Rockwell Automation/Allen-Bradley Company Inc.

The PLC manufacturer will provide details on installation in the hardware installation manual for their equipment. The National Electrical Code is an authoritative source for grounding requirements.

Grounding Guidelines

See Figure 17-10 for examples of ground connections. Grounding braid and green wires should be terminated at both ends with copper eye lugs to provide good continuity. Lugs should be crimped and soldered. Copper No. 10 or 12 bolts should be used for those fasteners that are used to provide an electrical connection to the single-point ground. This applies to device mounting bolts and braid termination bolts for subpanel and user-supplied single-point grounds. Tapped holes should be used rather than nuts and bolts. Note that a minimum number of threads are also required for a solid connection.

Paint, coatings, or corrosion must be removed from the areas of contact. Use external toothed lock washers (star washers). This practice should be used for all terminations: lug to subpanel, device to lug, device to subpanel, subpanel to conduit, and so on.

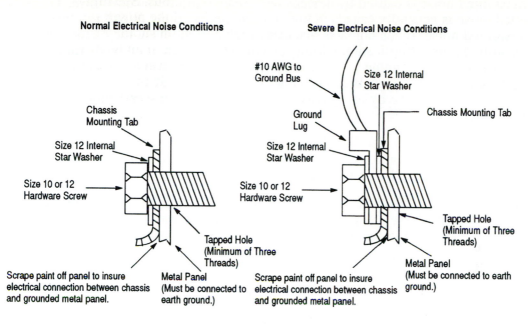

Figure 17-10. Ground connections. Courtesy Rockwell Automation/Allen-Bradley Company Inc..

Make sure to check the appropriate electrical codes and ordinances to assure compliance with minimum wire sizes, color coding, and general safety practices.

Handling Electrical Noise

Electrical noise is unwanted electrical interference that affects control equipment. The control devices in use today utilize microprocessors. Microprocessors are constantly fetching data and instructions from memory. Noise can cause the microprocessor to misinterpret an instruction or fetch bad data. Noise can cause minor problems or can cause severe damage to equipment and people.

Noise is caused by a wide variety of manufacturing devices. Devices that switch high voltage and current are the primary sources of noise. These would include large motors and starters, welding equipment, and contactors that are used to turn devices on and off. Noise is not continuous. It can be very difficult to find intermittent noise sources.

Noise can be created by power line disturbances, transmitted noise or ground loops. Power line disturbances are generally caused by devices that have coils. When they are switched off they create a line disturbance. Power line disturbances can normally be overcome through the use of line filters. Surge suppressors such as MOVs or an RC network across the coil can limit the noise. Coil-type devices include relays, contactors, starters, clutches/brakes, solenoids, etc.

Transmitted noise is caused by devices that create radio frequency noise. Transmitted noise is generally caused in high current applications. Welding causes transmitted noise. When contacts that carry high current open, they generate transmitted noise. Application wiring carrying signals can often be disrupted by this type of noise. Imagine wiring carrying sensor information to a control device. In severe cases, false signals can be generated on the signal wiring. This problem can often be overcome by using twisted-pair shielded wiring and connecting the shield to ground.

Transmitted noise can also "leak" into control cabinets. The holes that are put into cabinets for switches and wiring allow transmitted noise to enter the cabinet. The effect can be reduced by properly grounding the cabinet.

Ground loops can also cause noise. These are the noise problems that are often difficult to find. These are quite often intermittent problems. They generally occur when multiple grounds exist. The farther the grounds are apart the more likely the problem. A potential can exist between the power supply earth and the remote earth. This can create unpredictable results, especially in communications.

Proper installation technique can avoid problems with noise. There are two main ways to deal with noise: suppression and isolation.

Noise Suppression

Suppression attempts to deal with the device that is generating the noise. A very high voltage spike is caused when the current to an inductive load is turned off.

Inductive devices include relays, solenoids, motor starters, and motors. Suppression is even more important if the inductive device is in series or parallel with a hard contact such as a push-button or switch.

This high voltage can cause trouble for the device PLC. Lack of surge suppression can contribute to processor faults and intermittent problems. Noise can also corrupt RAM memory and may cause intermittent problems with I/O modules. Note that many of these problems are hard to troubleshoot because of their sporadic nature. Excessive noise can also significantly reduce the life of relay contacts. Some PLC modules include protection circuitry to protect against inductive spikes. A suppression network can be installed to limit the voltage spikes. Surge suppression circuits connect directly across the load device. This helps reduce the arcing of output contacts.

Figure 17-11 shows how AC and DC loads can be protected against surges. Noise suppression is also called *snubbing*. Snubbing can be used to suppress the arcing of mechanical contacts caused by turning inductive loads off (see Figure 17-12). Surge suppression should be used on all coils.

An RC or a varistor circuit can be used across an inductive load to suppress noise (1000 ohm, .2 microfarad). These components must be sized appropriately to meet the characteristic of the inductive output device that is being used. Check

Chapter 17: Installation and Troubleshooting

the installation manual for the PLC that you are using for proper noise suppression. A diode is sufficient for DC load devices. A 1N4004 can be used in most applications. Surge suppressors can also be used.

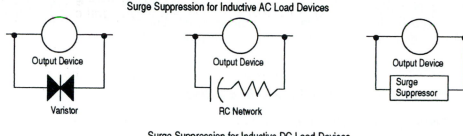

Surge Suppression for Inductive AC Load Devices

Output Device

Varistor

Output Device

RC Network

Output Device

Surge Suppressor

Surge Suppression for Inductive DC Load Devices

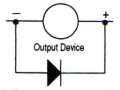

Output Device

Diode (A surge suppressor can also be used.)

Contact Protection Methods for Inductive AC and DC Output Devices

Figure 17-11. Surge suppression methods for AC and DC loads. Courtesy Rockwell Automation/Allen-Bradley Company Inc.

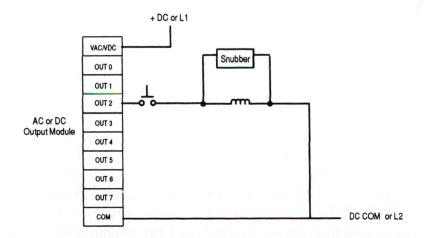

Figure 17-12. Example of snubbing. Courtesy Rockwell Automation/Allen-Bradley Company Inc.

Noise Isolation

The other way to deal with noise is isolation. The device or devices that cause trouble are physically separated from the control system. The enclosure also helps separate the control system from noise. In many cases field wiring must be placed in very noisy environments to allow sensors to monitor the process. This presents a problem especially when low voltages are used. Shielded twisted-pair wiring should be used for the control wiring in these cases. The shielding should only be grounded at one end. The shield should be grounded at the single-point ground.

How to Reduce the Effects of Noise:

> *Properly mount the PLC within a suitable enclosure*
>
> *Properly route all wiring*
>
> *Properly ground all equipment*
>
> *Install suppression to noise-generating devices*

INDUSTRIAL CONTROLLER MAINTENANCE

Industrial controllers are designed to be very reliable devices. They are used in automated systems that are very costly to operate. Downtime is very costly in industry. The technician will be expected to keep the systems in operation and downtime to a bare minimum. Industrial controllers have been designed to be low maintenance devices but there are some tactics that will help reduce downtime.

It is important, however to keep them clean. Industrial controllers are normally mounted in enclosures. Enclosures are typically air cooled. This means that a fan is mounted in the wall of the enclosure to circulate fresh air through the cabinet and cool the components. This can lead to the accumulation of dust, dirt and other contaminants. These can cause short circuits or intermittent problems in electronic equipment. One must make sure that the fans for such enclosures have adequate filters and that the filters are cleaned regularly. One should have a preventive maintenance schedule which includes a check of the enclosure to make sure the inside is free of contamination. This check should also include an inspection for loose wires or termination screws that could cause problems later. Vibration can loosen screws and cause intermittent or permanent problems. Modules should also be checked to make sure they are securely seated in the backplane, especially in high-vibration environments.

Many control devices have battery backup. Long-life lithium batteries are usually used that can have lifetimes of 2 or more years. One can be sure, however, that the batteries will fail eventually. They will probably fail when least convenient. Their failure can cause unnecessary system downtime which is often measured in hundreds or thousands of dollars. For this reason it is recommended that a regular schedule of battery replacements be made. This should be a part of the preventative maintenance schedule. Replacing the batteries once a year is a cheap invest-ment to avoid costly downtime and potentially larger problems.

Companies must maintain an inventory of spare parts to minimize downtime. With the high cost of downtime there is no time to spend repairing boards or other components. The technician must be able to find and correct the problem in a minimum of time. This means fault isolation and component replacement. It is a wise idea to keep approximately one spare each ten devices used. This applies to each type of input and output module and also CPU and other communication and special-purpose modules. Spare parts must also be maintained for sensors, drives, and other crucial devices. In many cases modules may be returned to the manu-facturer for repair. A spare one assures that production can resume while the defective one is returned for repair. A reasonable selection of spare parts can drastically reduce downtime, fire fighting, and frustration.

PLC TROUBLESHOOTING

The first consideration in troubleshooting and maintaining systems is safety. When you encounter a problem, remember that less than one-third of all system failures will be due to the PLC. Most of the failures are due to input and output devices (approximately 50 percent).

A few years ago, a technician was killed when he isolated the problem to a defective sensor. He bypassed the sensor and the system restarted with him in it. He was killed by the system he had fixed. You must always be aware of the possible outcomes of changes you make.

Troubleshooting is actually a relatively straightforward process in automated systems. The first step is to think. This may seem rather basic, but many people jump to improper, premature conclusions and waste time finding problems. The first step is to examine the problem logically. Think the problem through using common sense first. This will point to the most logical cause. Troubleshooting is much like the game 20 Questions. Every question should help isolate the prob-lem. In fact, every question should eliminate about half of the potential causes. Remember, a well-planned job is half done.

> *Think logically.*

> *Ask yourself questions to isolate the problem.*

> *Test your theory.*

Next use the resources you have available to check your theory. Often the error-checking present on the PLC modules is sufficient. The LEDs on PLC CPUs and modules can provide immediate feedback on what is happening. Many PLCs have LEDs to indicate blown fuses and many other problems. Check these indicators first.

The usual problem is that an output is not turning on when it should. There are several possible causes. The output device could be defective. The PLC output that turns it on could be defective. One of the inputs that allow, or cause, the output to turn on could be defective. The sensor or the PLC input that it is attached to could be defective. The ladder logic could even be faulty. It is possible that a ladder can be written that performs perfectly the vast majority of the time but fails under certain conditions. Again the module I/O LEDs provide the best source of answers (see Figures 17-13 and 17-14). If the PLC module output LED is on for that output, the problem is probably not the inputs to the PLC. The device is defective, the wiring is defective, or the PLC output is defective.

The next step is to isolate the problem further. A multimeter is invaluable at this point. If the PLC output is off, a meter reading should show the full voltage with which the device is turned on. If the output is used to supply 115 V to a motor starter, the meter should show the full 115 volts between the output terminal and common on the PLC module (see Figure 17-14).

If the PLC output is on, the meter should read zero volts because the output acts as a switch. If we measure the voltage across a switch, we should read zero volts. If the switch is open, we should read the full voltage. If we read no voltage in either case, the wiring and power supply should be checked. If the wiring and power supply are good, the device is the problem. Depending on the device, there may be fuses or overload protection present.

Output Troubleshooting Chart

Output Condition	Output LED	Status in Ladder	Probable Problem
on	on	true -⊛-	None
off	off	true -⊛-	Bad fuse or bad output module
off	off	false -O-	None
off	on	true -⊛-	Wiring to output device or bad output device

Figure 17-13. How to isolate a PLC problem by comparing the states of inputs/outputs, indicators, and the ladder status.

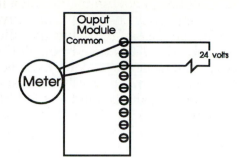

Figure 17-14. How to test an output. This will help isolate the problem for the technician.

Now pretend that the output LED was not turning on. Now we must check the input side. If the input LED is on, we can assume that the sensor, or other input device, is operational (see Figures 17-15 and 17-16). Next we must see if the PLC CPU really sees the input as true. At this point a monitor of some type is required. A hand-held programmer or a computer is often used. The ladder is then monitored under operation. *Note*: Many PLCs allow the outputs to be disabled for troubleshooting. This is the safe way to proceed. Check the ladder to see if the contact is closing. If it is seeing the input as false, the problem is probably a defective PLC module input.

Input Troubleshooting Guide

Actual input condition	Input module LED status	Ladder Status		Probable problem
off	off	false ⊣ ⊢	true ⊣/⊢	None
off	on	true ⊣█⊢	false ⊣/⊢	Short in the input device or wiring or a bad input module
on	off	false ⊣ ⊢	true ⊣/⊢	wiring/power to I/O module or I/O module
on	on	false ⊣ ⊢	true ⊣/⊢	I/O module
on	on	true ⊣█⊢	false ⊣/⊢	None

Figure 17-15. Chart for troubleshooting inputs.

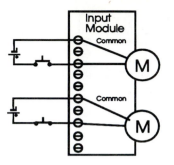

Figure 17-16. How to isolate input problems.

Potential Ladder Diagram Problems

Some PLC manufacturers allow output coils to be used more than once in ladders. This means that there can be multiple conditions that can control the same output coil. I have seen technicians who were sure that they had a defective output module because when they monitored the ladder the output coil was on but the actual output LED was off. In this case the technician had inadvertently programmed the same output coil twice with different input logic. The rung he was monitoring was true, but one farther down in the ladder was false, so the PLC kept the output off.

The other potential problem with ladder diagramming is that the problem can be intermittent. Timing is crucial in ladder diagramming. Devices are often exchanging signals to signify that an action has occurred. For example, a robot finishes its cycle and sends a digital signal to the PLC to let it know. The programmer of the robot must be aware of the timing considerations. If the robot programmer just turns it on for one step the output may only be on for a few milliseconds. The PLC may be able to "catch" the signal every time, or it may occasionally miss it because of the length of the scan time of the ladder diagram.

This can be a tough problem to find, because it only happens occasionally. The better way to program is to handshake instead of relying on the PLC seeing a periodic input. The programmer should have the robot output stay on until it is acknowledged by the PLC. The robot turns on the output and waits for an input from the PLC to assure that the PLC saw the output from the robot.

Summary

The installation of a control system must be carefully planned. People and devices must be protected. Hardwired switching should be provided to drop all power to the system. Lockouts should be provided to assure safety while maintenance is being performed. Proper fusing to assure protection of individual devices is a must. The cabinet must be carefully selected to meet the needs of the application environment. Control and power wiring should be separated to reduce noise. Proper grounding procedures must be followed to ensure safety.

Troubleshooting must be done in a logical way. Think the problem through. Ask questions that will help isolate the potential problems. Above all, apply safe work habits while working on systems.

Questions

1. What is a NEMA enclosure?

2. Why should enclosures be used?

3. Describe how an enclosure is chosen.

4. What is a fusible disconnect?

5. What is a contactor?

6. What is the purpose of an isolation transformer?

7. What is the major cause of failure in systems?

8. Describe a logical process for troubleshooting.

9. Describe proper grounding techniques.

10. Draw a block diagram of a typical control cabinet.

11. Describe at least three precautions that should be taken to help reduce the problem of noise in a control system.

12. A technician has been asked to troubleshoot a system. The output device is not turning on for some reason. The output LED is working as it should. The technician turns the output on and places a meter over the PLC output: 115 V is read. The device is not running. What is the most likely problem? How might it be fixed?

13. A technician has been asked to troubleshoot a system. The PLC does not seem to be receiving an input because the output it controls is not turning on. The technician notices that the input LED is not turning on. The technician notices that the output indicator LED on the sensor seems to be working. A meter is placed across the input with the sensor on: 24 V is sensed, but the input LED is off. What is the most likely problem?

14. A technician is asked to troubleshoot a system. An input seems to be defective. The technician notices that the input LED is never on. A meter is placed across the PLC input and zero volts is read. The technician removes and tests the sensor. (The LED on the sensor comes on when the sensor is activated.) The sensor is fine. What is the most likely problem?

15. A technician is asked to troubleshoot a system. An output device is not working. The technician notices that the output LED seems to be working fine. A meter is placed over the PLC output with the output on. Zero volts is read. Describe what the technician should do to find the problem.

16. What items should be included in system documentation?

Appendix A

This is a listing of web sites that may be helpful. Addresses may change rapidly. A search may provide the new addresses.

Allen-Bradley Company   *www.ab.com*

ASI bus   *www.as-interface.com/*

Bitbus   *http://vigna.cimsi.cim.ch/beug/Wbitbus.html*

CAN: Controller Area Network   *www.nrtt.demon.co.uk/can.html*

Control Technology Corporation   *www.control.com*

DeviceNET   *www.odva.org*

Event Technologies Inc.   *www.industry.net/event.technologies*

Field Bus Comparison   *www.synergetic.com/compare.htm*

Field Bus Links   *www.infoside.de/infida/wissen2.htm*

Field Bus Standards   *http://homepages.iol.ie/~readout/fieldbus*

Field Bus Tutorial   *http://rolf.ece.curtin.edu.au/~clive/Fieldbus/fieldbus.htm*

GE Fanuc   *www.industry.com/gefanuc.automation*

LonWorks   *http://www.lonworks.echelon.com/*

Omron Corporation   *www.omron.co.jp*

PLC Direct by Koyo   *www.plcdirect.com*

Profibus   *www.profibus.com*

Seriplex   *www.seriplex.org/wp.html*

SoftPLC Corporation   *www.softplc.com*

Steeplechase Software Inc.   *www.steeplechase.com*

Wonderware   *www.wonderware.com*

Z-World Inc.   *www.zworld.com*

Appendix B

Input Device Symbols

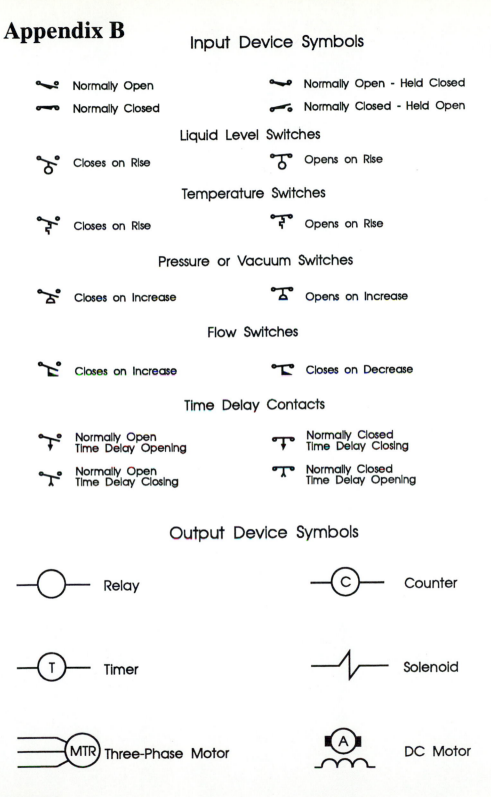

Normally Open	Normally Open - Held Closed
Normally Closed	Normally Closed - Held Open

Liquid Level Switches

Closes on Rise	Opens on Rise

Temperature Switches

Closes on Rise	Opens on Rise

Pressure or Vacuum Switches

Closes on Increase	Opens on Increase

Flow Switches

Closes on Increase	Closes on Decrease

Time Delay Contacts

Normally Open Time Delay Opening	Normally Closed Time Delay Closing
Normally Open Time Delay Closing	Normally Closed Time Delay Opening

Output Device Symbols

Relay	Counter
Timer	Solenoid
Three-Phase Motor	DC Motor

Glossary

A

AC input module: This is a module that converts a real-world AC input signal to the logic level required by the PLC processor.

AC output module: Module that converts the processor logic level to an AC output signal to control a real-world device.

Accumulated value: Applies to the use of timers and counters. The accumulated value is the present count or time.

Accuracy: The deviation between the actual position and the theoretical position.

Actuator: Output device normally connected to an output module. An example would be an air valve and cylinder.

Address: Number used to specify a storage location in memory.

Ambient temperature: Temperature that naturally exists in the environment. For example, the ambient temperature of a PLC in a cabinet near a steel furnace is very high.

Analog: Signal with a smooth range of possible values. For example, a temperature that could vary between 60 and 300 degrees would be analog in nature.

ANSI: American National Standards Institute.

ASCII: American Standard Code for Information Interchange. A coding system used to represent letters and characters. Seven-bit ASCII can represent 128 different combinations. Eight-bit ASCII (extended ASCII) can represent 256 different combinations.

Asynchronous communications: Method of communications that uses a series of bits to send data between devices. There is a start bit, data bits (7 or 8), a parity bit (odd, even none, mark, or space), and stop bits (1, 1.5, or 2). One character is transmitted at a time. RS-232 is the most common.

B

Backplane: Bus in the back of a PLC chassis. It is a printed circuit board with sockets that accept various modules.

Baud rate: Speed of serial communications. The number of bits per second transmitted. For example, RS-232 is normally used with a baud rate of 9600. This would be about 9600 bits per second. It takes about 10 bits in serial to send an ASCII character so that a baud rate of 9600 would transmit about 960 characters per second.

BEUG (BITBUS European Users Group): BEUG is a nonprofit organization devoted to spreading the BITBUS technology and organizing a basic platform where people using BITBUS can share application experiences.

Binary: Base two number system. Binary is a system in which ones and zeros are used to represent numbers.

Binary-coded decimal (BCD): A number system. Each decimal number is represented by four binary bits. For example, the decimal number 967 would be represented by 1001 0110 0111 in BCD.

Bit: Binary digit. The smallest element of binary data. A bit will be either a zero or a one.

BITBUS: BITBUS was created by Intel in 1983. It is one of the most widely used field-buses. It was promoted as a standard in 1990 by a special committee of the IEEE (standard IEEE-1118 1990).

Boolean: Logic system that uses operators such as AND, OR, NOR, and NAND. This is the system that is utilized by PLCs, although it is usually made invisible by the programming software for the ease of the programmer.

Bounce: This is an undesirable effect. It is the erratic make and break of electrical contacts.

Branch: Parallel logic path in a ladder diagram.

Byte: Eight bits or two nibbles. (A nibble is 4 bits.)

C

Cascade: Programming technique that is used to extend the range of timers and counters.

CENELEC: European Committee for Electrotechnical Standardization. It develops standards which cover dimensional and operating characteristics of control components.

Central processing unit (CPU): Microprocessor portion of the PLC. It is the portion of the PLC that handles the logic.

Color mark sensor: Sensor that was designed to differentiate between two different colors. They actually differentiate on the basis of contrast between the two colors.

Complement: The complement is the inverse of a digital signal.

CMOS (complementary metal-oxide semiconductor): Integrated circuits that consume very little power and also have good noise immunity.

Compare instruction: PLC instruction that is used to test numerical values for equal, greater than, or less than relationships.

Contact: Symbol used in programming PLCs. Used to represent inputs. There are normally open and normally closed contacts. Contacts are also the conductors in electrical devices such as starters.

Contactor: Special-purpose relay that is used to control large electrical current.

CSA (Canadian Standards Organization): Develops standards, tests products and provides certification for a wide variety of products.

Current sinking: Refers to an output device (typically an NPN transistor) that allows current flow from the load through the output to ground.

Current sourcing: Output device (typically a PNP transistor) that allows current flow from the output through the load and then to ground.

Cyclic Redundancy Check (CRC): A calculated value, based on the content of a communication frame. It is inserted in the frame to enable a check of data accuracy after receiving the frame across a network. BITBUS uses the standard SDLC CRC.

D

Dark-on: Refers to a photosensor's output. If the sensor output is on when no object is sensed, it is called a dark-on sensor.

Data highway: This is a communications network that allows devices such as PLCs to communicate. They are normally proprietary, which means that only like devices of the same brand can communicate over the highway. Allen Bradley calls their PLC communication network Data Highway.

Data table: A consecutive group of user references (data) of the same size that can be accessed with table read/write functions.

Debugging: Process of finding problems (bugs) in any system.

Diagnostics: Devices normally have software routines that aid in identifying and finding problems in the device. They identify fault conditions in a system.

Digital output: An output that can have two states: on or off. These are also called discrete outputs.

Distributed processing: The concept of distributed processing allows individual discrete devices to control their area and still communicate to the others via a network. The distributed control takes the processing load off the "host" system.

Documentation: Documentation is descriptive paperwork that explains a system or program. It describes the system so that the technician can understand, install, troubleshoot, maintain, or change the system.

Downtime: The time a system is not available for production or operation is called downtime. Downtime can be caused by breakdowns in systems.

E

EEPROM: Electrically erasable programmable read only memory.

Energize: Instruction that causes a bit to be a one. This turns an output on.

Examine-off: Contact used in ladder logic. It is a normally closed contact. The contact is true (or closed) if the real-world input associated with it is off.

Examine-on: Contact used in ladder logic programming. Called a normally open contact. This type of contact is true (or closed) if the real-world input associated with it is on.

Expansion rack: A rack added to a PLC system when the application requires more modules than the main rack can contain. A remote rack is sometimes used to permit I/O to be remotely located from the main rack.

F

False: Disabled logic state (off).

Fault: Failure in a system that prevents normal operation of a system.

Firmware: A series of instructions contained in read-only memory (ROM) that are used for the operating system functions. Some manufacturers offer upgrades for PLCs. This is often done by replacing a ROM chip. Thus the combination of software and hardware lead to it being called firmware.

Flowchart: Used to make program design easier.

Force: Refers to changing the state of actual I/O by changing the bit status in the PLC. In other words, a person can force an output on by changing the bit associated with the real-world output to a 1. Forcing is normally used to troubleshoot a system.

Frame: Packet of bits that will be transmitted across a network. A frame contains a header, user data and an end of frame. The frame must contain all the necessary information to enable the sender and receiver(s) of the communication to decode the user's data and to ensure that this data is right.

Full duplex: Communication scheme where data flows in both directions simultaneously.

G

Ground: Direct connection between equipment (chassis) and earth ground.

H

Half duplex: Communication scheme where data flows in both directions but in only one direction at a time.

Hard contacts: Physical switch connections.

Hard copy: Printed copy of computer information.

HDLC (High-level Data Link Control): Standard protocol of communication oriented in message transmission (frames). The user's data field in an HDLC-frame can be of a free number of bits. The SLDC is a subset of the HDLC that defines the whole protocol in more detail and is byte-oriented.

Hexadecimal: Numbering system that utilizes base 16.

Host computer: One to which devices communicate. The host may download or upload programs, or the host might be used to program the device. An example would be a PLC connected to a microcomputer. The host (microcomputer) "controls" the PLC by sending programs, variables, and commands. The PLC controls the actual process but at the direction and to the specifications of the host.

Hysteresis: A dead band that is purposely introduced to eliminate false reads in the case of a sensor. In an encoder hysteresis would be introduced in the electronics to prevent ambiguities if the system happens to dither on a transition.

I

IEC (International Electrotechnical Commission): Develops and distributes recommended safety and performance standards.

IEEE: Institute of Electrical and Electronic Engineers.

Image table: Area used to store the status of input and output bits.

Incremental: This term typically refers to encoders. Encoders provide logic states of 0 and 1 for each successive cycle of resolution.

Instruction set: Instructions that are available to program the PLC.

Intelligent I/O: PLC modules that have a microprocessor built in. An example would be a module that would control closed-loop positioning.

Interfacing: Connection of a PLC to external devices.

I/O (input/output): Used to speak about the number of inputs and outputs that are needed for a system, or the number of inputs and outputs that a particular programmable logic controller can handle.

IP rating: Rating system established by the IEC that defines the protection offered by electrical enclosures. It is similar to the NEMA rating system.

Isolation: Used to segregate real-world inputs and outputs from the central processing unit. Isolation assures that even if there is a major problem with real-world inputs or outputs (such as a short), the CPU will be protected. This isolation is normally provided by optical isolation.

K

K: Abbreviation for the number 1000. In computer language it is equal to two to the tenth, or 1024.

Keying: Technique to ensure that modules are not put in the wrong slots of a PLC. The user sets up the system with modules in the desired slots. The user then keys the slots to assure that only a module of the correct type can be physically installed.

L

Ladder diagram: Programmable controller language that uses contacts and coils to define a control sequence.

LAN: *See* Local area network.

Latch: An instruction used in ladder diagram programming to represent an element that retains its state during controlled toggle and power outage.

Leakage current: Small amount of current that flows through load-powered sensors. The small current is necessary for the operation of the sensor. The small amount of current flow is normally not sensed by the PLC input. If the leakage is too great a bleeder resistor must be used to avoid false inputs at the PLC.

LED (light-emitting diode): A solid-state semiconductor that emits red, green, or yellow light or invisible infrared radiation.

Light-on sensor: This refers to a photosensor's output. If the output is on when an object is sensed, the sensor is a light-on sensor.

Linear output: Analog output.

Line driver: A line driver is a differential output driver intended for use with a differential receiver. These are usually used where long lines and high frequency are required and noise may be a problem.

Line-powered sensor: Normally, three-wire sensors, although four-wire sensors also exist. The line-powered sensor is powered from the power supply. A separate wire (the third) is used for the output line.

Load: Any device that current flows through and produces a voltage drop.

Load-powered sensor: A load-powered sensor has two wires. A small leakage current flows through the sensor even when the output is off. The current is required to operate the sensor electronics.

Load resistor: A resistor connected in parallel with a high-impedance load to enable the output circuit to output enough current to ensure proper operation.

Local area network (LAN): A system of hardware and software designed to allow a group of intelligent devices to communicate within a fairly close proximity.

Lockout: The placement of a lockout device on an energy isolating device, in accordance with an established procedure, to ensure that the energy isolating device and the equipment being controlled cannot be operated until the lockout device is removed.

Lockout device: A device that utilizes a positive means such as a lock, either key or combination type, to hold an energy isolating device in the safe position and prevent the energizing of a machine or equipment.

LSB: Least significant bit.

M

Machine language: Control program reduced to binary form.

MAP (manufacturing automation protocol): "Standard" developed to make industrial devices communicate more easily. Based on a seven-layer model of communications.

Master: The master on a network is the device that controls communication traffic. The master of a network usually polls every slave to check if it has something to transmit. In a master-slave configuration, only the active master can place a message on the bus. The slave can reply only if it receives a frame from the master that contains a logical token that explicitly enables the slave to reply.

Master control relay (MCR): Hardwired relay that can be deenergized by any hardwired series-connected switch. Used to deenergize all devices. If one emergency switch is hit it must cause the master control relay to drop power to all devices. There is also a master control relay available in most PLCs. The master control relay in the PLC is not sufficient to meet safety requirements.

Memory map: Drawing showing the areas, sizes, and uses of memory in a particular PLC.

Microsecond: A microsecond is one millionth (0.000001) of a second.

Millisecond: A millisecond is one thousandth (.001) of a second.

Mnemonic codes: Symbols designated to represent a specific set of instructions for use in a control program. An abbreviation given to an instruction: usually an acronym that is made by combining the initial letters or parts of words.

MSB: Most significant bit.

N

NEMA (National Electrical Manufacturers Association): Develops standards that define a product, process, or procedure. The standards consider construction, dimensions, tolerances, safety, operating characteristics, electrical rating and so on. They are probably best known for their rating system for electrical cabinets.

Network: System that is connected to devices or computers for communication purposes.

Node: Point on the network that allows access.

Noise: Unwanted electrical interference in a programmable controller or network. It can be caused by motors, coils, high voltages, welders, and so on. It can disrupt communications and control.

Nonretentive coil: A coil that will turn off upon removal of applied power to the CPU.

Nonretentive timer: Timer that loses the time if the input enable signal is lost.

Nonvolatile memory: Memory in a controller that does not require power to retain its contents.

NOR: The logic gate that results in zero unless both inputs are zero.

NOT: The logic gate that results in the complement of the input.

O

Octal: Number system based on the number 8, utilizing numbers 0 through 7.

Off-delay timer: This is a type of timer that is on immediately when it receives its input enable. It turns off after it reaches its preset time.

Off-line programming: Programming that is done while not attached to the actual device. For example, a PLC program can be written for a PLC without being attached. The program can then be downloaded to the PLC.

On-delay timer: Timer that does not turn on until its time has reached the preset time value.

One-shot contact: Contact that is only on for one scan when activated.

Operating system: The fundamental software for a system that defines how it will store and transmit information.

Optical isolation: Technique used in I/O module design that provides logic separation from field levels.

OR: Logic gate that results in 1 unless both inputs are 0.

P

Parallel communication: A method of communications where data is transferred on several wires simultaneously.

Parity: Bit used to help check for data integrity during a data communication.

Peer-to-peer: This is communication that occurs between similar devices. For example, two

PLCs communicating would be peer-to-peer. A PLC communicating to a computer would be device-to-host.

PID (Proportional, integral, derivative) control: Control algorithm that is used to closely control processes such as temperature, mixture, position, and velocity. The proportional portion takes care of the magnitude of the error. The integral takes care of small errors over time. The derivative compensates for the rate of error change.

PLC: Programmable logic controller.

Programmable controller: A special-purpose computer. Programmed in ladder logic. It was also designed so that devices could be easily interfaced with it.

Pulse modulated: Turning a light source on and off at a very high frequency. In sensors the sending unit pulse modulates the light source. The receiver only responds to that frequency. This helps make photo-sensors immune to ambient lighting.

PPR (Pulses per revolution): This refers to the number of pulses an encoder produces in one revolution.

Q

Quadrature: Two output channels out of phase with each other by 90 degrees.

R

Rack: PLC chassis. Modules are installed in the rack to meet the user's need.

Radio frequency (RF): Communications technology in which there is a transmitter/receiver and tags. The transmitter/ receiver can read or write to the tags. There are active and passive tags available. Active tags are battery powered. Passive tags are powered from the RF emitted from the transmitter. Active tags have a much wider range of communication. Either tag can have several K of memory.

RAM (random access memory): Normally considered user memory.

Register: Storage area. It is typically used to store bit states or values of items such as timers and counters.

Repeatability: The ability to repeat movements or readings. For a robot it would be how accurately it would return to a position time after time. Repeatability is unrelated to resolution and is usually 3 to 10 times better than accuracy.

Resolution: A measure of how closely a device can measure or divide a quantity. For example, in an encoder resolution would be defined as counts per turn. For an analog to digital card it would be the number of bits of resolution. For example, for a 12-bit card the resolution would be 4096.

Retentive coil: A coil that will remain in its last state, even though power was removed.

Retentive timer: Timer that retains the present count even if the input enable signal is lost. When the input enable is active again, the timer begins to count again from where it left off.

Retroreflective: Photosensor that sends out a light which is reflected from a reflector back to the receiver (the receiver and emitter are in the same housing). When an object passes through it breaks the beam.

RF (radio frequency): *See* radio frequency.

ROM (read-only memory): This is operating system memory. ROM is nonvolatile. It is not lost when the power is turned off.

RS-232: Common serial communications standard. This standard specifies the purpose of each of 25 pins. It does not specify connectors or which pins must be used.

RS-422 and RS-423: Standards for two types of serial communication. RS-422 is a balanced serial mode. This means that the transmit and receive lines have their own common instead of sharing one like RS-232. Balanced mode is more noise immune. This allows for higher data transmission rates and longer transmission distances. RS-423 uses the unbalanced mode. Its speeds and transmission distances are much greater than RS-232 but less than RS-422.

RS-449: Electrical standard for RS-422/RS-423. It is a more complete standard than the RS-232. It specifies the connectors to be used also.

RS-485: Similar to the RS-422 standard. Receivers have additional sensitivity which allows for longer distances and more communication drops. Includes some extra protection for receiver circuits.

Rung: Group of contacts that control one or more outputs. In a ladder diagram it is the horizontal lines on the diagram.

S

Scan time: Amount of time it takes a programmable controller to evaluate a ladder diagram. The PLC continuously scans the ladder diagram. The time it takes to evaluate it once is the scan time. It is typically in the low-millisecond range.

SDLC: Serial Data Link Control, subset of the HDLC used in a large number of communication systems like Ethernet, ISDN, BITBUS, and others. This protocol defines the structure of the frames and the values of a number of specific fields in these frames.

Sensitivity: Refers to a device's ability to discriminate between levels. If it's a sensor it would relate to the finest difference it could detect. If it were an analog module for a PLC, it would be the smallest change it could detect.

Sensor: Device used to detect change. Normally it is a digital device. The outputs of sensors change state when they detect the correct change. Sensors can be analog or digital in nature. They can also be purchased with normally closed or normally open outputs.

Sequencer: Instruction type that is used to program a sequential operation.

Serial communication: Sending of data one bit at a time. The data is represented by a coding system such as ASCII.

Slave: On a master-slave configured network, there is usually one master and several slaves. The slaves are nodes of the network that can transmit informations to the master only when they are polled (called) from it. The rest of the time a slave never transmits anything.

Speech modules: Used by a PLC to output spoken messages to operators. The sound is typically digitized human speech stored in the module's memory. The PLC requests the message number to play it.

T

Tagout: The placement of a tagout device on an energy isolating device, in accordance with an established procedure, to indicate that the energy isolating device and the equipment being controlled may not be operated until the tagout device is removed.

Tagout device: A prominent warning device, such as a tag and a means of attachment, which can be securely fastened to an energy isolating device in accordance with an established procedure, to indicate that the energy isolating device and the equipment being controlled may not be operated until the tagout device is removed.

Thermocouple: A thermocouple is a sensing transducer. It changes a temperature to a current. The current can then be measured and converted to a binary equivalent that the PLC can understand.

Thumbwheel: Device used by an operator to enter a number between 0 and 9. Thumbwheels are combined to enter larger numbers. Thumbwheels typically output BCD numbers to a device.

Timer: Instruction used to accumulate time until a certain value is achieved. The timer then changes its output state.

TOP (technical and office protocol): Communication standard that was developed by Boeing. Based on the contention access method. The MAP standard is meant for the factory floor and TOP is meant for the office and technical areas.

Transitional contact: Contact that changes state for one scan when activated.

True: This is the enabling logic state. Generally associated with a "one" or "high" state.

U

UL (Underwriters Laboratory): Organization that operates laboratories to investigate systems with respect to safety.

User memory: Memory used to store user information. The user's program, timer/counter values, input/output status, and so on, are all stored in user memory.

V

Volatile memory: Memory that is lost when power is lost.

W

Watchdog timer: Timer that can be used for safety. For example, if there is an event or sequence that must occur within a certain amount of time, a watchdog timer can be set to shut the system down in case the time is exceeded.

Word: Length of data in bits that a microprocessor can handle. For example, a word for a 16-bit computer would be 16 bits long, or two bytes. A 32-bit computer would have a 32-bit word.

Index

A

Absolute encoders   145
Access methods   356
Accuracy   161
Active-high   11
Actuator Sensor Interface (ASI) Bus   372
Allen Bradley PLC-2 counters   116
Allen Bradley PLC-2 timers   98
Analog   547
Analog input modules   199
Analog output   27
Analog output modules   201
Analog sensors   139
AND   75
ANSI   547
Applications   35
ArcNet   357
Area control   350
Arithmetic instructions   226
ASCII   14, 345, 547
ASCII modules   205
ASI   359, 372
Asynchronous communications   547

B

Backplane   547
Bar-code modules   210
Baseband   354
Baud rate   547
Beckoff Light Bus   357
Binary   45, 548
Binary-coded decimal   47, 548
Bipolar module   199
Bit   548
BITBUS   359, 373
Bleeder resistor   192
Boolean   548
Bounce   548
Branch   548
Branching   75
Broadband   354
Burden current   160
Byte   47, 548

C

CAN   376
Capacitive sensors   156
Cascade   548
Cascading timers   114
Cell level   343
CENELEC   548
Central processing unit   8, 548
Channel   19
Closed-loop position control   206
CMOS   548
CMOS-RAM   11
Coaxial cable   353
COIL   59
Color mark sensor   144, 548
Communications modules   203
Compare instructions   229
Compensation   162
Contact   548
Contact sensor   138
Contactor   549
Contacts   58
Controller Area Network (CAN)   376
ControlNet   386
Convergent photo sensor   144
Counters   115
CPU   19, 89
CSA   549
CSMA/BA   358
CSMA/CD   357
Current ratings   27
Current sinking   549
Current sourcing   549
Current specifications   195

D

Dark sensing   141
Dark-on   549
Data   340
Data highway   549
DDE   395
Debounce   25
Decimal   44
Delay-off timing   96
Delay-on   96
Device bus   371, 372

Device level 342
DeviceNET 358, 377
Diffuse 141
Digital input modules 190
Digital output 27, 549
Digital output modules 190, 195
Digital sensors 139, 140
Distributed processing 549
Documentation 18
Down counters 116
Drum controller 280
Dumb terminal 14
Dynamic Data Exchange (DDE) 395

E

EEPROM 13, 550
Electrical noise 8, 533
Electronic field sensors 152
Embedded controllers 482
Enclosures 524
Encoders 145
Energize 550
Examine-off 550
Examine-on 550

F

Fail-safe 63
FBD 317
FDDI 355
Fiber optic 202, 354
Fiber-optic sensors 144
Field device 19, 27, 370
Field sensors 152
Fieldbus 383
Flow diagram 128, 468
Flow diagram symbols 128
Flowchart 550
Flowchart programming 436
Force 550
Frequency-division multiplexing 354
Full duplex 550
Function block diagram 317
Function blocks 310
Functions 311
Fuzzy logic 217, 294

G

Gate sensor 180, 182
GE Genius I/O 357
Gould Modicon counters 122
Ground 550
Grounding 531
Grounding guidelines 532

H

Half duplex 550
Hall effect sensor 168
Hand-held programmer 16
Handshaking 26
Hardwired control 3
Hexadecimal 49, 551
High-speed counter modules 193
High-density modules 199
Host level 362
Host-link modules 204
Hysteresis 154

I

I/O image table 10
I/O modules 190
IEC 551
IEC 1131-3 310, 454
IEEE 551
IL 322
Image table 10, 551
Immediate instructions 79
Incremental encoders 147
Inductive sensors 152, 153
Industrial automation controller 462
Industrial bus 357
Industrial terminals 15
Input modules 25
Input section 19
Installation 524
Instruction list programming 322
InterBus-S 357, 374
Isolation 25, 551

L

Ladder diagram 2, 5, 58, 60, 320, 552
Laser sensors 144
Latch 552
Latching instructions 81

Leakage current   159, 192, 552
Least significant bit   47
LED   140
Levels of plant communication   340
Light sensing   141
Line-powered sensors   160, 552
Linear Variable Displacement Transformer
    (LVDT)   171
Load   552
Load-powered sensor   158, 552
Local area network   350, 553
Lockout procedure   517
Lockout/tagout   507
LonWorks   358, 375

M

Magnetic reed sensors   168
Man-machine interface   394
Manufacturing automation protocol   359
Manufacturing message specification   361
MAP   553
Master control relay   83, 553
Memory   10
Memory map   11, 553
Microcomputer   17
MMI   395

N

NEMA   524, 553
Nibble   47
Noise   8, 533, 554
Noise isolation   536
Noise suppression   534
Nonretentive timer   98, 554
Nonvolatile memory   488, 554
Noncontact sensors   138
Normally closed   58, 65
Normally closed contact   62
Normally open   58
NPN   161

O

Object-oriented   446
Octal   48, 554
Off-delay   141
Off-delay timer   554
Off-line   17

Off-line programming   554
Omron counters   123
Omron timers   108
On-delay   141
On-delay timer   554
On-line   17
One-shot contact   554
Open-loop position control   205
Operating system   554
Operating system memory   10
Optical isolation   25, 554
Optical sensors   140
Output image table   27
Output modules   27
Output section   26

P

Parity   554
PC-based control   432
Peer-to-peer   554
Peer-to-peer modules   204
Photodetector   141
PID   555
PID modules   210
PLC   6
PLC applications   35
PLC installation   524
PNP sensor   160
Polarizing photo-sensors   142
Position control modules   205
Power supply   19
Precision   161
Preset value   97
Pressure sensors   176
Primitive communication   344
Process bus   371, 383
Profibus   359, 384
Programmable controller   3, 6
Programming devices   14
Pseudocode   129

Q

Quadrature   149

R

Rack   19, 21, 555
Radio frequency   355, 555

Radio-frequency modules   218
RAM   555
Random access memory   11
Read-only memory   10
Reflective sensors   141
relays   27
Remote I/O modules   202
Repeatability   161
Resolution   163, 199, 200
Resolvers   174
Response time   158
Retentive timer   98, 556
Retroreflective   142, 556
ROM   556
RS-232   344, 346, 556
RS-422   346, 556
RS-423   346, 556
RS-449   347, 556
RS-485   347
RTD   165
Rung   556

S

SCADA   348, 394
Scan   61
Scan cycle   61
Scan time   88, 556
SDS   358, 380
Sensing distance   154, 156
Sensoplex   359
Sensor   25, 58, 138, 180, 556
Sensor applications   180
Sensor installation   178
Sensor wiring   157
Sequencer instructions   282
Sequential control   280
Sequential function chart   325
Sercos   357
Serial communication   557
Seriplex   357, 381
SFC   325
Shift resister programming   285
Single-board controllers   482
Sinking   161, 549
Slots   19
Smart Distributed System (SDS)   380
Snubbing   534

Sourcing   160
Speech modules   220, 557
Square D counters   124
Square D timers   111
ST   312
Stage programming   286
Start/stop circuit   77
State language   462
State logic   302
Step programming   291
Strain gage   169
Structured text programming   312
Supervisory control and data acquisition   394
System wiring 526

T

Tagout   512
Temperature sensors   167
Thermistor   167
Thermocouple 162, 164, 557
Thru-beam   143
Thumbwheel   47, 557
Time base   97
Time-division multiplexing   354
Timer   96, 557
Token passing   356
Topology   351
Transitional contact   80, 557
Troubleshooting   537, 5, 16, 18
Twisted-pair   202, 353

U

Ultrasonic sensors   150
Unipolar modules   199
Up counters   116
Up/down counter   116
Upload/download   17
User memory   10, 558

V

Vision modules   208
Visual logic controller   436
Volatile memory   558

W

Watchdog timer   558
Wiring guidelines   531
Word   47, 558